The Ringtone Dialectic

The Ringtone Dialectic

Economy and Cultural Form

Sumanth Gopinath

The MIT Press
Cambridge, Massachusetts
London, England

MIT Press books may be purchased at special quantity discounts for business or sales promotional use. For information, please email special_sales@mitpress.mit.edu or write to Special Sales Department, The MIT Press, 55 Hayward Street, Cambridge, MA 02142.

Set in Stone Sans and Stone Serif by Toppan Best-set Premedia Limited. Printed and bound in the United States of America.

Library of Congress Cataloging-in-Publication Data

Gopinath, Sumanth S.
The ringtone dialectic : economy and cultural form / Sumanth Gopinath.
pages cm.
Includes bibliographical references and index.
ISBN 978-0-262-01915-6 (hardcover : alk. paper)
1. Ring tones—Social aspects. 2. Cell phones—Social aspects. 3. Cell phone services industry. 4. Mobile games industry. 5. Internet entertainment industry. 6. Digital media—Social aspects. I. Title.
HE9713.G67 2013
384.5′35—dc23
2012045906

10 9 8 7 6 5 4 3 2

for my parents, and for Michael and Hazel

Contents

Acknowledgments

Many thanks to Danielle Kuntz, research assistant and copy editor—the project would not have come to completion without her tremendous help; she also did much of the primary research for chapters 8 and 9. Thanks to Brian Schmidt, research assistant and copy editor, who was very helpful during the project's final stages.

Thanks to the readers of the manuscript for their perceptive comments and suggestions: Timothy Brennan, Michael Cherlin, Michael Denning, Eric Drott, Walter Everett, Robert Fink, Gabrielle Gopinath, Teresa Gowan, Anahid Kassabian, Henry Klumpenhouwer, George Lewis, Ralph Locke, John MacKay, Fred Maus, Martin Scherzinger, and the anonymous reviewers.

Many thanks to Doug Sery, Paul Bethge, Katie Helke Dokshina, and Pamela Quick at the MIT Press.

A thousand thanks to the many contributors of ideas, references, source materials, contacts, and help (and apologies to those unintentionally omitted): Matthew Arndt, Arved Ashby, Christopher Atkins, Michael Austin, Trevor Bača, Matthew Belinkie, Leslie Blasius, Mark Borrello, Chesa Boudin, Bill Bahng Boyer, Soniya Brar, Bruce Braun, Robert Brenner, Seth Brodsky, Stacey Burns, Jalylah Burrell, Sian Butcher, Lisa Campbell, Thomas Campbell, Amanda Ciafone, Juliette Cherbuliez, Paul Clabby, Susan Cook, James Currie, Scott Currie, Ursula Dalinghaus, Maria Damon, Stephen Dembski, Nandini Deo, Tracey Deutsch, David Dies, James Dillon, Hanna Dorn, Clare Eng, Michael Ethen, Kale Fajardo, Sujatha Fernandes, Nicolas Fogelholm, Michael Friedmann, Keya Ganguly, Michael Gaudio, Madhura Gopinath, Shamin Gopinath, Sudhir Gopinath, Claudia Gorbman, David Grayson, Matthew Hadley, Dan Hancox, William Helmcke, George Henderson, David Hesmondhalgh, Barry Hill, Karen Ho, Robert Hullot-Kentor, Brian Hyer, Christopher Isett, Roman Ivanovitch, Neelima Joshi, James Katz, Matt Katz, John Kim, Karen Kinoshita, Michael Klein, Colin Kloecker,

Charles Kronengold, Diyah Larasati, Emily Lechner, Richard Leppert, Henry Liu, Justin London, Silvia López, Tonglin Lu, Patrick Macey, Noriko Manabe, Lev Manovitch, Wayne Marshall, Joseph Matson, Shanai Matteson, Patrick McCreless, Matthew McDonald (especially for his Freudian reading of Crazy Frog), Jason McGrath (especially on Jia Zhangke), Dilip Menon, Carlos Miranda, Ali Momeni, August Nimtz, Mark Nowak, Mark Nye, Sean Nye, Joanna O'Connell (on the PQNTC phenomenon), Adam Olson, Dara Orenstein, Ian Maghinay Padios (on the "Hello Garci" scandal), A. Naomi Paik, Marc Perlman, Wayne Marshall, Peter Rachleff, Ronald Radano, Nora Renka, Steven Rings, Gil Rodman, Ron Rodman, Marynel Ryan, G. S. Sahota, Arun Saldanha, Justin Schell, Matt Schley, Anna Schultz, J. B. Shank, Tiffany Shelton, W. Anthony Sheppard, Robert Slifkin, Susannah Smith, Terri Smith, Hoon Song, Juliana Spahr, Michael Spencer, Gustavus Stadler, Tyler Staples, Jonathan Sterne, Aaron Stob, Jordan Stokes, Teresa Sutton, Sara Swain, Marion Traub-Werner, Pam Trudeau, Antonio Vázquez-Arroyo, Michael Veal, Ann Waltner, Holly Watkins, Alexander Weheliye, Dana Whitco, and Dag Yngvesson.

Thanks to students in my "Ringtone Culture" graduate seminars who read early drafts of some chapters, offered helpful comments, and provided a critical space to think through many of these ideas: Aaron Biren, Colin Holter, Jennifer Karon, David McCarthy, Ross McPherson, Kalani Michell, Lauren Shell, Ryan Thompson.

I am grateful to the University of Minnesota for a McKnight summer fellowship, Imagine Fund grants, one semester leave (through the College of Liberal Arts), and a semester residency at the Institute for Advanced Study. Many thanks to all academic departments/schools and institutions that gave this work a hearing and generously supported it; these include the University of Wisconsin School of Music; the Eastman School of Music; Macalester College; Carleton College; Northwestern University (Weinberg College Freshman Seminar and Department of Music); the Yale University Initiative on Race, Gender, and Globalization; the "New Directions in Musical Aesthetics" conference sponsored by the University of Texas at Austin and the Donald F. Harrington Foundation; the Walker Art Center's "Opening the Field" lecture series; and the Library Foundation of Hennepin County (People's University lecture series). Additionally, presentations on this work were given at the Society for Ethnomusicology, Working Class Studies Association, and Music and the Moving Image conferences, and in lecture series or classroom talks in the University of Minnesota Departments of Asian Languages and Literatures, Communication Studies, English, Geography, the School of Music, and the Markets in Time collaborative.

Thanks to *First Monday* and Edward Valauskas for publishing my earliest work on ringtones ("Ringtones, or, The Auditory Logic of Globalization," *First Monday* 10/12, December 2005, www.firstmonday.org).

Many thanks to all primary contacts, including composers/artists giving me permission to use their work, composers or performers sending recordings, writers on ringtones, and interviewees from the mobile entertainment industry: David Baker, Eric Bernatchez, Jennifer Bird, Andy Booth, Andrew Bud, Chia-Chun Chen, Julie Corwin, Nathan Davis, Henry Dillard, Billy Dixon, Yasmin Eid, Paul Freeman, Jordan Gray, Eric Gundersen, Evan Henshaw-Plath, Luke Holden, Michael Jo, Martin Kearns, Maneesh Kenia, Bernd Kremling, Golan Levin, Vincent Lo, Arne Lucas, Bill Luoma, Kevin Mullen, Alex Noppe, Keith Nowak, Vesa-Matti Paananen, Elena Pereira, Jocelyn Pook, Thomas Dolby Robertson, Brice Salek, Emily Turrettini, Aaron Vandermeer, Katrin Verclas, Alex Vlassopulos, Darren Vogel, Erik Wernquist, and Brad Zutaut.

Emily Turrettini's ringtonia.com blog was an indispensable source for this project.

Thanks to rights holders for granting permission to reproduce copyrighted work, including Chester Music, Edizioni Musicali Rai Trade, Vincent Lo, Nokia Corporation, Brice Salek, Theodore Presser, and They Might Be Giants.

Many thanks to my colleagues at the University of Minnesota College of Liberal Arts and School of Music, particularly those in the theory division: Matthew Bribitzer-Stull, Michael Cherlin, David Damschroder, and James Dillon.

Jason Stanyek has been an important collaborator and influence on my thinking during the writing of this book. Thanks to the contributors to our co-edited volume, the *Oxford Handbook of Mobile Music Studies*.

This project originated in 2004 in Michael Denning's Working Group on Globalization and Culture at Yale University, from a group presentation on the commodity chain of the mobile telephone. Thanks to all participants in my two years in the group, including Nikhil Anand, Jenny Carrillo, Amanda Ciafone, Rossen Djagalov, Andrew Friedman, Daniel Gilbert, Myra Jones-Taylor, Nazima Kadir, Christina Moon, A. Naomi Paik, Laura Trice, and Kevin Woods.

Many thanks to all participants in the other research groups informing this project, including the Working Group on Marxism and Cultural

Theory at Yale, and the University of Minnesota's Dialectics and Society Workshop and Music and Sound Studies Initiative. Thanks, too, to the tireless efforts of the Graduate Employees and Students Organization (GESO), the Federation of Hospital and University Employees (FHUE), and the Connecticut Center for a New Economy (CCNE)—all of which have influenced my work profoundly.

Many, many thanks to my family, friends, and loved ones, without whom this book would not have been written. Thanks to the past and present members of the Things Themselves, Quarterstance, the Gated Community, Saam & Dev, and Beth Hartman for making music with me during the writing of this book.

A very special thanks to Beth Hartman, who was closely involved with the project, offered many perceptive comments, ideas, and references, and read and critiqued my writings numerous times.

I would like to dedicate this book to my parents, Sudhir and Madhura Gopinath, for their love and guidance during and long before this project; I feel very fortunate to have been raised by the two people whom I admire most. I would also like to dedicate this book to Michael Denning and Hazel Carby, whose ideas, generous mentorship, and political commitment shaped its contents for the better.

A multimedia website including numerous examples discussed in the text is available at www.theringtonedialectic.com. Unless it has been changed, the password should be economyandculturalform. If you have questions, please contact me at sumanth.s.gopinath@gmail.com.

Introduction

The task of this book is, to cite Blake, to see a world in a grain of sand. Its eccentric, even perverse ambition is to examine the ringtone—the customizable, often musicalized ringer signal on mobile phones, typically several seconds long and first popularized in the late 1990s—as a means of understanding a spatial and temporal totality, a world-system and a global conjuncture. The relationship between part and whole, however, is not exactly one of microcosm and macrocosm, in which the minuscule element expresses and reflects the vast system in all its features; nor is the exercise to divine the future of the world from the present of the grain. Rather, for a period of time the ringtone itself *became* a world—a multi-billion-dollar micro-industrial economy and a novel cultural form—that both mimicked and shaped much of the world writ large; thus, by grasping the history of the former, one captures an historical semblance of the latter. Moreover, the ringtone is fundamentally a *digital* commodity and as such is a part—a central part, I argue—of the momentous transformations in the music industry and the consumer's experience of music resulting from the digitization of sound and from its articulation with mobile devices.[1] Blake's sand in "Auguries of Innocence," it would seem, was refined and processed into a silicon wafer, the substrate for a microchip in a now long-discarded flip phone.

It is fair to say that the world conjured by the ringtone has now ended, with less a bang than a whimper.[2] About ten years ago, in many of the world's metropoles and other urban centers, the ringtone seemed to be ubiquitous, the tinkling (and often individualized) treble frequencies of beeping mobile phones differentiating the soundscape of that moment from the then-recent past. The creation of a worldwide fad and opportunity for booming profits was predicated on the thinnest of offerings when compared to other new content available on the market. In the words of Andrew Bud, co-founder and executive chairman of the mobile billing firm mBlox (and Global Chair of the Mobile Entertainment Forum), the essential problem of the ringtone could be summed up in the question "How much can you charge people for things that are more or less worthless?"[3] Indeed. The ringtone was typically an expensive digital commodity and played a formative role in producing the mobile entertainment business, which Bud himself assessed as being worth $25 billion worldwide in mid 2008 (roughly at the peak of ringtone sales in the largest markets) and is said to have been worth $33 billion by 2011.[4]

The centrality of the ringtone to the formation of the mobile entertainment business, including the technological forms and labor practices particular to its production, warrants the description of an entire system of production: a global ringtone *industry*.[5] In this book I view the development of that industry rather critically, but my goal, rather than to produce an unnecessary polemic against a waning entity, is to examine the material logics and marketing strategies of the industry and to study its effects on cultural production in a variety of aesthetic and geographical contexts. I attempt to both elucidate and circumvent one of the primary marketing ideologies that governed it: *personalization*. Discussed in greater depth in chapter 1, marketing personalization essentially means that advertising and exchange should be customized to the individual consumer, producing an idealized marketplace for that consumer. Although such a marketplace is difficult to produce in physical space, it is far more feasible in the digital portals of the wired world, with recommendation algorithms, collaborative filtering, and, more recently, shifting hierarchies on search engines helping to tailor the online experience to individual users—and with online marketers hoping to maximize sales profits and to commodify attention spans.[6] As recommendations become more accurate, the benefits of consumer customization are apparent to sellers and, perhaps, to buyers. But there are dangers. In addition to the extraordinary growth of the consumer surveillance trade, Eli Pariser has cogently argued that personalization produces "filter bubbles"—i.e., that one's access to digital media confirms one's prejudices and interests, producing a kind of "autopropaganda" and potentially preventing the development of collective civic responses to large-scale societal problems.[7] The ringtone industry, however, goes one discursive step further: it makes personalization the voluntary task of an individual, and transforms a marketing rubric into language about customizing one's own commodities. Although commodity customization has a long history, one that poses consumer practices against the interests of sellers and marketers (and authorities representing them), here personalization is encouraged by sellers themselves as a means for a consumer to express his or her identity, in the process mystifying the broader marketing agenda at work.[8] Aiding and abetting that agenda is the mass of journalistic prose dedicated to demonstrating connections between one's ringtone and one's personality, crystallized in the discursive trope that Meghan Daum characterizes as "you are your ringtone."[9] Suspicion of this trope is warranted: your ringtone, or lack thereof, may say much *less* about you than you have been led to think.

The logic of personalization may have affected the priorities of ringtone scholarship, which has typically employed surveys, interviews, or other forms of sociological or ethnographic methodology as its primary means of examining how consumers use ringtones in their everyday lives, sometimes with an eye toward businesses' strategies for attracting consumers.[10] This work is consistently good and sometimes excellent, and did not seem in urgent need of replication within the geographic sites to which I had access. Like any scholarship, it has its limits; for example, it tends to focus on particular scenes of consumption and use (the choosing and sounding of ringtones in everyday contexts), and, while offering variable depictions of ringtones in action, it tends to privilege analysis at the level of the individual.[11] Sometimes broader societal claims are made; the most provocative of these can be found in the work of the French sociologist Christian Licoppe. Licoppe's basic argument, beyond the familiar demonstration of varieties of individual ringtone use and expressions of identity, is that there is an almost antinomial relationship between the ringtone as a signal (a summons that demands an answer) and the ringtone as a musical "treat" that offers pleasures of various kinds to the phone user. The ringtone offers owners of mobile phones a compensation of pleasure within the labor regime of contemporary capitalism that demands perpetual availability.[12] Although one could argue that this line of reasoning is based in social critique, the picture of society here is a largely functional and normative one, attuned to the maintenance of social harmony and characteristic of certain (functionalist and consensus-based) strains of academic sociology.[13] However, one could examine ringtones and related phenomena with greater and more overt societal conflict in mind; work by Wayne Marshall and Dan Hancox on the use of mobile phones as new, trebly, blaring quasi-boomboxes (called "sodcasting" in the United Kingdom) highlights conflicts of class, race, and age—particularly the conjunction of working-class, black, and "chav" youth cultures—and brings to mind classic cultural studies approaches to subcultures.[14]

It may be the case, however, that views of society in conflict and accord—or, more specifically, societies as comprising clashes between oppositional groups (some dominant, others subaltern) or relatively stable functional wholes—may come to some form of resolution or synthesis. Consider: in late 2005, a British inventor and entrepreneur, Howard Stapleton, designed a youth-repellent device called "the Mosquito," a high-frequency (17.4 kilohertz) noise device that emits sounds that are unpleasant to young people but are increasingly inaudible with age (as a result of normal hearing loss). First sold and tested in Wales, the device became increasingly popular among shop owners as a way to discourage loitering by young people, especially gang members. Taking advantage of the disparities in hearing between adults and youths, the tone was adapted for more limited phone frequency ranges and adopted as a ringtone by teenagers, who exchanged it via Bluetooth. Soon the ringtone could be heard (and go unheard) in Welsh classrooms. It quickly spread within and then beyond the British Isles. Ringtone companies got into the act, marketing versions of the tone as the "Mosquito Ringtone" or "Teen Buzz." Stapleton's own company, Compound Security, began to produce Mosquito-related ringtones and paraphernalia. The sound made its way into a film (*The Messengers*, 2007), a commercial for Kentucky Fried Chicken, and a Compound Security-sponsored dance track ("Buzzin'"). Complaining teachers eventually prevailed on a ringtone aggregator called Phat Tonez to stop selling its version of the sound.[15]

In short, a device intended to produce social stability was appropriated by youths and transformed into an oppositional, subcultural ruse. The valences of these terms can be easily flipped: one can say that the Mosquito disrupted youth congregation, and that the ringtone facilitated youth communication—making it clear that the previous position was taken from an adult perspective. But *other* adults appropriated the Teen Buzz ringtone, commodifying it in numerous ways and reintegrating it into normative society while not ameliorating and even exacerbating the conflicts it spawned, as demonstrated by the teachers' protests. Moreover, as Louis Menand viewed it, the episode provided a pretext for a meditation on adult health (through discussions of age-related hearing loss, or presbycusis), drawing attention away from the youth demographic initially spotlighted in the first place; even claims that the Mosquito is potentially harmful to one's hearing (including children's), instead of merely being a metric of hearing ability alone, reveal the concerns of adults (as parents and/or societal caretakers) rather than children themselves.[16] Geographic dispersion was also at work: the phenomenon originated in Wales, then spread around the world through adult-run English-language media, especially on the Internet. Hence, a phenomenon initially understood as a self-evident opposition between youths and adults can be interpreted as a conflict between different segments of adult populations divided along socioeconomic, professional, and geographical lines. Whether seeking to impose disciplinary control on youths or to construct lucrative youth-product markets, these adults participated in a contest of meaning over the definition of "youth"—as subversive or consumer—in accordance with their specific material interests.

What is at stake here has little to do with individual commodity personalization and identity and much more with a larger collective (although potentially identitarian) dynamic of age and class, one that quickly transformed into a media phenomenon and elicited a significant amount of commentary. An ethnographic study would have been useful in teasing out a wealth of details and information as to, for example, how youth actually used the Mosquito Tone in classrooms. But the phenomenon is now "over," and what remains are the memories of participants (who could be interviewed, even if participant observation would be impossible) and the information on the event as covered in a wide array of mass media, the latter of which is sufficient to generate a complex story needing interpretation and would not be likely to change in its broadest outlines after closer investigation. In addition, the method of interpretation illustrated above focuses on the large tendencies and processes as represented in media narratives of the Mosquito Tone. In principle, these tendencies could be dealt with statistically, although data on the phenomenon, outside of a few numbers (such as Stapleton's interesting claim that by May 2006 Compound Security had sold nearly 1,000 of his original Mosquito devices), are not readily available.[17] Instead, in recounting the story above the analysis focuses principally on qualitative movements and developments (the need for which, of course, statistics do not obviate), as well as the interpretive frames used to grasp them. Particularly when we take note of the changing of places and values of the oppositions child/adult, harmony/conflict, and disability/health, all of which point to the more general dynamics of bodies and age cohorts intersecting with a new technological form, this process suggests, in miniature and perhaps only haltingly, that a compelling way to understand such large movements is through the application of dialectical thought—one of the principal perspectives informing this book.[18]

Mention of the word "dialectics" may seem outdated in a study of a new technology and a form of new, digital media. Although a commonsense and not incorrect notion of the term, involving a contradictory unity of opposites as delineated especially by Hegel and his successors, has embedded itself in humanistic and social-scientific study, other reference points, metaphors, and methods are ascendant. In fact, we are said to occupy a period "after dialectics" whose origins have multiple points of determination—the emergence of language-based philosophies of structuralism and poststructuralism in the postwar period, the massive transformations in computation speed and technology in the past few decades, the series of defeats of the left since the 1960s up to and after the end of the Cold War and, with it, the disappearance of most examples of "actually existing socialism"—that have variously led to the premature discrediting of dialectical thinking and of Marxist thought.[19] In its stead, the sociologist Göran Therborn argues, a number of theoretical responses on the left have emerged, one of which involves a discourse of *networks* that is still in its intellectual "honeymoon" phase.[20] This discourse ranges from reticular metaphors to full-blown systems of analysis—the latter including various forms of quantitative network analysis, the "network society" theory of Manuel Castells, and the writings of Bruno Latour, John Law, and other figures in Science and Technology Studies.[21] At its most extreme, network discourse produces theoretical equivalences for the maximal production of difference, following a general trend favoring difference over similarity in the postmodern period, but its variable application makes it difficult to generalize about. But despite an apparent favoring of the network over dialectics in recent years, including in new media studies, engagements with the dialectic are by no means absent from research on digital media, perhaps the best known being Peter Lunenfeld's edited volume *The Digital Dialectic*, which invokes dialectics for the study of new media in part as a way of circumventing a "'science-fictionalized' discourse" not grounded by materiality.[22] However, dialectics may pervade scholarship on new media (and even network-based studies) while going unnamed. As Fredric Jameson argues, perhaps "something like the dialectic will always begin to appear when thinking approaches the dilemma of incommensurability, in whatever form; and that the dialectic henceforth seems to be the shift of thinking on to a new and unaccustomed plane in an effort to deal with the fact of distinct and autonomous realities that seem to offer no contact with each other."[23]

The most important dialectic of incommensurables in this book is the canonical relationship in Marxist theory between base and superstructure, often dismissed as undialectical.[24] In contrast with many cultural studies critical of this paradigm, the relationship between base and superstructure—interpreted here as *economy* and *cultural form*, respectively—defines the basic plan of the book, which begins with a study of the economy of the ringtone industry and then examines the cultural effects of that industry's rise and fall.[25] In doing so, the project takes seriously Raymond Williams' famous critique of this relationship, which sought to produce a "more active, more complicated and more contradictory" and ultimately processual understanding of the base, even if my treatment of the latter still focuses on economic and technological activities and does not encompass the entirety of the working life and the reproduction of the laborer as Williams recommended.[26] One way to better understand base-like processes would be to restrict the primary scale of analysis in duration and scope (down from "the economy" to sectors, industries, and individual firms). In doing so, a more concrete picture of the relationship between economy and culture materializes, involving the *components* of the latter that originate in the former—with the mobile phone handset and ringtone as found in musical compositions (as in chapter 4) being rather obvious examples of cultural components.[27] By focusing on a micro-industry over a roughly conjunctural duration, causal relationships between culture and economy may become relatively transparent, even "mechanical" in Jameson's sense.[28] Moreover, such a restricted focus does not avoid (and may exacerbate) risking the interrelated sins of economism and technological determinism,[29] for the direction of causality here still follows the traditional model—the dominant story of the base, in all its complexities, is one that determines the cultural forms that could be said to express it[30]—even if it is implicit that culture, writ large, dialectically makes the activities of the base possible in the first place. But to the extent that the direction from base to superstructure is privileged in what follows, this has to do with the specific selection of cultural objects, whose relationship to the ringtone industry is largely one of market-based heteronomy.[31] One aspect of the ringtone's industry's demise, we might speculate, was its very inability to become a medium of cultural production that gained enough economic success to produce a significant measure of autonomy—to the extent that such a process is at all possible in the age of neoliberalism. Even if some ringtone producers generated creative practices and forms immanent to mobile telephony itself (see chapter 7) and hence partly escaped the orbit of novelty production that characterizes the emergence of many forms of media, this failure may turn out to be of far greater significance than the economic decline of the ringtone industry itself.[32]

An accurate account of the base, at least where the ringtone is concerned, is a complicated affair, and necessitates an understanding of four different economic layers of increasing scale and value (the first of which is the primary focus of chapter 1):

- the "ringtone industry," or the intersecting economic interests and systems of production producing for-profit ringtones, often but not exclusively by mobile entertainment firms
- the mobile entertainment industry, or the firms and interrelated interests producing entertainment and customization products for mobile phones, including text messaging, various forms of ringtones, wallpaper, video games, "mobisodes" and other moving image forms for mobiles, and, most recently, apps (although a broader understanding of mobile entertainment views it as "any leisure activity" conducted via mobile, networked devices[33])
- the entertainment and telecommunications industries, the intersection of which produces mobile entertainment but whose priorities extend far beyond it
- the broader dynamics of the global capitalist economy, as typically expressed in national and regional markets and their interrelations

In order to make sense of these nested economic layers, I have drawn upon the work of Marxist political economists and world-systems theorists to interpret my less systematic findings and observations of economic trends in the ringtone industry. For example, from Robert Brenner's work I take both the primary emphasis on the three most powerful economic regions of the global economy—North America, Europe, and East Asia—and Brenner's observations on the speculation bubbles of the 1990s (tech) and the 2000s (assets) in relation to shifts in currency rates and export strength that affected the relative economic vibrancy of Europe relative to North America in the early 2000s; the ringtone's popularity in Europe may be symptomatic of these differences.[34] Of course, such explanations only go so far; consider that Japan's recession of the 1990s and the early 2000s coincided with the initial development of ringtones. Far more important was the role of the global economic crisis of 2008, which may have delivered the death blow to the ringtone's economic viability in many parts of the world. Yet of equal if not greater significance was the spectacular advancement of mobile telephony worldwide during the 1990s and the 2000s, which was undoubtedly the great success story of neoliberalism and which has inspired a formidable body of research.[35]

Despite the logical priority accorded to things economic, though, this study is ultimately one of cultural forms and expressions. The creation of ringtones was a cultural (and especially a musical) practice, and ringtones made their way into numerous media and genres, including advertisements, contemporary classical compositions, art installations and performances, popular songs, political movements, films, and television shows. Thus, a secondary dialectic is often at work in this book: that of hermeneutics or interpretation, in which subjective and objective understandings of cultural texts or objects are resolved, temporarily, in close readings. This characteristically humanist dialectic is quite distinct from the more sociological base-and-superstructure dialectic, although I hope the two dialectics work together productively if not always in perfect consonance.

Two emphases that were developed in the course of this study deserve mention. First, a great deal of attention was paid to issues of artistic labor, both in the ways that laborers actually make cultural forms (including ringtones) and, to a lesser extent, the ways that laborers are represented vis-à-vis their ringtones. I am inspired by Michael Denning's argument for a "labor theory of culture," which involves paying attention to the stratified labor forms of cultural production, avoiding reductive political readings of cultural forms, appreciating labor's potential as a powerful counterpoint to the possible mystifications of the aesthetic, and examining the cultural lives of laborers themselves—in short, a wide-ranging examination of the many possible meeting points of culture (especially as arts and leisure) and the work it requires and sustains.[36] Second, numerous types of cultural production were examined in this study, as mentioned above; but insofar as the ringtone is, at its base, a sonic form, there is a primacy given to sonic and musical materials that either directly register the impact of ringtones or supply the content of particular ringtones themselves. Thus, although numerous primary sources were employed in the course of this study, including news reports, interviews (many conducted by me), a small number of firsthand observations (especially at industry conferences), blogs, online message boards, videos, and images, the bulk of my attention was focused on musical recordings. In some cases I cite musical scores, but in many cases I provide my own transcriptions in staff notation as representations of my musical observations. Although I consulted a few music based precedents and examples (Karen Collins' study of video game music was a helpful model[37]), for the most part I rely on ad hoc and contextually specific ways of coming to terms with ringtone music and sound—ways that stem from my training as a music scholar.

The book has four parts. Part I (chapter 1) deals with the technical and economic structure of what I call the global ringtone industry, although in it I focus primarily on the United States and Europe, with some attention paid to Japan. In addition to providing historical background and ideological critiques of industry rhetoric, I discuss the way ringtones transformed from monophonic, single-line synthesizer files to polyphonic MIDI files to digital sound files and explain a concomitant change in the nature of capital and rent accumulation within the industry. Part II focuses on sociocultural practices that seemed to decline in the course of the ringtone's history as a result of these shifts: the craft of producing monophonic and especially polyphonic ringtones (chapter 2); the forms of notation and representation that accompanied early ringtone production, including their not unrelated favoring of Western classical music (chapter 3); and the creation of new classical and experimental compositions and art installations and performances using or quoting ringtones (chapter 4). Part III examines the dialectical reversals resulting from the appearance of the sound file ringtone, including the Crazy Frog fad (chapter 5), the development of new forms of political ringtones (chapter 6), and the growth and maturation of independent original ringtone composition (chapter 7). Part IV addresses two cases that do not fit this pattern of decline or reversal—instead, they span roughly the entirety of the ringtone's appearance—and that could be said to feature a particularization of what are claimed to be more "global" processes in the previous chapters, in the sense that they address English-language films and television shows (chapter 8) and popular music (chapter 9). Chapter 8 also contains an extensive treatment of the films of the Chinese director Jia Zhangke, whose extraordinary attention to ringtones and related sonic forms offers a striking counterpoint to the tendencies discussed in the rest of that chapter. Chapter 9 includes speculations about race and class in ringtone consumption, suggesting that African Americans in the United States and perhaps working-class whites in the United Kingdom contributed disproportionately to the ringtone industry's profits. The epilogue addresses the ringtone's relation to the notion of "mobile music," offers an interpretation of the ringtone phenomenon at the level of the conjuncture, and concludes with thoughts on the ringtone's relationship to the emergence of cloud computing.

The patterns of decline, reversal, and particularization charted in the following chapters define what I mean by "the ringtone dialectic"—a series of dialectical processes specific to the history of the ringtone's emergence, popularity, normalization, and waning, and involving content repurposing and cycles of nostalgic revival, the development of new aesthetic capacities, and harmonizations with specific ethnolinguistic communities and national or regional cultural practices. That this history took place over an economically and even politically coherent period, roughly 1998–2010, allows me to claim that the ringtone defined an entire international conjuncture, which I call "the ringtone conjuncture." The lineaments of that conjuncture encompass a major global economic crisis punctuating a longer, systemic downturn, minor shifts in state power within a persistent neoliberal frame, ever greater and naturalized digital surveillance of and voluntary labor by the consumer-citizen,[38] and steps toward more thoroughgoing transformations in the global tripolar economy involving a possible new weighting of its poles—features that are already familiar to most observers. The trick, however, is to *hear* the conjuncture by listening to a carefully curated playlist of ringtones both popular and obscure.

I The Rise and Fall of the Ringtone Economy

1 This Business of Ringtones: The Unstable Value Chain and Accumulation of Capital by Rent in the Global Ringtone Industry

"Let me be the first to say that the iPhone is [a] huge problem for the ringtone industry. In many ways it makes the ringtone industry completely obsolete. . . . In 3–5 years just about all of the newest phones will have easy syncing with your music library and the paid ringtone market will shrivel up and die."[1] This statement, made by the businessman blogger Blackbeard SEO, was picked up by ABC World News on January 10, 2007 and elicited more than 200 Web-posted responses within a few days. Several respondents countered that many mobile phones already allow the use of MP3 files as ringtones, and that Steve Jobs, chair of Apple Inc., initially decided that the iPhone would not allow easy conversion of iTunes selections into ringtones.[2] But the larger point of the blog—that the ringtone industry would not be sustainable in the medium term—was less controversial. Indeed, that this statement would be cited by a major news organ suggests, beyond the growing importance of the blogosphere in the United States, that the ringtone industry and fears about its decline were big news during the height of its success.

At first glance, the uproar seems entirely out of proportion to the trivial nature of ringtones—the purchasable musical/sonic fragments supplanting the functional ringer on mobile telephones. But in light of the international economic and cultural impact of the ringtone, it is rather fitting. A multi-billion-dollar business and the former centerpiece of mobile-entertainment sales, the *global ringtone industry* comprises a complex of small, specialized firms, multinational corporations crossing several sectors, and state-regulatory apparatuses that combine to service and stoke consumer demand for ringtones. The purpose of this chapter is to describe that industry by providing a history of its development and discussing its economic structure. In the historical narrative, technological advances—the changes from monophonic and then polyphonic synthesizer instruction files to digital sound files—take center stage, the ringtone developing in sonic complexity and fidelity each time it changes garb. These changes were made possible by the now-familiar world-historical transformation of the mobile phone handset—increasingly miniaturized, with greater processing power, memory integration, storage capacity, and uploadable software, to the point of becoming a pocket microcomputer—and data delivery systems across the global telecommunications network (especially the replacement of dial-up connections with broadband networks that soon merged with digitized telephony). Because it has already been explained by others, I do not fully recount this larger story below, although I allude to it often—for example, in my brief discussion of the smartphone or when invoking the telecommunications industry language of phone generations (1G, 2G, 3G, and so on).[3]

In my discussion of the industry's economic structure, the most attention will be paid to its byzantine scheme of capital accumulation, whereby numerous claims on the profit generated by a ringtone bespeak the presence of a complex set of interests and forces. Perhaps most important, the industry consists of an unstable mix of (sub-)sectoral entities whose efforts to forestall or mitigate destructive intersectoral competition held the industry together tenuously for some time before slowly unraveling. Specifically, the music industry's interest in repurposing its catalog to accumulate capital by *rent* meant pressuring the mobile phone handset industry to upgrade its file formats.[4] Both interests, and numerous others, collaborated on a project of "convergence," with content holdings accessible by networked devices, initially producing a limited-access "walled garden" of mobile entertainment in lieu of a fantasized "celestial jukebox." The project was marketed to consumers through the language of "personalization," with the fluidity of instant customization tethered ideologically to postmodern, consumerist constructions of identity. But as convergence progressed, phones would replicate file-sharing practices rampant on personal computers, making ringtones a part of "free" culture. This fundamental contradiction in the ringtone industry—between content and hardware—led to the demise of the ringtone as a leading-edge entertainment commodity, but it was by no means the sole factor in this development. The conjunction of misleading marketing campaigns (see chapter 5), the short half-life of fads (see the introduction to part II), the global recession of 2008, and the emergence of new mobile/digital commodity markets for the smartphone (particularly for the application, or "app") all contributed to the decline of the ringtone market.

The characterization of some set of practices and/or institutions as an "industry" is often not a neutral act. It has become a commonplace within academic and even journalistic prose to append the word "industry" to a noun that names the object of inquiry or criticism. Hence, industries of the most unexpected types are flourishing everywhere—for example, industries devoted to particular thinkers ("the Bourdieu industry," "the Deleuze industry") or even forms of thought itself ("the theory industry"), industries attached to perpetuating particular ideologies ("the Holocaust industry," "the 9/11 industry") and institutions ("the detox industry," "the think tank industry"), and even industries for academic degrees ("the PhD industry," "the MFA industry").[5] Uses of this construction inevitably hark back to Adorno and Horkheimer's polemic against the "Culture Industry" in *Dialectic of Enlightenment*, and claims about each of these industries would seem to confirm the "absolute power of capitalism," whose very pervasiveness *industrializes* all forms of human behavior.[6] Unlike the ostensibly neutral description of a manufacturing or service industry—even when, as in the cases of "the sex industry" and "the prison industry," the labels rub somewhat uncomfortably against expectations of what an industry should be—the term's metaphorical use encapsulates a critique by linking the object of study with the broader, determinate forces of the capitalist world economy and the capitalist mode of production. The industry metaphor, however, does more than simply delegitimize through guilt by association; it theorizes an entire system, of whatever size, of production and consumption that coordinates ideologies with material resources, sustains livelihoods, and wields political influence. In its very invisibility, such an industry gains power, and it is in naming and elucidating this system that we begin to comprehend its reality and devise strategies for addressing and transforming it. The industry metaphor is therefore particularly apt as an analytic and a rhetorical device, and it has produced powerful insights into a wide range of phenomena. Although my understanding of the global ringtone industry is more literal than some of the foregoing examples (I describe an actual set of production practices linked together in a global value chain), I hope that the following analysis retains their critical inflection. For the ringtone industry is, in the last instance, an elaborate smoke-and-mirrors game—a mere repackaging of existing content whose heyday has come to an end.

How did the ringtone transform from a functional component on a mobile phone into a device for playing back music? Understanding this transformation is no easy task, insofar as it crosses many ethnolinguistic boundaries—especially those between East Asia (especially Japan and Korea), central and northern Europe (especially Scandinavia), and English-speaking regions and international business cultures—and is situated in the context of competitive claims over industrial pre-eminence, representing in miniature the strategic positioning between the three dominant economic zones in the world economy: North America, Europe, and the Pacific Rim. What follows, then, is a provisional, somewhat speculative attempt at such a history.

Precursors to the Cell Phone Ringtone

The recent historical origins of the ringtone are found in the application of microelectronics technology to the land-line telephone. In the early 1980s, press releases in the United States and elsewhere announced the appearance of variable telephone ringer signals or bells, sometimes called "ringer tones," "ringing tones," or "ring tones."[7] In June 1981, for example, Bell Laboratories engineers patented an integrated circuit (IC) that facilitated two different kinds of ringer signals. "The integrated circuit," one report noted at the time, "has already been built into about 130,000 of the electronic telephones trademarked Touch-a-Matic."[8] Neo-Cold War competition with the Soviet Union extended to the domain of telephone technology, and in 1982, the British Broadcasting Corporation reported that the Lenin VEF Works in Riga had produced new phones that, in addition to having number memory and automatic redialing, included an electronic ringer signal.[9] By the mid 1970s, such devices had begun replacing the electromechanical "gong ringer" system in Europe, in which "a solenoid coil with a clapper . . . strikes a single or double bell."[10] Aside from the fact that it facilitated the possibility of customized ringer signals (in the United States long the province of electronics tinkers and experts[11]), the change had two economic motivations. One was that piezoelectric or loudspeaker signals from an IC used a much smaller percentage of the 40–150-volt signal (usually 20-hertz alternating current) sent via the phone jack that powered the ringer signal and was paid for by the local branch of the Bell System monopoly than a digital ringer powered by batteries or power lines at consumers' expense.[12] The second economic motivation was that the new technology represented the most current sort of telephonic modernity—sleeker, lighter, smaller, digital sounding—and thereby encouraged consumers to replace their newly obsolete telephones. Building on earlier developments in which telephony moved from electromechanical to electronic to microelectronic and digital switching and signaling technologies (most famously in the rise and refinement of "touch tone" dialing), the digital conquest of the telephone was part of a broader history of the profitable making and remaking of household and personal devices through

IC design that began in the 1970s.[13] Not surprisingly, most of these new developments were initially targeted at upper-income and business users (often the same people), who would be crucial consumer groups in the early days of cellular telephony (which more or less coincided with the land-line telephone's assimilation of the integrated circuit).

Most familiar to contemporary consumers in the form of little rectangular and square "chips" in computers and other devices, integrated circuits are essentially small electrical circuits etched directly into a semiconductor material such as silicon. The compact size of these circuits makes them far more efficient and cost-effective than traditional, multi-component circuits, and they quickly became essential to digital computation systems being used in a wide range of contexts. Although the IC was invented and designed in the United States—in part by Jack Kilby at Texas Instruments and Robert Noyce at Fairchild Semiconductor in 1959 and 1961, respectively, and even earlier by others[14]—device development quickly spread to companies in other countries, most notably in Japan and subsequently Western Europe. In the case of Japan, the IC appeared quite early on—for example, the Nippon Electric Company (NEC) claims to have begun developing them as early as 1960 and began digitizing the national telecommunications system within the decade.[15] Research and development in the microelectronics field merely constituted the next stage in the transistor-based electronics industry in postwar Japan, but the speed with which Japanese companies could competitively enter the field hinged upon the close relations between the state, manufacturing corporations, and banks that extended back to the Meiji Restoration modernization program.[16] Production of integrated circuits advanced greatly after the 1971 protectionist "Nixon shocks" of increased import restrictions and dollar depreciation and the 1973 OPEC oil embargo, both of which severely hampered the economies of the United States and Japan. Many companies, including NEC, Hitachi, Mitsubishi Electric, Sony, Seiko, Sharp, Casio, Toshiba, and Fujitsu, sought more markets for exports, primarily in Europe, revised their operations to incorporate new quality-checking measures, and greatly expanded their research and development programs in lighter industries such as semiconductor devices, calculators, digital watches, computers, satellites, fiber optics, home electronics, and mobile telephony.[17] Indeed, by the end of the 1970s the state of integrated-circuit production in Japan was so advanced that alarmist reports from the United States warned that, despite a global IC shortage that necessitated more overall production, Japanese competition—artificially enhanced by "dumping," or underselling—would have a deleterious effect on the American semiconductor industry.[18]

The IC market in Japan was responsible for one of the most important precursor inventions for the ringtone: the digital music synthesizer chip. Even more than the IC, the synthesizer chip was a transnational innovation. The Stanford University electroacoustic composer John Chowning developed the process of frequency-modulation (FM) synthesis in the 1960s and the early 1970s, patenting the technique in 1971. Frequency modulation, in which one signal varies the frequency of another signal (a simple example would be the rate and depth of vibrato) creates complex harmonic spectra, and hence timbres, by very simple means. The "extreme economy" of the system made it easy to implement, and engineers at Yamaha began working with Chowning's technique for use in a digital synthesizer.[19] For its part, Yamaha spearheaded the development of the IC-based music synthesizer, developing its own IC-production facilities and helping to advance the use of Very-Large-Scale Integration (VLSI) chips in musical instruments. The culmination of this research and development in combination with the newly developed Musical Instrument Digital Interface (MIDI) standard was the landmark DX-7 synthesizer, which first appeared in 1983. More than 200,000 DX-7 keyboards were sold in only a few years, and the instrument was ubiquitous in Anglo-American popular music of the mid 1980s—the song "Axel F" (1984) by Harold Faltermeyer, included in the soundtrack of the film *Beverly Hills Cop*, is one example.[20] Around the same time, the global musical instrument industry pursued development of digital keyboard samplers, the first of which was the Australian company Fairlight's CMI (Computer Music Instrument) from 1979 and which became mass-market items thanks to Akai, Korg, Roland, and other Japanese instrument companies.[21] Both digital synthesizers and samplers would be incorporated into various products such as video game consoles and PC sound cards, creating the kind of media "convergence" that would be crucial to the development of the ringtone.[22]

Meanwhile, a smaller and more modest type of music circuit was making its way into consumer devices in the late 1970s and the early 1980s. Instead of producing complex forms of sound synthesis, it merely sounded a simple single-oscillator signal. Such circuits combined a simple memory circuit storing pitch and duration data for melodies, pitch frequency and duration frequency divider circuits that would produce pitches and durations in accordance with the stored melodies, an address counter to store memory addresses for the melodies, and a speaker (usually a small piezoelectric buzzer) to convert electric signals into sound signals.[23] Various devices used this technology from the 1970s on, including alarm clocks and electronic phones. A portable precursor of the ringtone was the digital alarm watch, which, beginning around 1979, could play musical melodies of various sorts (folk songs, classical melodies, and sometimes popular songs still under copyright—perhaps used without permission). In 1980 the Japanese calculator firm Casio released the M12 and M321, perhaps the earliest well-known digital alarm watches.[24]

But the device that perhaps most directly mediates the musicalized cell phone and the twin forces of digitized global telephony and Japanese integrated-circuit manufacturing is the pocket pager or beeper. The pager was first developed in the immediate postwar years in the United States and was commercialized in the late 1950s. "Pager" was the name given to a device made by Motorola in 1959, "a small receiver that delivered a radio message individually to those carrying the device."[25] Pager technologies developed during the 1960s, and Nippon Telegraph and Telephone (NTT) introduced the pager in Japan in 1968. Pager use increased in the US in the 1970s with the introduction of the Motorola Pageboy I in 1974, and the US commanded a large share of the 3.2 million pagers in use globally in 1980.[26] In contrast, use increased in Japan primarily only after the NTT monopoly-controlled pager market was deregulated in 1987. But this economic shift set off a boom in pager use in Japan.[27] In the 1980s and the early 1990s, the Japanese pager—once a rather anonymous device used in business contexts and shared among groups of individuals—increasingly became a highly popular fashion device open to "personalization," especially among youth. Pager manufacturers added new features to pagers, including melodic ringers.[28] As early as 1988, a group of twenty companies entering the liberalized pager market mentioned that new pagers would include "a musical alarm instead of a beep." Soon a significant subset of pagers would play brief melodies chosen from preset options.[29]

Historical Development of the Ringtone Industry

The foregoing sketch would suggest that the ringtone, at some level of mediation, emerged from Japan. In various ways this appears to be true. But there are two conflicting accounts of the ringtone's development, each tending to the status of myth and each having ramifications for national-economic identity in the global marketplace. One narrative traces the ringtone's roots to Japan and the NTT subsidiary DoCoMo ("Do Communications over the Mobile Network"); another traces the ringtone's emergence to the Nokia Corporation in Finland. The first narrative identifies the pager as the source and direct inspiration for the ringtone, which appeared first in Japan in May 1996, when DoCoMo made preset ringtones available on handsets.[30] Shortly afterward, in September, IDO introduced the possibility of entering original ringtones through the handset keypad. But it was with the release of the *Keitai Chaku Mero Doremi Book* (*Cellphone Ringtone Do Re Mi Book*) in December 1997 that the ringtone phenomenon began to take off. With 3.5 million copies sold, the *Keitai Chaku Mero Doremi Book* was the ninth-best-selling book in Japan in 1998. Meanwhile, in 1997 Astel Tokyo had begun selling ringtones commercially for download, and in 1998 J-Phone followed suit.[31] An additional connection between the Japanese pager and mobile phone is the early existence of the vibrate mode, which was introduced in NTT pagers during the mid 1980s.[32] Ringtone sales, both legal and illegal, spread quickly within the East Asian market. As early as 1998, small phone shops in Hong Kong were selling pirated ringtones, "often charging [US]$10 for a 15-second ring."[33]

A competing narrative—one more prevalent in the Western media—claims that the Finnish corporation Nokia first developed the ringtone. In 1865 Nokia was a wood-pulp mill. By the early twentieth century, the name Nokia was used for a vertically integrated foresting concern, and also for the Finnish Rubber Works and the Finnish Cable Works, which produced telegraph and telephone cables. Characteristic of the cartelization of Finnish companies in the wake of World War I and the Finnish Civil War, the three industrial operations existed jointly under the control of the Rubber Works starting in the 1920s.[34] By 1967, the three firms had merged into the Nokia Corporation, with the new conglomerate expanding into many industries, particularly electronics and telecommunications. After the collapse of the Soviet Union and the massive recession that followed, Nokia took off as a global corporation, playing a major role in the widespread use of mobile telephones—particularly in Europe with Finland's accession to the European Union in 1995.[35] In contrast to the deregulation of state-supervised corporate telecommunications monopolies that took place with the Bell System in the United States and NTT in Japan, Finland's small telecommunications system had long been run by smaller companies, an outgrowth of the local telecommunications cooperatives of the late nineteenth century.[36]

The global rise of Nokia in the field of mobile telephony essentially coincided with the development of preset ringtones in that company's handsets. As early as 1994, phones in the 2100 series featured a number of preset ringtones, including one titled "Grande Valse," which would later become known as the Nokia Tune.[37] In March 1996, press releases appeared for the Nokia 8110 phone, which was to be released that summer and fall and among whose many features were "16 ringing tones," including "a few bars of well-known tunes."[38] But ringtone-wise, the crucial phone for Nokia was the 6110, which was announced in a press release from November 1997 and available in January 1998.[39] In particular, the development of phone modification and customization was advanced by this phone's ability to assign different ringtones to different callers.[40] According to one writer, the ringtone's first appearance in a Nokia product was accidental:

> The personal ring tone dates back to the early 1990s when a senior Nokia marketing executive, walking past the handset lab and hearing the ringer being tuned and tested for maximum penetration, commented that it almost sounded like real music. It was decided to risk putting musical ring tones on a mass-market handset. At the last minute, one of several tones had to be picked as the default; without much thought, a tone containing a phrase from an obscure waltz called "Gran Valse" was selected. This gave birth to what many people consider a Nokia-branded jingle, and it has become so ubiquitous that thrushes in Copenhagen trees are starting to imitate it with their beaks.[41]

Soon downloadable ringtones were available in the European regional market, thanks to the work of Vesa-Matti Paananen, a self-described "Mobility and Information Worker Evangelist" and the founder of Add2Phone, an early mobile marketing and media firm based in Helsinki.[42] Paananen's epiphany has been reported as having occurred in March 1998 and in December 1997. Paananen, the 26-year-old chief technology officer of the new media-oriented Yomi Group, had awoken with a terrible hangover "after a night of Koskenkorva vodka and beer." His Nokia 6110 had rung with the Grande Valse tune now familiar around the world. As he puts it, "I didn't want to hear 'de de de de deeeee' ever again. I wanted to hear Van Halen's 'Jump,' and I was willing to pay for it."[43] Inspired to develop a downloadable ringtone service, Paananen spent months pitching it to telecommunications firms before Radiolinja, a major Finnish wireless provider, agreed to host the service. Paananen's idea was predicated on the fact that Nokia had developed phones with the capacity to upload ringtones, which could be done through a binary variant of the messaging system called the Short Messaging Service (SMS)—the same system now used worldwide to send text messages.[44] As Paananen noted, by 2002, "30 per cent of all SMS messages in Europe [were] requests for downloadable ringtones. Back in 1998, no one could have imagined anything like that."[45] Harmonium, an SMS-based software system for programming ringtones developed by Paananen, was posted online as Freeware by its author for other individuals and companies (or "aggregators," which stock and sell digital content) to begin marketing monophonic ringtones on a huge scale—a generous gesture that helped to create the ringtone phenomenon and may have cost Paananen billions of dollars.[46]

As the phenomenon spread through the European mobile phone market—within which Nokia had, on the handset side, become dominant[47]—others began to develop the downloadable ringtone as a business opportunity. In July 1999, James Winsoar, a 23-year-old college graduate in Nottingham, England, took advantage of the new technology and began to compose new ringtones, first selling them to customers individually online (at a price of £4.95–5.99 per ringtone) and eventually automating the delivery system (which lowered the price to £3.00). Initially trading on Nokia phones' ringtone features by calling his company My Nokia, he changed the name to Phat Tonez, probably because numerous phone brands soon accommodated uploadable ringtones and, in his words, to "appeal more to our target [i.e., young, working-class, chav, black] audience." (Also see chapter 9.) Winsoar may have thereby helped to initiate a national phenomenon in the United Kingdom.[48] Around that time, numerous small companies there began entering the mobile-media market, in some cases switching over to ringtone production from providing phone-sex lines.[49] But perhaps the most important figure within the United Kingdom was Anthony Stonefield, CEO and founder of the company YourMobile, which gave ringtones away (along with personalized advertisements, the source of the company's revenue) and, as a result, ran into copyright conflicts with EMI, eventually settling out of court for $45 million. Having given away 4.6 million downloads of ringtones in three months (and having attracted the ire of Winsoar and other competitors), YourMobile seems to have played the biggest role in spreading the popularity (and knowledge) of ringtones in the English-speaking world (especially the United Kingdom) and beyond.[50]

Once the ringtone fad had become well established in Europe, claims to historical priority in ringtone entrepreneurship began to proliferate. In July 2004, Vesa-Matti Paananen received the first Special Recognition Award from the Mobile Entertainment Forum, a London-based global trade association for the mobile-entertainment industry, for "his pivotal role in the creation of the mobile ringtone industry." "The whole mobile entertainment industry," said Andrew Bud, vice-chair of the MEF's board, "owes a great deal to the ringtone, and to the man who started it all."[51] More informally, the blog Ringtonia.com, written by Emily Turrettini, dubbed Winsoar the "first ringtone provider" in 2003.[52] Perhaps the most unusual ringtone "first" was claimed in 2005, when Marco Luzzatto, son of the prominent Israeli patent lawyer Kfir Luzzatto, claimed that in the late

1990s he had developed a now-patented means of "recording and storing new tunes on cellular devices, and then playing them as ringtones," and that the patent had then been violated by Siemens, Ericsson, and other handset manufacturers. These firms settled out of court with Luzzatto for undisclosed sums, possibly in the millions of shekels.[53] Although the story actually concerns a later moment in the development of ringtones (sound-file ringtones), it also could have been viewed as another claim for priority in originating the ringtone.[54]

The ringtone-origination stories emerging from Finland and the United Kingdom both feature Nokia phones. If they took place after the Japanese developments discussed above, might there be transnational connections worth exploring? One answer can be found in the saga of the Nokia Corporation itself, which began to downsize in the late 1980s and the early 1990s. In 1992, the new CEO, Jorma Ollila, proposed that Nokia focus solely on wireless communications. With that new strategy, Nokia would concentrate on developing fewer products—primarily mobile handsets—in global markets, tailoring its products to local demands, focusing on product design and customer service, and absorbing external technological developments through business partnerships.[55] In the 1990s, as a part of this process, Nokia had developed strategic ties with Japanese firms and had gained entry to Japanese markets, supplying cell phones to DoCoMo in the early part of the decade and later developing phones targeted at Asian markets.[56] One might reasonably assume, then, that sufficient numbers of Nokia executives were aware of the melodic alarms in pagers and, perhaps later, mobile ringtones, and that they adopted such features in order to maintain Nokia's presence in the Japanese market. But ultimately it may not be possible to ascertain whether the specific development of ringtones was borrowed wholesale from Japanese firms or whether it was a relatively independent innovation in Finland.

Ringtone Technologies: Monophonic, Polyphonic, Sound File

So far, I have been discussing a particular type of ringtone: the monophonic ringtone. This type is the simplest, involving a somewhat high-pitched single-line melody (or sound effect), having a limited frequency range, and producing a tinny, somewhat unpleasant, sometimes piercing sound. Like almost all ringtones, monophonic ringtones are typically limited in length to 30 seconds or less, and on account of the small piezo speakers on most cell phones they have a limited frequency range—from about 300 to 3,000 hertz—and little bass response.[57] As one commentator noted, "Considering that the technology used to generate ring tones—typically a monophonic square wave played through a small transducer or 'buzzer,' allowing for a simplistic electronic rendition of a known tune—is so primitive, it's quite astonishing how popular downloadable ring tones have become."[58] Indeed, the "primitive" monophonic ringtone is based on a now-old-fashioned beeping oscillator that had appeared on integrated circuits for more than 20 years in digital alarms on clocks and watches, in the earliest personal computers, in hand-held video games, and, of course, in pagers and digitized land-line telephones, especially but by no means exclusively in Japan.

Spurred by the earning potential of ringtones, handset manufacturers and musical instrument and sound software firms conspired to create a polyphonic ringtone. Now phones used synthesizer circuits of varying complexity to create multi-voiced music, with more than one pitch sounding at a time. Whereas monophonic ringtones used simple text languages for encoding instructions to an elementary waveform generator,[59] polyphonic ringtones involved sending more elaborate coded instructions to a relatively complex in-built synthesizer, eventually using one of several variants of the MIDI protocol.[60] Polyphonic ringtones produced music that straddled the line between video-game music and elevator music. The story of the polyphonic ringtone's development is, again, a story of competition between Europe and East Asia, with a kind of "technology transfer" from the United States to those regions looming in the background. The tale begins, again, in August 1999 in Japan, with the "personal handy-phone system" provider Astel Tokyo offering ringtones that could sound three pitches (or "voices," in musical terminology) simultaneously, perhaps as part of a strategy to counter decreasing subscriptions.[61] The decisive event, however, was the entry into the market, in September 1999, of the Yamaha MA-1 chip, a more complex, four-voice, 128-timbre MIDI-based synthesizer similar to the Chowning-inspired, FM synthesis-based DX-7 and sound card chips described above. The Yamaha MA-1 and its successors used the Yamaha Synthetic Music Mobile Application Format (SMAF), which conveniently allowed for the combination of "FM synthesis data, MIDI, samples, wavetable synthesis data, and other media, such as graphics, into one convenient format."[62] Several Japanese mobile handset manufacturers utilized the chip, which in turn both made polyphonic ringtones ubiquitous in the Japanese market and raised the cost of cell phones slightly. Advances in that market continued: in June 2000 a 16-voice phone-based synthesizer chip (the Yamaha MA-2) appeared, and in 2001 a synthesizer chip with 40 voices (the MA-3) was unveiled.[63]

In the European market, a few phones had begun to include the capacity to play polyphonic ringtones by 2000. However, these were primarily Japanese phones that were being sold in Europe, such as the Panasonic GD92, which provided the ability to e-mail and could play polyphonic ringtones—described as "a first" by a *Guardian* reviewer—as well as short sound samples recorded from the phone's microphone.[64] But, fearing a loss in competitive, profit-making advantages in price, most European companies balked at the notion of switching to polyphonic ringtone technologies by adding a costly chip to their phones.

A cheaper option—creating a polyphonic ringtone service without the use of synthesizer chips—was developed by Thomas Robertson. Robertson is better known as Thomas Dolby, the synth-pop musician whose classic record *The Golden Age of Wireless* (1982) included the hit song "She Blinded Me with Science." In 1996, Robertson launched Beatnik, Inc., a successor to his earlier company Headspace. Based in San Mateo, California, Headspace had been involved in online audio software (particularly music-player software) and had been moderately successful during the tech-speculation boom of the 1990s. One of Beatnik's innovations was the Rich Music Format (RMF), which provided an efficient file format for combining MIDI instructions and digital audio files. Some other Beatnik innovations were the Rich Music Markup Language (a language in which to edit such files), Beatnik (a software package that could edit such files interactively, in real time, with great speed and efficiency), and the Beatnik Audio Engine (a sound-synthesis program).[65] In the last example, Robertson's basic maneuver was "to apply the technology of the MIDI synthesizer to PC sound creation," a strategy which directly influenced ringtone development.[66] With the growing popularity of hand-held video games and mobile phones, Robertson began to develop a smaller variant of the Audio Engine that could be used on such mobile applications. In late 2001, Beatnik announced that its new mini-Beatnik Audio Engine (mBAE) would be appearing on select Nokia phones for the production of polyphonic ringtones, mobile video-game sounds, and multimedia messaging.[67] In addition, Beatnik played a major role in developing new standardized formats for the use of MIDI in mobile and miniaturized contexts—Scalable Polyphony MIDI or SP-MIDI, which allowed for prioritizing the most important lines or parts in ringtones and other polyphonic RMF files, so that phones with different degrees of polyphonic capability would all be able to use the same ringtone file.[68] In 2002, Nokia—which had, by that point, almost 40 percent of the global handset market—incorporated these technologies into its mass-market phones.[69] Soon enough, the polyphonic ringtone—

particularly Beatnik's variant of it—would spread steadily and ultimately replace the monophonic ringtone. In November 2000, only 9.2 percent of the handsets in the world could produce polyphonic ringtones; by November 2003, 66.1 percent had that capacity, according to a survey conducted by the NPD Group.[70] And by 2007, Beatnik's ringtone software would be available on more than 750 million phones, by Robertson's own estimate.[71]

The 1990s brought about the MP3, a format for compressing sound files that helped to spread digital music beyond the limits of specific physical media. In retrospect it may seem inevitable that the mobile telephone would have "converged" with this technological development. Indeed, the present (and final) form of the ringtone is the sound-file ringtone, in which a digital sound recording is played back on a cell phone's speakers. Partly because of marketing campaigns, the sound-file ringtone has accrued a plethora of names—ring song, ringtune, mastertone, master ringtone, truetone, realtone, superphonic ringtone.[72] As early as 2000, the German company Siemens had developed a phone for release in 2001 that could also play MP3 files—yet it could play only monophonic ringtones.[73] Sound-file ringtones would not appear on the market until more than a year later.

The development of the sound-file ringtone in Europe and Japan proceeded at almost exactly the same pace, marking one small area in which the latter lost its sizeable technological lead on developments in Europe and in North America. In Japan, the sound-file ringtone (*chakuuta*) was first offered by KDDI's au brand in December 2002, through its "EZchakuuta" service. This ringtone capability was connected to the adoption of the third-generation (3G) mobile-telephone standard determined in 1999 by the International Telecommunications Union in its International Mobile Telecommunications program known as IMT-2000.[74] The 3G system allowed broadband or high-speed wireless data transfer, thereby making mobile Internet service and Global Positioning System service available on cell phones for the first time. Third-generation services were first offered on the mass market in Japan in October 2001 by NTT DoCoMo, and have been dominant there. By late 2006, 3G services had been taken up far more enthusiastically in Japan (two-thirds of subscribers) than in the United Kingdom (14 percent) or the United States (8 percent).[75] Despite DoCoMo's early adoption of the 3G standard, however, it was only in 2002 that 3G became significantly popular in Japan, with KDDI/au's sound-file ringtone service providing further impetus for switching to 3G in late 2002 itself.[76] For these ringtones, KDDI used a low-grade variant of the Advanced Audio Coding (AAC) compression format—designed as a successor to the MP3 format and famously used by Apple for its iTunes files—misleadingly called AACplus, encoding sound-file ringtones about 100 kilobytes in size.[77]

Anticipating the development of sound-file ringtones while working on polyphonic ringtones, Beatnik co-authored the eXtensible Music Format (XMF) with Sun Microsystems and Nokia. Like the SP-MIDI specification, XMF was supported by the MIDI Manufacturers Association. Able to accommodate both complex polyphonic ringtones with samples and pure audio sound files,[78] the XMF format seems to be roughly analogous to the AACplus format. Its 30-second audio ringtone files add up to 100 kilobytes. In April 2002, Robertson predicted that at the end of that year high-end devices would begin to include sound-file ringtones with the XMF specification, and that mass-market phones with this capability would appear in 2003.[79] One example of the latter would be Nokia's 3300, which, the company noted in a press release, was "the first Nokia device for the GSM markets supporting True tones."[80] Phones able to play MP3 files as ringtones, such as the Motorola V300, were now being released. By 2004, designer phones such as the Nokia 7280 could play ringtones in both the MP3 and the AAC format.[81]

"New" Ringtones and Mobile Music

As the ringtone fad and the ringtone market began to expand in the early 2000s, handset manufacturers sought new means by which phones could become customized through music and sound. In part these manufacturers were also responding to a new degree of regionalization within the ringtone market. Ringtones had spread throughout the European Union, and were particularly popular in the United Kingdom, in Germany, and throughout Scandinavia. Regional development of a distinctive sort took place in East Asia, particularly with the rise of South Korea as a major contributor to the telephone music phenomenon. Two cases will suffice to demonstrate: ringback tones and mobile music downloading.

The ringback tone, originally, is a ringer cadence heard by the caller after he or she has dialed and before the called phone has been picked up. Hence, it marks the presence of a potential telephone connection and the signaling process itself. Different countries have different ringback signals associated with them—for example, the famous "beep beep" of the UK system (notably appearing at the end of Pink Floyd's song "Young Lust"), or the North American warbling two-tone (440 Hz + 480 Hz), two-seconds-on-and-four-seconds-off signal.[82] Perhaps as early as April 2002, the South Korean mobile telecommunications provider SK Telecom began to offer the Color Ring Back Tone (CRBT), which had been developed the previous year by the Korean firm Witcom.[83] Essentially, the service allowed for alteration of the conventional, state or monopoly-determined ringback cadence, so callers would now hear distinctive sounds or selections of music piped to them while calling—adapting the elevator/Muzak principle of telephonic hold music for personal, novelty entertainment. In contrast to the public display function of the ringtone—largely dependent on the portable, ubiquitous cell phone—the ringback service is not inherent to the mobile telephone. Yet, in view of the ringtone precedent, it is not surprising that it emerged within the context of mobile telephony. Becoming hugely popular and generating millions of dollars in revenue, the service soon spread throughout East Asia, with Japan's NTT DoCoMo offering an analogous service in September 2003.[84] Ringback tones (sometimes called "caller tunes" in India, where they are popular) ultimately appeared in Europe and North America in 2004. Despite journalistic quasi-advertising and analysts' optimism, they have not proved particularly lucrative in those regions.

More significant is the development of mobile music downloading, in which cell phones merge with portable audio file players. The first such service in the world appears to have been founded in South Korea. SK Telecom launched a "Music-on-Demand" (MoD) sales program (and a video-on-demand program) in November 2002, initially relying on lower-grade sound files and updating the service in early 2004 once Korean handsets could function as MP3 players.[85] Responding to Japanese consumers' use of sound-file ringtones to test music selections, KDDI introduced the *chaku-uta full* service for downloading music to mobile phones in November 2004. It became wildly popular, rivaling the sales figures of ringtones.[86] In Europe and North America, however, mobile music was initially slow to take hold, in part because of the popularity of the iPod as a vehicle for mobile music listening and the relatively limited capacities of mobile phones in those regions at the time.[87] Indeed, some market researchers proclaimed that the success of *chaku-uta full* in Japan was not likely to be repeated in the United States or in Europe, on account of differences in Internet access and use between East Asia and the other two core economic zones. Although this prediction may have been borne out at least somewhat in the United States, it's difficult to tell from sales figures alone.[88]

The extremes to which commodified sonico-musical phone customization has gone with the development of alert tones and video ringtones (among other tones one might apply to one's phone) should be noted. In alert tones, we have sounds that can effectively customize virtually any activity or action (such as opening a particular menu, receiving an SMS text message, or receiving a voice-mail message). In video ringtones, video clips are displayed as a ringtone sounds. It seems, though, that alert tones, like most other ringtones, are usually "sideloaded" (or freely uploaded from personal computers, from the Internet, or by means of Bluetooth), and that video ringtones don't perform the same novelty-display function any better than a ringtone might (since few would initially see the video clip on the small phone screen).[89] For these reasons, these forms point primarily to the *limits* of phone customization—the exhausting of a capitalist paradigm—and perhaps the limits of atom-splitting commodification itself.

Accumulation in the Global Ringtone Industry

The preceding histories of what is essentially a trivial novelty might seem excessive were it not for the fact of the massive investments of time and resources on the part of companies small and large and the substantial returns on those investments through ringtone sales. The numbers in the global ringtone industry are, for a microcommodity such as this, nothing short of staggering. Profits were estimated by Jupiter Research in 2006 to be as much as $6.6 billion, and estimates for 2007 and beyond were much more generous (see below).[90] The major markets for ringtones have, to date, been in East Asia and Europe, with the North American market only subsequently catching up and even surpassing them. Although they are partial figures and probable underestimates, one can get a sense of the industry's profits in the three core zones of the capitalist world-system by looking at the three largest national markets for mobile music (sound-file ringtones and full-track downloads to phones), which happen to map directly onto the three core zones. In 2007, according to the International Federation of the Phonographic Industry (IFPI), these top national markets were Japan ($516 million), the United States ($458 million), and the United Kingdom ($42 million).[91] Though it seems incontrovertible that ringtones have become a billion-dollar global industry, we should be cautious in approaching these numbers. Ascertaining the ringtone industry's actual profits is not easy. First of all, the quickly changing nature of the industry and the overlaps between these digital commodities make data keeping inconsistent between sources and even from year to year. The casual observer must contend with monophonic and polyphonic ringtones often being separated from master/sound-file ringtones in industry counts because they provide different rent rates to different interests, as well as the sometime inclusion of ringtones, ringback tones, mobile music downloading, and even mobile entertainment generally in published accounts of profits. Second, the sources of data on the industry come from a broad range of sources, each with its own interests. Whose figures should be used? The types of sources themselves can be classified as mobile music and entertainment aggregators (companies selling ringtones, mobile video games, wallpaper, etc.), wireless service providers, music industry/media firms and their recording labels, music-industry trade associations (copyright collection organizations, such as ASCAP, BMI, and SESAC, or the Harry Fox Agency, in the United States, and the various national and international recording industry associations, such as the Recording Industry Association of America, the Recording Industry Association of Japan, the British Phonographic Institute, or the IFPI), and mobile consulting and measurement companies (such as M:Metrics, Ovum, Jupiter Research, Informa Telecoms & Media, Gartner Research, etc.). Some of these parties, such as the aggregators, have become less relevant in the counting of ringtone profits, particularly with master ringtones giving a much larger stake in the ringtones market to the music industry. In addition, wireless companies have not necessarily had the time or the resources to track detailed information on ringtone sales, and credit-based systems of payment—in which consumers buy credits for ringtones or mobile media generally—lead wireless firms to see only the amounts paid for the credits themselves and not the amounts paid for downloaded content.[92] Publishing agencies and trade associations are more meticulous about counting ringtone sales, insofar as they already engage in the vast and tedious work of tabulating copyright royalties, but they do have a tendency to underemphasize and possibly undercount ringtone profits so as to keep media attention trained on decreasing CD sales and illegal file sharing.[93]

Mobile-media consulting and measurement firms occupy a special position in the firmament of the ringtone industry. Serving a niche role within the mobile marketing world, and acting in parallel to other "advice industries" (such as management consulting firms, which promoted and rationalized the corporate restructurings of the 1980s and the 1990s[94]), mobile-media consulting and measurement companies exist because of the inherent difficulties in measuring ringtone sales, which was a particularly anarchic process in the industry's earliest days before its monopolistic/oligopolistic consolidation and which remains difficult because of the sheer volume of transactions and the complexity of apportioning profits. Seizing upon this new need for information, the members of this "consultocracy" do two things: they act as cheerleaders for the mobile-entertainment industry by forecasting exaggerated possibilities for profits, priming the requisite optimism of mobile-media capitalists themselves, and they promote their own growth as a para-industry, making their knowledge proprietary and charging hundreds or even thousands of dollars for research reports.[95] Keeping the loop tightly closed between themselves and mobile-entertainment firms, mobile consulting and measurement firms reproduce and expand their role within the broader mobile-media marketplace.[96] Hence, in terms of their calculations, these firms tend to work in just the opposite way from the music-industry publishers and organizations, though under the guise of credibility some firms avoid habitually inflating sales figures and predicting an abundance of opportunity in the marketplace. Still, the tendency is pronounced, leading one writer to marvel at the disparities of ringtone sales figures for 2006: according to the blogger Jared Moya ("soulxtc"), Gartner Research claimed that mobile music downloads (including ringtones) earned $9 billion, whereas the IFPI claims making only about $1 billion on mobile music that year.[97] Apparently this sort of problem is characteristic of the mobile-entertainment industry as a whole. Reporting a discussion with Steve Mayall of the mobile consulting firm Music Ally, the reporter Adam Webb writes:

> Nobody within the mobile business anticipated the demand for either ringtones or text messages in the first place, indicating the market's unpredictability, says . . . Mayall. And since the industry is rife with unrealistic predictions and hype, that makes it incredibly difficult to assess. "Somebody said yesterday (at Monaco's Mobile Entertainment Market conference) that mobile content is a $40bn industry, but it's not," says Mayall. "It's not even half that. It's not bigger than the entire music or film industry, and that kind of hyperbole is in danger of hurting the whole business."[98]

Figure 1.1 illustrates the mobile consulting industry's vague method of knowledge creation.[99] In January 2007, competing estimates of mobile music revenues (including ringtones) were offered by the IFPI and by Gartner Research. The IFPI claimed that mobile music revenues (including digital downloads and master ringtones/ringbacks, etc.) were $1 billion in 2006; Gartner claimed the revenues amounted to $3 billion. Furthermore, Gartner projected revenue growth through 2010 (at rates averaging 74 percent per year), with revenues for mobile music at that time said likely to be $26 billion—and if revenue from old-fashioned monophonic and polyphonic ringtones (which were projected to decline slowly) is added, the number reaches an astounding $32 billion by that year.[100] In comparison, the IFPI claimed 100 percent growth from 2005 to 2006, though starting from a much lower number ($500 million in 2005). In 2007, as the ringtone industry's prospects began to sour for various reasons, Gartner's projections became vulnerable to more modest counter-projections

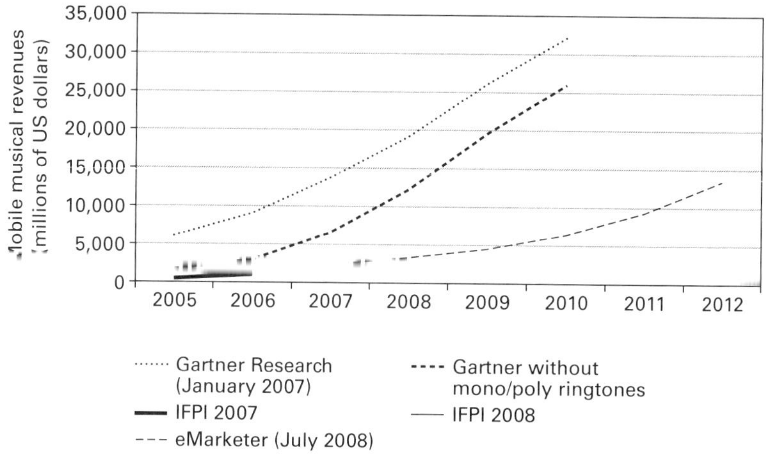

Figure 1.1
Overestimation and correction of profit predictions in the ringtone industry.

by other mobile consulting firms. By January 2008, when the IFPI released new revenue figures for mobile music, they were substantially lower ($1.45 billion) than Gartner's predicted number for 2007 ($6.6 billion). With the IFPI declaring a slower 40 percent mobile music sales growth in 2006–07, it was clear that the earlier predictions by Gartner Research had been far too optimistic. In July 2008, eMarketing produced a new set of figures somewhat closer to the IFPI's numbers, claiming $2.43 billion in mobile music income in 2007. However, upholding the mobile consulting firm's role as an industry booster, eMarketing nonetheless counted 2007 revenues as being $1 billion higher than the IFPI figures and projected increasing rates of revenue growth (averaging 41 percent per year), forecasting $13.4 billion in global mobile music revenue by 2012 and arguing that advertising on mobile phones will aid growth. In addition to this intrasectoral dialectic of industry boosterism and revenue underreporting, it is also worth noting that the significant global revenue of monophonic and polyphonic ringtones—which were still lucrative in 2006 and accounted for the overwhelming majority of mobile music sales ($6 billion in Gartner's probable overestimation)—were never counted by the IFPI and not included in the press release for the eMarketing report. Although this can partly be explained by the ongoing supersession of these earlier ringtone forms by the sound-file ringtones and full-track mobile downloads, they also demonstrate the music industry's success in framing the discourse on its own terms.[101]

It would seem, then, that concrete statements about the state of capital accumulation in the global ringtone industry are vexed terrain. But even if it is difficult to ascertain specific figures on ringtone profits, it is possible to get a sense of how those profits are made. To do so, we must first acquaint ourselves with the numerous parties involved in a mobile-entertainment transaction. In the lingo of the industry, "content providers" are those companies and institutions that offer the entertainment substance of what is being sold. For most ringtones, this is both the actual ringtone digital file itself and the popular song, piece of music, or other proprietary sonic pattern on which it is based. These content-providing firms can include ringtone and mobile-media wholesale firms, ringtone sales companies, publishing companies, and record labels, and the role and presence of these different types of firms depends on historical, regional, and ringtone-technological factors. Next in the ringtone commodity chain are the so-called content aggregators, who sell ringtones to consumers—and these aggregators are often mobile-entertainment companies that sell ringtones, ringback tones, video games, cell phone wallpaper, and other digital commodities for mobile phones. Types of such companies may include independent ringtone or mobile-entertainment sales firms (such as Phat Tonez or Ringtone.com), mobile-media branches of wireless carriers (such as the "download shop" of the UK wireless company Orange), or branches of publishing companies, record labels, or music-industry conglomerates (such as Sony BMG's Music Box).[102] Other participants in the mobile transaction include providers of wireless service, wireless billing companies (such as VeriSign, the company that created secure Internet transactions, facilitating online shopping), and other providers of specialized services and applications, who may sell software or protocols for various aspects of the production and distribution process.

The flow of activity within the chain begins with the act of consumption: a customer initiates the process of purchasing a ringtone, usually by text message and paying with pre-purchased credits from the aggregator or through the wireless provider itself (perhaps as part of monthly charges). The ringtone must then be delivered by the aggregator, which keeps a large database of ringtone files ready for sale and secures copyright clearances if that is necessary. The production of the ringtone itself involves either the arrangement of some preexisting music or the composition of new music or sound, either produced by in-house composers and arrangers or supplied by a ringtone wholesale firm. The music is encoded into a proper file format and then transmitted to an individual phone via SMS (for monophonic ringtones) or Wireless Application Protocol (WAP) (for polyphonic and sound-file ringtones). The wireless billing company takes care of the payment arrangements in tandem with the wireless provider, ensuring that payment is secured and confirmed efficiently.[103]

Owing to the numerous economic interests making claims on each ringtone sale, the process of accumulation with the industry is rather complicated. First, the aggregator selling the ringtone necessarily receives some percentage of the net ringtone income, some of which may go to the wholesale firm supplying the ringtones if necessary. Second, if the ringtone makes use of preexisting copyrighted material, which is typically the case, some percentage of the price of the ringtone must go to the music publisher who owns the copyright of the song or musical selection in question. Third, the wireless service provider takes a cut of the ringtone sale, which it can do easily since payments are made through the mobile phone's wireless service. Finally, the wireless provider contracts one or more separate billing companies to take care of billing arrangements and verification of sales. Each participant in the ringtone market receives a certain proportion of ringtone sales that varies with the type of ringtone sold, with the particular companies involved, and of course with the economic and legal structures in different countries. With this in mind, I should emphasize that the following discussion focuses on the United States and does not necessarily represent what goes on elsewhere, though the parties involved and their respective proportions of a ringtone's net income probably would remain similar. Indeed, it appears to be more the case that pricing, rather than the network of payment relations, varies quite widely from country to country, reflecting different marketing strategies. For example, Noriko Manabe notes that in Japan master or sound-file ringtones are sold for around 100 yen (around $1), whereas full-track and album downloads cost more—making ringtones promotional tools for mobile and online music sales (which is quite different from the American and European strategies of higher pricing for ringtones than for downloaded single tracks).[104]

In the case of polyphonic ringtones, the mobile-entertainment companies providing the ringtones usually get the largest share of the income: 35–65 percent. The wireless carrier receives a significant proportion (10–40 percent), and various billing companies receive only 10–16 percent. Music publishing firms receive a relatively standardized proportion of revenue, which hovers slightly above the minimum fees for mechanical (reproduction) and (public) performance copyright payments.[105] In the magazine *Billboard*, Scott Banerjee explains:

> The performing-rights organizations have different rates for ringtone use. BMI collects 2.5% of gross; ASCAP collects 2%. SESAC fees vary depending on the number

of ringtones sold. . . . On the mechanical side, some publishers have made direct deals for ringtones and are collecting a minimum of 10 cents or 10% of gross per track, whichever is higher. This exceeds the statutory mechanical rate of 8.5 cents. When a direct deal is not in place, the Harry Fox Agency handles the mechanical licensing. HFA collects 10 to 12 cents per download, plus an initial "fixation" fee of up to $50 to include the composition in a ringtone service.[106]

Banerjee also posits an average, perhaps ideal-typical version of these payment proportions for polyphonic ringtone arrangements of copyrighted material (monophonic ones not being substantially different here): 58 percent, 20 percent, 10 percent, and 12 percent of the $1–$2 cost of the ringtone would go to the aggregator, the carrier, the billing firm, and the publisher, respectively.[107] Securing the rights to arrangements can be very complicated with popular songs—particularly with sample-based, multiply composed hip-hop tracks. "For example," Jason Ankeny notes, "in the case of a P. Diddy single built on melodies, basslines and beats sampled from other recordings, there can be dozens of different composers credited with the finished product, and [the aggregator] must clear 100% of a song's rights before it can create a ringtone. Also, not all songs are even up for grabs: Some artists refuse on principle to license their music for ringtone consumption."[108]

With sound-file ringtones, the situation is somewhat different. Ringtones of this kind favor the recording industry, since legally sound files fall under the more lucrative category of sampling rather than covering or arrangement. For a sound-file ringtone, sampling rights allow the record company or label that released the original sound recording to claim 40–50 percent of these more expensive ringtones (about $2.50–$3.00). Sound-file ringtones have delivered even higher returns, and one might initially attribute pricing and sales increases to their ability to accurately reproduce current radio hits and their relative ease of production (clipping and conversion/reformatting of a preexisting sound file). In fact, however, the shift in the balance of competing interests has led content owners to have a much greater slice of the overall pie, forcing aggregators to accept a lower percentage of an individual ringtone's revenue. Banerjee's alternate ideal-typical arrangement reveals a 40 percent–20 percent–18 percent–12 percent–10 percent subdivision of profits for the record label, the wireless carrier, the aggregator, the music publisher, and the billing company, respectively. With the costs of the publishers being essentially fixed and the sound-file ringtone creating no substantial differences for carrier and billing company, one can see that record labels are primarily competing with aggregators. In fact, as Banerjee notes, "When dealing in master ringtones, . . . the aggregators can get squeezed, typically down to 15% to 20%, but sometimes the cut can drop all the way down to zero."[109]

The price increases that accompanied the shift in ringtone technology are mystifying. After all, the costs of production and labor involved in making sound-file ringtones are much lower than those for polyphonic ringtones, so in theory one would not expect price increases as high as 300 percent.[110] The increases are due to rent and the proportion of value that it claims in a ringtone sale. Unlike the multitude of music publishers found in the marketplace, the vast majority of published music recordings are owned by one of the Big Four music conglomerates (Universal, Sony, Warner, and EMI), and that allows them to engage in cartel pricing and thereby to secure sizable monopoly rents, to the detriment of the only productive components of the commodity chain—content-producing aggregators and mobile content wholesale firms. The overwhelming proportion of rent in this equation—which can be understood as well over 50 percent, if the carrier is the digital equivalent of transportation, the wireless billing firm is tantamount to a sales-machinery firm (providing the digital equivalents of cash registers, like NCR of old[111]), and the aggregator is merely the point of sales markup for wholesale-produced goods—is further exacerbated by the fact that productive or non-rent-based accumulation plays such a small role in the commodity chain of the ringtone (at most 20 percent, much less if wholesale arrangements are involved), thereby threatening the profitability of the industry as a whole. Such concerns have loomed large in the ringtone industry, and with good reason. High rent rates can push up prices, but only to a point given the realities of consumer demand, leading to the (often temporary) profit-rate equalization characteristic of monopoly accumulation.[112] If prices are inflated in excess of consumer demand, an industry can become vulnerable to new competition (though this can be staved off by internal dumping, or temporary price reduction to eliminate competition) or can become depressed or even be destroyed. With the prices of ringtones rising in Europe around 2004 (up to as much as £4 in the United Kingdom, for example) just as downward pressure was being exerted on sales of digital music files, Simon Dyson at the consulting firm Informa bluntly stated: "The reseller is really between a rock and a hard place. They are torn between raising the price or keeping it steady in the hopes of establishing a market. Demanding such high percentage rates by the record companies could certainly lead to the market being depressed."[113] The reseller, or aggregator, whose only reason for being is the ringtone market, faces by far the greatest risk of economic failure in this situation.

The Music Industry, Sound-File Ringtones, and Declining Aggregators

The tendency toward the aggregator's destruction is not an accidental outcome guided by the "invisible hand" of the marketplace; it has been the music-industry oligopoly's unstated goal, aided by handset manufacturers in their fervent updating of mobile phones' ringtone technologies, to squeeze out the larger ringtone providers that once dominated regional ringtone sales.[114] To this end, the music industry has pursued several concurrent strategies aimed at eliminating competition and maximizing profits. First, music-industry conglomerates sought to bypass mobile-media aggregators by partnering directly with wireless companies, with a wave of such partnerships taking place with the rise of the sound-file ringtone, such as BMG and EMI with Verizon Wireless in the United States or Sony and Universal with T-Mobile in Europe.[115] Second, music labels took advantage of the reduced costs of sound-file ringtone production by creating online sales portals for their own proprietary content. Third, the music industry pushed for greater shares of profits within the existing ringtone payment structure—ostensibly by demanding higher percentages of ringtone revenue itself and by promoting music owned by industry publishers such as BMG Music Publishing and Universal Music Publishing.[116] A particularly cynical ploy on the part of the industry concerns its desire to have ringtones legally classified as sales instead of licensed goods, thereby ensuring that contractual royalty rates for music sales (say, 15 percent) would go to artists instead of the 50 percent owed to them as licensed goods.[117] The situation apparently came to a head in April 2006, when several law firms representing a number of older or defunct recording acts, including the Allman Brothers Band, Cheap Trick, and the Youngbloods, sued BMG Music for violating contractual obligations over digital music sales—though the initial hearings in the case resulted in the plaintiffs' being granted leave to clarify their arguments.[118] A similar case was argued involving Eminem's music. The court ruled in favor of the defendant recording firms, although an appeal may be in process.[119] Additionally, some publishers in the monophonic and polyphonic days secured favorable rates in excess of standard contractual royalty rates, sometimes as much as 50 percent of ringtone income, and battles over these royalty rates in the new digital music marketplace continue to rage in various theaters of conflict, including in testimony before the US Copyright Royalty Board (CRB).[120]

There also have been conflicts between record companies and music publishers, the primary components of the ringtone industry—particularly over whether or not the mechanical licensing of a ringtone is subject to compulsory license rates, which means that "someone wanting to make part of a cover record available as a ringtone can 'go compulsory' if they are unable to obtain a negotiated license from the music publisher."[121] The lower the compulsory rate paid to music publishers, the more money ringtone producers (which, in the era of sound-file ringtones, are often the record conglomerates or vehicles/suppliers for them) make. Set in 2006 at the same rate as the statutory license rate for any other cover recording (9.1 cents per song, to oversimplify), the rate was increased and set to 24 cents (favoring publishers) after a month of testimony before the CRB.[122]

Internal legal battles notwithstanding, it appears that a critical component of the music industry's efforts to have ringtones legally understood as sales is the *normalization of the ringtone*, by which I mean eradicating public perceptions that a ringtone sale is substantively different from any other musical purchase, digital or otherwise. Aiding in, or perhaps enlisted for, this effort, the Mobile Entertainment Forum created a sales chart in 2004 for ringtones sold in the United Kingdom, in the manner of a singles chart.[123] Later that year, *Billboard* adopted the idea for the North American market, creating the Hot Ringtones top-20 list for sales of polyphonic ringtones.[124] However, owing to the differential appeal of purchasable ringtones among various populations, polyphonic ringtone charts had a strange appearance, often combining purchases uniquely suitable for and popular as ringtones such as the "Super Mario Brothers theme song" video-game music and ringtone adaptations of classic hit sound recordings such as Waylon Jennings' "Good Ol' Boys." The more recent "Hot Ringmasters" charts for sound-file ringtone sales, on the other hand, tended to reproduce the R&B charts, demonstrating the importance of ringtone use in African American communities. Nonetheless, for a few years the sales of sound-file ringtones had outpaced the sales of single recordings of the same song,[125] and music-industry officials were doing all they could to ensure that ringtone and digital sales would help them recover their losses from declining sales of CD singles and albums. Although it appears improbable in retrospect, the sound-file ringtone at one time was itself seen as a means of "saving" the music industry, whose alleged profit losses caused partially by file sharing are often discussed.[126] Yet in 2007, with sound-file ringtones still constituting 40 percent of digital music sales, the music industry continued to promote them until ringtone profits began to decline irrecoverably.[127]

Of course, the music industry wasn't alone in seeking to advocate sound-file ringtone hit charts and sound-file ringtone sales more generally.

Tabulating hits and legitimizing sound-file ringtone sales was in the interest of all parties involved in the ringtone industry, leading even the targeted-for-death aggregators to rationalize the move from polyphonic to sound-file ringtone technologies.[128] In 2004, Fabrice Grinda, former CEO of the mobile-entertainment aggregator Zingy, said "The [profit] margins will go down, but the price points are increasing, so the dollar profit usually remains the same."[129] Famous last words, one might be tempted to add. Only three years later, Zingy closed down its ringtones division and renamed itself Vindigo after a company that it had bought out—a company that specialized in mobile applications useful in outdoor navigation (MapQuest Mobile) and in nightlife entertainment (MovieGoer, Vindigo City Guide). The company itself was bought in 2004 for $36.5 million by the Japanese mobile-content firm For Side, which made unsuccessful efforts to sell the company in 2006 and which still appears to hold Vindigo as a subsidiary at present.[130]

Although buyouts of this type have been common to the advanced capitalist economies since the 1980s and are reminiscent of the purchases and failures of 1990s Internet start-up firms, the closures of ringtone divisions and the buyouts of sometime failing mobile-entertainment companies suggests that the music industry's sound-file ringtone strategy has been largely successful.[131] There are many tales of indebted and defunct mobile-entertainment aggregators and their buyouts, and though their meaning is not always clear—particularly for buyers—the trend appears to mark the presence of failing or underperforming ringtone sales firms *and* of aggregator consolidation within the industry. In a recent example of the latter, W3i, sold Ringtone.com to New Motion (better known as Atrinsic) in July 2008 for $6.25 million, leading the growing mobile-entertainment firm to capitalize possibly on the smaller company's primary asset—its domain name, which surely aided in securing a putative 1,000 subscribers per day.[132] Of course, some mobile-entertainment entrepreneurs made out quite well during this phase of buyouts. Simon Buckingham, who secured the domain name ringtones.com for £20 in 1997, formed a highly profitable aggregator firm Mobile Streams and, after reaping "hundreds of millions" of pounds in profits, ended up selling the URL for $750,000 in December 2010—arguably at just the right time. The funding was, in Buckingham's words, "a good way to raise some investment capital" for his new "cloud-based app store Appitalism."[133]

Perhaps one of the more lurid of examples of mobile-entertainment buyouts is that of MonsterMob, a mobile-content firm based in the United Kingdom. The story begins in 2000, when the British firm was founded by Martin Higginson, who quickly became known as "Lord of the Ringtones," a not-infrequently granted moniker in the global ringtone industry on account of the popular Tolkien-based film series.[134] MonsterMob was well positioned during the early explosion of uploadable monophonic ringtones and was quite successful at the time, distinguishing itself from other mobile service aggregators through its *Mob Magazine*, an award-winning publication targeting youth readers and featuring a large number of advertisements for ringtones.[135] However, early on the company seems to have focused significantly on a supplementary strategy—mergers with and buyouts of other mobile-media aggregators. At first, the buyouts were nationally based. In 2001 the company merged with "the mobile technology m-Quest to create a [£60 million] business."[136] After pursuing various content ventures (including an effort to put classic Commodore games on mobile phones, in tandem with Jester Interactive[137]) and updating itself for the polyphonic ringtone market, in late 2003 the company began seeking to raise £10 million in capital by placing itself on the Alternative Investment Market of the London Stock Exchange, with the express purpose of "expand[ing] the company through its existing operations, which include three million UK customers, and *through acquisitions*."[138] Acquiring a great deal of capital in a few months, MonsterMob soon revealed its insatiable appetite for other mobile-entertainment firms, quickly penning a distribution-and-promotion deal with the UK electrical retailer Dixons Group in February 2004 and taking over the UK mobile-content firm Mediaprom in March 2004 for £10 million.[139] After the June takeover of Phunky Phones for £730,000, MonsterMob made its first global foray in the acquisitions business by buying out, for some $3 million in cash and later deferred payments, the shares of the Denver-based company 9 Squared, Inc., which was founded in 2001–02 by a college dropout named James Eberhard in an effort to corner the US ringtone market, then perceived as an untapped gold mine.[140] With 9 Squared's substantial resources ready to hand, MonsterMob unveiled a new ringtone and mobile-entertainment subscription service with China Mobile's network of 160 million mobile subscribers (in 2004).[141] Subsequent buyouts followed, including the Malaysian company UnrealMind Interactive (November 2004), Upper Mobile Limited (owner of the Philippines' Information Gateway) and Creative Minds Pte of Singapore (March 2005), ATOP Century Ltd of China (August 2005), the Russian mobile-content provider Mobicon (December 2005), the Chinese mobile games company M Dream (January 2006), and the Chinese mobile-content firm W-Infinity Communications (April 2006).[142] In view of the ephemeral and unpredictable nature of the mobile-content industry, it would seem that MonsterMob's solution to this problem was to buy out a large swath of its present and future competition, as well as claiming a sizable proportion of the mobile-content business in the Pacific Rim. But the buyouts predictably turned out to be difficult to manage, and with further difficulties marked in part by layoffs, the firing of Higginson as CEO, and mobile-content clampdowns by the Chinese government, the company began accumulating losses on account of some of its firm purchases and soon was itself targeted for a buyout. In February 2007, the Spanish mobile and entertainment content giant LaNetro Zed succeeded in doing so, in the process bringing Higginson back to run its newly acquired, and failing, subsidiary.[143] Although Zed had planned to continue the growth-hungry tendencies of MonsterMob, a report noting that it raised capital for mergers and acquisitions through bank loans suggested accumulative difficulties instead of potential—a notion that was, perhaps unintentionally, reinforced by the company's tepid press statement that "the market continues to support Zed's growth policy despite the tough economic climate."[144] By early 2009, Higginson was promoting mobile gaming and TV rather than music—prospects that may have stabilized its mobile-content business, though since the 2008 crisis the economic climate surely has not favored consistent or predictable growth.[145]

Of course, if the MonsterMob saga is a single, extreme case exemplifying ongoing independent ringtone aggregator decline—constituting an apparent success for the music industry—the shifting realities of the industry understood on a comprehensive, firm-by-firm basis are simply outside the scope of this study, though we might note that widespread industry consolidation on a global scale is undeniable.[146] Of course, those companies that have not been bought out and/or swerved into more lucrative mobile-content areas are left to languish and disappear, with the only traces of these once-profitable aggregators and wholesalers being inactive websites still elicited or referenced in Google searches—Winsoar's Phat Tonez appears to be one of these, to judge from the site, as is the UK wholesale firm 3rd Ear Productions, formerly one of the main suppliers of polyphonic ringtones to Orange and T-Mobile.

Ringtones and the Rhetoric of Personalization

The great irony in this flurry of financial activity to make ringtones increasingly profitable is that, initially, they were not predicted as potential means of accumulation by mobile sector capitalists, let alone music publishers or music-industry conglomerates. The technology was essentially conceived as a novelty, but young consumers creatively used ringtones and similar mobile entertainments to "personalize" their phones. Of course, with the industry's rising profitability, marketers have embraced the notion of personalization itself, almost treating the concept as a dogma within the world of wireless accessories—and trade industry journalists have predictably reproduced the idea with alacrity. Scholarly articles, such as an ethnographic study by Heikki Uimonen, must now address the prevalence of the idea and its effect on social behavior. Researching ringtones in Finland, Uimonen interviewed 19 teenagers about their use of ringtones. He describes the numerous ways in which the youths strove to make their ringtones unique.[147] In *Always On*, Naomi Baron takes for granted that "downloaded ring tones are an audible form of personalization."[148] Manuel Castells, Mireia Fernández-Ardèvol, Jack Linchuan Qiu, and Araba Sey emphasize the role of personalization as part of a broader autonomization process within "mobile youth cultures," noting that the proximity of the mobile phone to the body in part lends itself to its transformation into a fashion item. However, these writers also emphasize that the "whole concept of individuality may also have some limitations," and they note that young consumers of mobile media prefer "prepackaged individuality" over features that allow users creative control over personalization, such as phone-based ringtone composition programs.[149]

Thus, the phenomenon of personalization isn't as simple as the paradox of "the mass appeal of individuality,"[150] as some have suggested. From another perspective, it looks like the promotion of an apparent individuality to a demographic segment most powerfully molded by peer and marketing pressure. Indeed, the marketing strategy of personalization, or tailoring marketing to a single individual, has quite a long history, extending at least as far back as the history of direct marketing (i.e., marketing emphasizing reduced mediation between seller and buyer, and the use of statistical information to target potential buyers) and its postwar transformation into direct mail, telemarketing, infomercials, and other means of bypassing distributors and other retail systems.[151] Another important concept in postwar consumer-information-based marketing was what Michael Weiss has described as "clustering," which correlated ZIP codes, census data, and consumer surveys to create statistically significant consumer groups making targeted marketing possible.[152]

"Personalization," then, ostensibly moved the target market from the small group to the individual, relying on a database of information for each consumer and thus logically being linked to the strategy of "database marketing," which widely available computing power and data storage made feasible in the 1980s.[153] By creating the possibility of automated customer discrimination and the rewarding of privileged or preferred customers, database marketing engendered loyalty marketing. "In the 1980s and 1990s," Joseph Turow notes, "as database-driven loyalty marketing emerged, financial and leisure firms and elite retailers began to adopt the logic of segmenting their customers and pursuing the 20 percent who supposedly accounted for about 80 percent of sales."[154] Turow's equivocations suggest that regardless of the realities of database marketing's success—even if its prioritizing of statistical analysis makes calculable marketing success a central goal—"personalization" became a buzzword within the corporate world and thereby an ideologeme—that is, a coherent unit of ideological meaning, in Fredric Jameson's sense.[155] Moreover, as Turow argues, the rise of Walmart served as an impetus for personalization by a large range of competing firms who could not match the extremely low prices offered by the behemoth retailer firm.[156]

If computational database marketing was one significant element in the spread of personalization, another was the rise of the Internet, which could create the possibility of fully digitized sales customized to a particular consumer. As Kim Veltman argues, computer personalization technologies were first developed in the 1980s by Charles River Analytics, who applied artificial intelligence to personal computer use in their Open Sesame! software package, which would customize a computer on the basis of repetitions demonstrating personal habits.[157] By the 1990s, numerous Internet-based firms began using these customer-relationship management (CRM) technologies, creating a number of business-to-business software applications for particularizing customer interactions, including Andromedia and Macromedia's collaborative filtering technology, as well as similar products by Net Perceptions, Allaire, and the previously mentioned Open Sesame.[158] The discursive relationship between personalization and the Web was cemented during the remainder of the decade, and the year 2000 witnessed the birth of an industry-funded Personalization Consortium, whose goals were to promote personalization technologies and to build a set of "best practices" for Internet-based consumption and data collection, in what one reporter described as "the latest of various industry attempts at self-regulation on privacy matters."[159] Despite examples of "success" in online customer personalization, such as Amazon.com's product recommendation algorithm, the combination of dampening enthusiasm for personalization technologies, the bursting of the tech-speculation bubble, and the post-9/11 conjuncture's new relationship to information and security—in which state and corporation more freely traded consumer information and which led to an ongoing civil-libertarian public backlash on the issue—appeared to temporarily displace Internet-based marketing personalization as a prominent theme in discourse on the US economy, before returning with a vengeance through Google's personalized browsing system and other websites.[160]

However, the term "personalization" has had a more consistent home in the world of mobile content. Claiming that personalization is "evident on four levels 1) sales leads, 2) news and content, 3) experience, and 4) learning," Veltman goes on to emphasize that in the United States "personalization is a euphemism for mass customization, which in turn is closely connected with the vision of agile manufacturing [and] a rhetoric of creating a customer-centric or customer-driven market."[161] Sometime in the 1990s, the discourse of personalization was applied to *content* instead of *delivery*—not a surprising transposition, insofar as, in the fantasy world conjured by marketing, delivery is supposed to be merely a means of providing desired content, rather than a way of "managing consumers," as the title of one database marketing book puts it. The pioneering spaces for this form of personalization were news and information webpages, such as My.Yahoo.com, that facilitated the transmission of customized content to online readers. Around the same time, the idea of personalization also drifted into the new, burgeoning market for mobile phones, the notion of content itself becoming increasingly relevant to that commodity. Hence, in the English-language press we find discussions of new "smartcards," which could be used as "personalization equipment" for mobile phones as early as 1995.[162] As with the ringtone phenomenon generally, the discourse of mobile phone and device personalization took hold in Japan and East Asia roughly at that time. But whereas the smartcards were intended to provide forms of customer and news service personalization (including business news for corporate owners), young girls placing photograph stickers on their pagers were also eliciting analogous attention in Japanese scholarship.[163] In August 1998, Nokia described its 5190 series as including "'personalization' features including a selection from more than 35 ringing tones with melody lines from popular songs, a directory that stores up to 255 names and numbers and three language options (English, French, and Spanish)."[164] A month later, Nokia introduced the sale of painted mobile phone covers produced by company-recommended painters, again under the rubric of personalization.[165] From then on, it would seem, personalization became increasingly associated with mobile entertainment—an association that persists more than 10 years later, even though Internet-based personalization has remained a topic of interest in the press.

Even today, the target markets driving cell phone personalization are the young adult and teen/youth markets. In 2004, before Crazy Frog and the subsequent barrage of unwanted ringtone subscriptions negatively affected spending on mobile entertainment, teens' spending on consumer electronics (especially cell phones and iPods) and accessories was so high that it had significantly eaten into clothing consumption in that demographic segment.[166] Still, marketing executives seem unable to predict exactly what will become the next fashionable form of mobile entertainment, and instead companies have been vague in their promotion of newer technologies, such as 3G phones, that offer a host of accessories and devices.[167] Thus, the ceaseless appearance of new gadgets and devices on phones (from screen "wallpaper" to video games to cameras to video recorders to the surfeit of ringtone derivatives) is driven by that familiar dialectic of youth consumption desires and consumer marketing strategies. The meeting of these two forces has, to this point, resulted in the increased consumption of both new handsets and digital mobile-entertainment files. But with the rise of notions of user-generated content that have become central to present-day use of the Internet, the mobile phone, lagging behind and reproducing the Internet, appears also to be having its own "2.0" moment, Mobile 2.0 discourse having becoming rampant as of late.[168] Perhaps, then, we ought to demarcate a different process, wherein cultural entertainment and leisure consumption are progressively *outsourced* to consumers themselves—first the labor of making a plethora of micro-level decisions of taste and preference for a wide array of prefabricated content, and eventually the labor of constructing the content itself.

The Celestial Jukebox and Convergence

The digital utopia of the perfectly integrated, portable entertainment device is the limit toward which many phones currently point, and has loomed on the cellular horizon for some time. The competing visions of the cell phone as music player and integrated entertainment device constitute somewhat different goals for the phone of the future. The *Wireless Review* and Allmusic.com writer Jason Ankeny perhaps best articulated the former perspective with the phrase "the celestial jukebox."[169] Borrowed in 2003 from Shawn Conahan, then president of the Los Angeles-based mobile-entertainment firm Moviso (which was bought by InfoSpace in 2003, then sold to FunMobility in 2007[170]), the phrase encapsulates a vision of complete musical access available instantaneously through a cell phone. Indeed, similar outlooks on the future of music have at various points been articulated by music-industry artists (e.g. Jay Sean), copyright experts (e.g. Lawrence Lessig), and online companies (e.g. Kazaa)—with the differences between them hinging on the nature and amount of payment required by music consumers.[171] But Conahan has articulated an ideal that extends beyond music alone: "my vision is that someday you'll be able to reach into the sky and pull down entertainment content, no matter what the device."[172] Perhaps unknown to Ankeny, the term "celestial jukebox" had been in use since the late 1980s as part of "a rhetorical effort to gain legislative support for digital-performance rights on the Internet."[173] Currently, the phrase is used by media executives to describe a total media device, in which every accessed cultural good is paid for by an individual transaction or a proprietary subscription plan.[174] Such a perspective on the future of the mobile phone seems to be corroborated by the rapid growth of multimedia messaging services (MMS) and multimedia broadcast messaging services (MBMS), WAP, 3G, and broadband mobile Internet access. Through these services, different kinds of media—music, video games, photos, text messaging, e-mail, and video—have been made readily available on mobile phones.

Multimedia content distribution also became central to the industry: Nokia's 2004 announcement that it would offer "visual radio," in which mobile phone users would receive FM radio broadcasts directly on their phones, along with synchronized text and images, might serve as an early example.[175] In the mid 2000s, one of the primary goals of the media industry was the transmission of TV broadcasts on cellular networks. At that time, at least four handset manufacturers (NEC, Nokia, Samsung, and Toshiba) set about designing phones for that purpose, and Sprint and other wireless carriers began to offer streaming TV to their subscribers.[176] Although mobile TV is somewhat popular in Japan and South Korea and is growing popular in Europe, with the rise of "mobisodes" or 3–5-minute programs being developed for the mobile TV medium itself, as of 2008 it was a niche service elsewhere in the world.[177] Moreover, such services are still hindered by the physical capacities of batteries (which cannot yet power such devices for more than a few hours), by the memory and processing power of the devices, by the infelicities of watching TV on small display screens, and by a market increasingly flooded with phones that redundantly reproduce features on other portable hand-held gadgets (the iPod, portable video game devices, portable film and TV viewscreens, and so on).

But despite these obstacles, a number of newer devices were developed in the mid 2000s that moved the cell phone closer to some version of the cell phone's theoretical endpoint. Motorola's decision to sell MP3-playing cell phones was an important development. Working in tandem with Apple Inc., the handset manufacturer announced that a mass-market Motorola music phone would appear in 2005 that would hold from a half-dozen to a few dozen songs, depending on the model.[178] Disabusing any notions that the designer and vendor of the iPod was working against its own economic interests, Apple argued that the ability to purchase and play a few songs on a cell phone would encourage phone users to purchase the iPod—and the venture retrospectively paved the way for Apple's unveiling of the iPhone in 2007.[179] On the software end of such projects, a company called MusicGremlin attempted to create programs that could connect its own song database directly to cell phones, bypassing computer-based websites such as Apple's iTunes site.[180] Moreover, with Qualcomm and other companies increasingly successful in their attempts to create high-quality three-dimensional audio sound from cell phones using only two speakers or headphones, public music playback via phone—as a kind of celestial boombox—has become a popular pastime on London buses.[181]

Apple's increasing involvement in the mobile-entertainment business—spurred by efforts to expand and consolidate its successes with the iPod and the iTunes store, which as of 2008 is reportedly the largest music retailer in the United States (surpassing Walmart)[182]—led the company to develop its own mobile telephone, the iPhone. Unveiled in January 2007 and first sold in June of that year, the iPhone combined the mobile phone with the iPod and a hand-held computer with Internet access, all facilitated by an elegant, touch-sensitive screen interface. Described as a "game-changing" device that "changes the rules in the cellphone market," the device received a tremendous amount of advance press coverage (some 11,000 articles and 69 million Google hits *before* being released). It was estimated that about two-thirds of the population of the United States had heard of the device before it was sold.[183] Consumers all over the country stood in long lines before and on the day of the phone's release, and the iPhone quickly sold out at many Apple stores.[184] Sold with exclusive contracts to select wireless carriers in different national markets—AT&T in the United States, O2 (Telefónica) in the United Kingdom, T-Mobile in Germany—Apple "locked" its phones into these arrangements and likewise discouraged independent software being uploaded on the phones. But within two months of its release, the iPhone's closed system was hacked by George Hotz, a 17-year-old from New Jersey. Soon a global gray market for the unlocked iPhone emerged. In February 2008 it was estimated that more than one-third of the 3.7 million iPhones sold at that time were not registered on its official wireless networks in the United States and Europe.[185] Although Apple gradually expanded the official global market for the iPhone—for example, in the summer of 2008 Apple's partner telecommunication firms announced its arrival in India, Estonia, and the Czech Republic[186]—the device was not unique or even particularly novel in its combination of features, but it inspired imitation by various handset manufacturing firms (LG, Samsung, Nokia, Motorola) and hence pushed the market one step further along the path of creating an all-purpose technological fantasy-device. The iPhone *is*, in a sense, the Celestial Jukebox.

But of course it isn't. As Patrick Burkart and Tom McCourt emphasize, what is fundamentally at stake in the jukebox imaginary is the proprietary control of content by media conglomerates. According to Burkart and McCourt, Universal, Sony, Warner, and EMI control about 85 percent of recorded music.[187] The fantasies inspired by the iPhone and any similar mobile phone—whether understood as "black box" or as a "general purpose remote control device"[188]—tend to mask the fact that media content is by no means entirely coterminous with the technological advances and the "technocratic ideology" that undergird those fantasies. For Burkart and McCourt, the Celestial Jukebox is based on the application of personalization technologies to music sales, which can maximize profit by customized product delivery, and through digital-rights management (DRM) technologies that can prevent or reduce file sharing, thereby maximizing profit (in principle) by restricting content use and access.[189] In view of the long history of the music industry's opposition to various recording technologies, the Celestial Jukebox seems to be a compensatory strategy of distribution control.

If "the Celestial Jukebox" is an ideological term of content, the crucial ideological term of delivery is "convergence," which refers to the unification of disparate media forms into a single mechanism. It originated in largely scientific and mathematical discourse and transformed into a Cold War-era term signaling the increasing overlap of the capitalist and communist worlds, a process debated by Isaac Deutscher, Herbert Marcuse, John Kenneth Galbraith, Daniel Boorstin, and (most importantly for the prophet of convergence, Henry Jenkins) Ithiel de Sola Pool.[190] By the 1980s, discussions of political convergence, appeared less frequently, and the Gorbachev-era Glasnost and the effective convergence of the 1989 liberal revolutions and their aftershocks rendered them moot. Meanwhile, an apparently depoliticized variant of technological convergence had become ascendant within corporate discourse. Discussions of the merging of computers and communications networks were quite common in the 1980s.[191] The hoopla surrounding the deregulation-authorizing 1996 Federal Telecommunications Act provided a crucial moment in the feeding of convergence dreams, which were tantamount to lucrative tech-business opportunities.[192] Upon its deregulated transformation into a commercial technology, the Internet would serve as the foundation for the ideology of convergence, but the union of mobile devices with Internet access has magnified its scope and tangibility. As Veltman notes, "at present there are three main networks: 1) a broadcast network (of television and radio); 2) a communications network (of telephones), and 3) the Internet," and the Internet appears to be swallowing up and becoming intertwined with the others to such a degree that Hiroshi Fujiwara, CEO of the Japan-based Internet Research Institute, claimed as early as 2000 that the Internet would fully subsume the other two by 2010, at least in Japan.[193] However, the visible melding of once-monofunctional devices may partly obscure the fact that economic consolidation and integration are conditions of possibility for convergence, which requires competing industries either to merge or to act in concert. As Jenkins puts it, "digitization set the conditions for convergence; corporate conglomerates created its imperative."[194]

But corporate conglomerates are not coherent monoliths, nor do they consistently span the various industries involved in bringing convergence and the Celestial Jukebox closer to consumerist reality. Hence, new mobile digital devices continue to flood the marketplace, with relatively newer ones like the iPod or mobile phone both creating new needs and supplanting older devices such as the Walkman or pager. Likewise, smartphone purchases have more recently cut into iPod sales, particularly as the phones have reduced in size and weight.[195] If Jenkins avoids addressing this problem in depth by focusing on the ability of audiences to navigate the fractured mediascape—convergence, in his argument, being both a technical *and* a social relation—the cleavages between content and delivery industries, which are often, but not always, housed in different divisions of the same conglomerate (for example, the Sony Corporation), seem to create a structuring contradiction that cannot be simply resolved by industry coordination and cartel pricing.[196] Indeed, the contradiction between the content and delivery industries can be felt in the threats to the very existence of the ringtone industry.

Threats to Accumulation

From the quotation of "Blackbeard SEO" at the opening of this chapter, and from the responses it elicited, it appears clear that the increasingly easy cost-free creation and uploading of ringtones helped to destroy the ringtone market altogether. The costs of such a loss for the global music industry have been momentous—as early as 2003, ringtones accounted for approximately 10 percent of global music sales (which totaled $32.2 billion in 2003), and as of 2007 sound-file ringtones still accounted for the vast majority of mobile music sales.[197] Although this decline appears to be relatively recent, its roots extend at least as far back as the creation of the sound file ringtone itself and are inherent to the fungible, flexible character of the digital audio file, which can be very easily converted into different file formats, accessed by playback equipment, and distributed via the Internet. A good starting point would be the formation of Xingtone, a Los Angeles-based company founded in early 2003 that produced a downloadable computer program allowing an individual to transform any sound file (from a CD or an MP3, for example) into a sound-file ringtone. After paying a one-time fee of $15 for the software, a consumer can easily produce ringtones without any further cost. Upon registering a phone with a uniquely purchased copy of Xingtone, the software allows you to edit the ringtone sound file; upon completion, the file is sent from a computer via the Internet to a company file server and then back through the phone subscriber's wireless Internet service to the particular phone. Brad Zutaut, co-founder and former CEO of the small company, has stated repeatedly (with the support of the Recording Industry Association of America) that the program falls under the domain of fair use in copyright law. Beginning his enterprise from the impulse of wanting to make ringtones that were not commercially available, Zutaut has argued that ringtones, which are merely data transfers, should not be so expensive, and that "ring tones are not going to save the music industry."[198] The company seems to have been successful and has pioneered music promotion deals with record labels like Disney's Hollywood Records and the independent Artemis Records (whose band Sugarcult released a single from its album via ringtone in partnership with Xingtone). In 2004, Xingtone received financial support from Siemens to expand its operations, allowing it to become, according to one industry observer, "a viable company in the mobile music market."[199] In an interview in May 2004, Zutaut stated that the company had a window of 3–5 years, and although the Xingtone website (xingtone.com) was active as of the summer of 2008 its latest press release dates from June 13, 2006.[200] It is not clear whether Xingtone was bought out by a major media or entertainment conglomerate. Although the music industry's interests conflict with those of the company, making it at one time a target of the industry, Xingtone's software is no longer unique.[201] ToneThis (also from Los Angeles) followed Xingtone's lead in 2003 and produced similar software packages.[202]

In subsequent years, a large number of free ringtone-creation programs (such as SmashTheTones in 2005) and, later, phone apps became available, and freeware audio editors such as Audacity or WavePad became capable of creating ringtones. Making this possible was the fact that newer phones could use familiar sound-file formats for ringtones (especially MP3 files); that new, often free, software could help mobile phone users manage files on their phones; that multiple ways of transferring files to and from phones became available; and that the mobile music industry increasingly encouraged phone owners to upload songs on their phones, in the hopes that they would be more likely to download paid-for music on those phones. All of these trends helped to break down what is known as the "walled garden" of mobile information use and commerce. The notion of the walled garden, at least within information and telecommunications technologies, extends back to the mid 1990s, when educational institutions sought to limit Internet access for their students, in the process creating a mediating entity between the end-to-end use principle that governs the structure of the Internet.[203] Toward the end of the 1990s, the term became applied to a recent development in the mobile communications industry, the wireless application protocol (WAP). The goal of WAP (first announced as a joint effort by Ericsson, Motorola, Nokia, and Unwired Planet in June 1997) was to create a "refined, license free protocol" that could provide the basis for Internet access and data exchange on mobile phones.[204] Forming the WAP Forum in early 1998, these four companies progressively released the protocol's architecture and specifications online, and worked with wireless service providers, in part through those already working with Unwired Planet (which had already developed a form of mobile Internet service), to ensure that handsets with WAP capabilities would be put to use.[205] Phones such as the Alcatel One Touch were made available on a test basis to French subscribers in late 1998 and early 1999, but WAP services went online starting in late 1999: one report mentioned that about 10,000 WAP pages were available in December 1999 and that around 2.4 million were available by June of the following year.[206] Slightly predating the development of WAP, NTT DoCoMo introduced its "i-mode," offering an HTML-based mobile Internet service (in this way, rather unlike WAP) in Japan starting in February 1999.[207]

During the early days of WAP's presence in the mobile telecommunications market, a number of critics emphasized its structural deficiencies. WAP is essentially a parallel scheme of accessing information content on the Internet via mobile phones and is comprised of its own text-only browser and a protocol-conversion "gateway" that mediates between the phone and the accessed website. The impact of limited access for users immediately gives them the not entirely accurate impression of a "walled garden," as advocates for WAP have emphasized.[208] Moreover, as many critics noted, wireless service providers could further limit or at least control access to the Internet by privileging or only allowing specific pages to be accessed by their phones—Orange and Vodafone in the United Kingdom, for example, initially only allowed access to their own respective WAP pages.[209] Despite gaining traction within the wireless service industry, WAP's success was strikingly limited, with criticism continuing into the new millennium. The computer scientist Dan Duchamp argued in September 2002 that the Internet was being transformed into what he described as "the Discrete Internet," with numerous separate networks accessed through various intermediaries; despite lamenting the decline of genuine end-to-end accessibility online, Duchamp also took an accommodationist view by arguing that "the original homogenous Internet model is gone forever" and devising new models for managing and recovering a semblance of that earlier model's benefits.[210]

Not exactly living up to its promise—and in this way, showing poorly with respect to DoCoMo's less restrictive i-mode, which was already transforming Japanese public culture shortly after its release[211]—WAP appeared to be a failure within the mobile industry. However, the rise of the ringtone industry—particularly the emergence of polyphonic ringtones—saved WAP's fortunes, validating it as a commercially viable technology. Al Russell, head of mobile Internet and content services at Vodafone, vividly describes this moment in the industry's history:

> I joined the GSM story in 2000, the year the dotcom bubble burst and WAP came in for some criticism. Revenue was being generated primarily from ringtone advertisers in the back pages of tabloid newspapers and men's magazines. In November 2002, something changed. Sharp and Vodafone created a genuinely integrated experience between a phone and a mobile portal that was launched across Europe. Customers suddenly had one-click access to a portal that used colour icons. They could also buy polyphonic ringtones and Java games through a simple one-click billing mechanism. I remember the afternoon it went live. We saw an immediate volume of traffic that we found hard to believe. I witnessed my first inflexion point in the mobile data story.[212]

Russell's ascending "inflexion point" marked the speedy rise of profitability within the ringtone industry, but the countervailing tendencies unleashed by the sound-file ringtone and new file-transfer protocols for mobile phones within just a few years facilitated ringtone sideloading: the uploading to phones of ringtones not purchased from ringtone aggregators. Like the practice itself, the term "sideloading," applied equally to ringtone and full-track uploads, has become extremely common worldwide. Surveying the main North American and European markets in November 2007, M:Metrics reported that "sideloading accounts for 83 percent of mobile music usage across France, Germany, Italy, Spain, the United Kingdom, and the United States" and that the use of musicphones (phones capable of playing back all major sound files and having at least 47 megabytes of storage space) was partly to account for this trend, with some 63.8 million in use in those markets.[213] Although the mobile music industry attempted to spin this information as a positive sign, since it could have led to greater mobile music consumption, this hypothetical compensation for declining ringtone revenue was surely cold comfort for mobile music capitalists. In 2007, Andrew Bud, then vice-chairman of the Mobile Entertainment Forum and executive chairman of mBlox, put it succinctly: "I think the ringtone business is in peril now because the operators have allowed into the market mobile phones which can sideload MP3s and use them as ringtones."[214] The reporter Adam Webb characterizes the "contradictions" of the mobile-content industry as follows:

> Sideloading highlights some of the mobile market's inherent contradictions. The facility is essential for handset manufacturers such as Nokia and Ericsson, which are in effect competing head-to-head with Apple in the MP3 player market. But that is not necessarily in the vested interest of the networks that want consumers to buy content over the air, preferably within their walled garden.[215]

Another factor making free ringtone distribution even easier is the presence of Bluetooth technology on mobile phones. Bluetooth, a short-distance wireless device communications and file-transfer protocol that effectively substitutes for USB ports and cables, was invented by the Swedish telecommunications firm Ericsson and named after Harald I (Harald Bluetooth), a tenth-century Danish king who unified conflicting Danish tribes under his kingdom. Developed in 1998–99 with the intent of similarly unifying different devices, Bluetooth has become increasingly common on mobile phones, allowing them to connect freely with computers and other mobile phones—in some cases, too freely, leading to the phenomena of "Bluesnarfing" (information/file theft) and "Bluejacking" (uploading unwanted information, including text messages, video, etc.) through open Bluetooth connections.[216] Nonetheless, with a large number of phones all over the world becoming outfitted with Bluetooth—in the United States, nearly 80 percent of phones sold in the first quarter of 2008 were Bluetooth-capable, for example[217]—mobile phone users are increasingly trading ringtones within their friend and family networks. Casual mention of such exchange has become common within the global English-language press. For example, one story of a "howling dog" ringtone, which made its way around parts of South Africa in the summer of 2008 and sparked bouts of (possibly racialized) ringtone rage, seems to have been traded in just this sort of manner.[218] Moreover, to judge from interviews, ringtone exchange via Bluetooth appears to be exceedingly common, perhaps the norm, in many parts of the world.[219] Also, Bluetooth has been used as part of promotional ringtone giveaways—for example, a British band called Elbow did just that at the 2005 Reading Festival.[220] And Bluetooth has become integral to a new means of ringtone sales: interactive electronic kiosks selling mobile content. Though they extend back at least to 1997 in the United Kingdom and in South Korea, these kiosks have become popular throughout East Asia and elsewhere, selling print pictures from digital image files on mobile phones, ringtones, wallpaper, full music tracks, and more.[221]

The growth of mobile-content sales outside of the wireless carrier's "walled garden" created multiple means of purchasing or freely downloading ringtones, leading to greater pressure on the industry. In the United States, which for years lagged far behind East Asia and Europe in its mobile communications amenities, "off-deck" mobile-media sales—those occurring outside the boundaries of the wireless carrier's sanctioned services—greatly affected the ringtone market. Making these off-deck sales more common were the multiple forms of information transfer, such as the previously mentioned Bluetooth, soon common on mobile phones. As Antony Bruno noted in 2008, "while the worldwide ringtone market is largely flat, off-deck sales are booming," which suggests that the once-dominant wireless service-supported aggregators were losing revenue at a precipitous rate, as were the carriers themselves.[222] The shifting revenue figures for mobile content tell this tale clearly. In 2005, North American on-deck vs. off-deck income was $1.1 billion vs. $381 million, whereas in 2007, the revenue numbers were at $2 billion vs. $1.5 billion, respectively, with the latter probably passing the former as early as 2008. Part of the story here is one of wider product availability off-deck. Regarding one such off-deck aggregator, Thumbplay, Bruno states that "on the carrier deck, the top 20 ringtones represent around 90 percent of the revenue made from the category. [CEO Are Traasdahl] says Thumbplay's 20 best-selling ringtones only make up 17 percent of its revenue. The other 83 percent comes from Long Tail catalog and niche-audience titles."[223]

One of the most important forms of information transfer enabling these consumption shifts is mobile Internet access, which, while common in Europe and East Asia, took off later in the United States, with the rise of the iPhone's (and iPhone-inspired) haptic Internet browsing being a crucial motivating factor. Internet access is a critical line of demarcation within the digital music business: when it comes to full-track downloads, consumers are most likely to buy them online and download them to PCs first, in contrast to the largely impulse-based ringtone purchases that happen most often through SMS text messages or on-deck WAP pages.[224] Hence, mobile Internet access, especially that which bypasses the garden walls of mobile carriers, has opened the ringtone and mobile music markets to a wider range of competing aggregators and significantly depressed profitability in digital music retail, wherein margins already appear to be "razor-thin."[225]

With the growth of the mobile Internet, we may also see the continued prevalence of another means of scaling the garden walls: ringtone piracy. The problematic notion of ringtone piracy has been associated both with users copying ringtones illegally and with small aggregators failing to pay for copyrighted material used in monophonic or polyphonic ringtones. Both forms have been widespread all over the world, particularly throughout Asia, where music and media piracy are common more generally.[226] Industry-produced piracy statistics on developing markets reached the misleadingly lurid extremes produced by the multinational media industries —for example, the claim in 2004 that an estimated 90 percent of ringtones in Malaysia were pirated—and the recording industry has continued protecting its developed world markets and pricing by eradicating what it refers to as a piracy "epidemic," through legal action against rogue aggregators and using DRM and secured websites to limit the accessibility and exchangeability of content.[227] But when it comes to user-based "piracy," the phenomenon appears to be universal. One study by Frost & Sullivan of European mobile phone use in 2006 claimed that 80 percent of mobile phone content (the majority of which was ringtones) "has been hacked or downloaded onto mobile phones, illegally from websites, or through the use of Bluetooth, memory cards and other storage devices" and that the cost by that year's end would be around 2.7 billion euros (about US$3.5 billion at the time).[228] The Recording Industry Association of Japan (RIAJ) conducted a survey of 1,054 Japanese music listeners, probably in late

2007 or early 2008, "which found that 37.1 percent said they use illegal sites for mobile music downloads, and 14 percent said they upload files to the sites. The RIAJ estimates that more than 399 million mobile tracks were illegally downloaded in Japan last year."[229] One factor making illegal ringtones particularly easy to obtain was the presence of websites that were not "secured"—in which the sound-file excerpts of ringtones were easily downloaded and converted into ringtones themselves—leading to what was called "ringtone shoplifting."[230] One online and wireless billing firm, Qpass, identified this situation in 2005, noting that two-thirds of ringtone websites offered unsecured music excerpts. Certainly, ringtone and mobile music piracy are increasingly inseparable from the general realities of digital music file sharing worldwide, which is deeply ingrained among younger, tech-savvy consumers and which the global music industry interprets as lost sales revenue—the IFPI complained in its 2008 annual report that piracy cost American record companies $3.7 billion in 2007.[231] Bracketing the likelihood of inflated "substitution effect" rates and ignoring of the "countervailing benefits" of piracy on profits (for example, hearing a pirated ringtone may encourage legal purchase of the same one), the ringtone adds an extra layer of irony to alarmist claims of individual users' music piracy.[232] Specifically, music-based ringtones often, if not always, represent musical tastes or interests of listeners at a given moment, and it is likely that they either already possess the music in some form or find it widely available on radio broadcasts. In many of these cases, then, the ringtone is almost certainly a "fair use" of a music file that itself may have been "legally" or "illegally" obtained. Characterizing the problem as one of ringtone piracy, then, appears not to be the most accurate or even effective way of combating it, were that desirable.

The notion of fairness in the context of ringtones—which many consumers and commentators feel should not have to be purchased—has motivated writers in various media organs to offer a pedagogy of free ringtone creation in the apparent service of consumer's rights. Among the many explanations of ringtone creation software, sound editing, file extensions, and file storage on mobile phones, some of the most striking are those advocating free ringtone creation on the iPhone, which as noted earlier, despite its "locked" protocol has provided many with the hopes of free and easy ringtone use. On the vanguard of such efforts has been *Wired* magazine, whose "radical" consumerist and entrepreneurial self-image was brilliantly savaged by Keith White in the magazine *The Baffler* during the height of the 1990s tech-economy boom.[233] Indeed, *Wired*, like many other publications, published a self-consciously transgressive article on how to make free ringtones around when the sound-file ringtone became dominant in the United States (2005–06).[234] Just a few weeks after the iPhone was successfully hacked, Scott Gilbertson of *Wired* wrote a follow-up article describing how to make free ringtones on the iPhone.[235] The article's most useful observation was that whereas the version of iTunes (7.4) used first on the iPhone stored music files in Apple's AAC format with an .m4a extension, its AAC-file ringtones were labeled with an .m4r extension, making the conversion between the two simple enough; the article also mentioned the purchasable programs iToner and iPhoneRingtoneMaker which combine sound-file editing and conversion, further simplifying the process. However, Gilbertson also noted that Apple made subtle coding corrections to iTunes to prevent easy music file-to-ringtone conversion, and to counter this *Wired* devised a user-written and user-edited "how-to Wiki" revealing hackers' successes in circumventing Apple's iPhone system and software updates.[236] The article includes a disclaimer about the legality of the process—noting that some claim that a ringtone "does *not* constitute fair use and may be grounds for copyright infringement"—and cheekily proffers mobile style advice: "And before you go crazy making a dozen ringtones, we can't help but suggest you consider another ring option: the ever-popular vibrate setting. Like a timeless black suit or little black dress, vibrate is classy, demure and doesn't annoy the hell out of everyone within 20 feet of you."

Wired's pretensions to consumer radicalism notwithstanding, the magazine's article demonstrates the simplicity of making ringtones from sound files. It is a small gesture within a complex set of technological and social forces making the free ringtone a permanent fixture in the world of digital music and thereby forcing the music-industry-owned aggregators and wireless carriers to reduce ringtone prices. Indeed, although Apple's iPhone did not make ringtones freely available by linking them without cost to the iTunes library, Apple did, as the article notes, reduce the price of the sound-file ringtone to a maximum price of $2 in the United States (down from the once-standard price of $2.50 or $3.00). Moreover, the price of the ringtone itself is effectively halved, in that the full track sound file still costs $1 when purchased through iTunes, whether on a computer or a mobile phone, and a ringtone based on the sound file costs an additional $1. Likewise, full-track downloads to mobile phones, which once cost as much as ringtones—and are still more expensive than full-track downloads to computers, since most wireless carriers charge by the size of the file transferred—have come down in price, Sprint's reduction from $2.49 to iTunes' standard $.99 in 2007 being indicative of a broader trend.[237]

But while reduced ringtone pricing in the United States and Europe is a relatively recent phenomenon, other pricing models have been extant for sometime elsewhere in the world. In Japan, where the i-mode protocol offers greater access to the Internet than the more restrictive WAP, the forces impinging downward on ringtone accumulation long ago led to a different pricing standard, in which ringtones cost a fraction of full-tracks and could be used as ways for consumers to sample or try out a song, as mentioned earlier.[238]

Oligopoly Consolidation, the Long Downturn, and the Death of the Ringtone Industry

The prognosis for the ringtone industry is not good, and while digital music sales are increasing, it is not clear that the technological forces facilitating the easy, cost-free distribution of digital music files are likely to be transformed anytime soon, if ever. Indeed, despite earlier, gloomy assessments of the music industry's DRM clampdown on file sharing, the recent trend has been for the industry to reduce its use of DRM.[239] The British Phonographic Institute put a good face on the process:

> While the industry is still not entirely unified on DRM (Digital Rights Management), common ground is steadily being arrived at. While several labels were happy to offer DRM-free music from the start, many others were more cautious. EMI's deal with iTunes in April 2007 was significant and others have all dipped toes deeper and deeper into the waters. DRM will, of course, remain for subscription services, but the move towards DRM-free à la carte looks set to finally deliver on what many consumers want—interoperability.[240]

The "of course" regarding subscription services appears more nervous than assured, though, and the music industry may continue lurching toward some sort of license or fee-based means of accumulation, requiring payment by Internet providers to cover the indiscretions of their clients.[241] The industry had once hoped that ringback tones, directly controlled by the carrier and therefore impossible to acquire freely by file transfer, would be the "savior of the music industry" (Multimedia Intelligence claimed the ringback tone would be worth \$4.7 billion by 2012), but one wonders how long consumers will pay for choosing the near-equivalent of "hold music."[242]

The history of the ringtone industry might be recapitulated as follows: Emerging in the late 1990s as a small boutique industry servicing a novelty feature of the second-generation mobile telephone, ringtones soon became profoundly lucrative and led handset manufacturers to upgrade ringtone technologies from monophonic to polyphonic and then sound-file formats. The ringtone industry "took off" with the emergence of the polyphonic ringtone, whose specialized labor demands (involving careful musical transcription by skilled workers) created a new division of labor and process of firm specialization within the burgeoning industry, which was worth more than a billion dollars by the early 2000s. However, the move to the sound-file ringtone allowed the global recording industry to enter decisively into the ringtone market and cash in on the ringtone craze. In doing so, the recording industry demanded extreme cartel-level rates in North America and Europe—and less so in Japan and other markets—both to maximize short-term gain and to eliminate the presence of independent ringtone aggregators. However, the impact of piracy/free exchange, downward pricing, and consumer anger in response to manipulative ringtone subscription services had a negative impact on the market, particularly in Europe. Despite market potential for ringtones in the developing world (especially China and India, and perhaps especially in countries abandoned by the multinational recording conglomerates, such as Kenya[243]), lower prices reflecting lower living standards and exchange rates set an upper limit on the profitability of those markets, leading ultimately to the gradual decline of its overall profitability (with continued but reducing growth rates). The global economic crisis of 2008 greatly reduced disposable incomes for consumers worldwide—yet another nail in the coffin of the for-profit ringtone business. Add to that the decreasing fashionability of the ringtone, which by 2008 had existed in different guises for a decade—bringing to mind Benjamin's claim that the immediately preceding generation's fashion is "the most radical anti-aphrodisiac imaginable"—and it isn't difficult to see why the profitable ringtone would soon disappear.[244]

The window of opportunity that the ringtone industry represented was made possible through a fortuitous alignment of particular global economic interests—primarily the media and entertainment industry, on the one hand, and the telecommunications industry, on the other. On each side of the divide we find interests that both collude and compete with one another—music publishers and recording labels in the music industry and wireless carriers and handset manufacturers in telecommunications—and an independently owned mobile-content industry's attempt to insert itself into this already crowded space has been only marginally successful. To the extent that this is the case, we can point to two separate developments enabling the existence of that space. First, we can point to the emergence of digital capitalism, which entailed a project of massive neoliberal privatization and finance capital investment and in which the largely US-owned and developed technology of the Internet (with the support of the US computer industry more generally) overtook the existing US telecommunications network, favoring data transfer over voice traffic and mobile telephony over land lines.[245] In this narrative, the US underwrote the global expansion of Internet technology, partly as a weapon in the broader propagation of neoliberal economic policies and partly as a means of fueling the tech-sector bubble that was artificially bolstering the US economy during the overcapacity-driven post-1970 downturn. The second was the seesaw effect on the tripolar global economy since the onset of that downturn, which was produced by periodic alternations of currency deflation and resulting export sales between the United States, on the one hand, and Europe and Japan, on the other. If development of ringtones in Europe and East Asia (and its relative absence in the US) points both to the specific tactical failures on the part of the US telecommunications industry to hew to a single transmission standard (GSM or CDMA) in particular, the industry's rise and fall more generally correlates with the 1990s US speculation bubble's jumpstarting of the international economy (particularly after the disastrous 1997–98 East Asian downturn), which magnified the strength of US consumer demand and facilitated what amounted to a massive, credit-fueled wealth transfer into the other two poles of the global economy.[246] Moreover, following Peter Golding's useful oversimplification that, during the 1990s, in "Europe the activity has been mainly in telecommunications, in the United States in the audio-visual sector, and in Japan in electronics," we can see that something of a global division of sectors—and possibly a transnationalized "sectoral clash"—favors leading-edge productive interests in Europe and East Asia and rentier interests in the United States, a tendency that seems to align with the economic forces necessary to produce and then quash the ringtone industry, while also hinting at the possible prognosis of US economic decline.[247] Indeed, the previously discussed story of technology transfer that marked the move of polyphonic ringtone technologies from the US to Europe may be one small example of a larger pattern.

The Death of an Industry?

To judge from a few economic reports, the remaining life span of the ringtone is rather short. The established publishing firm BMI claimed that 2007—the first year in which ringtone sales in the United States would decrease after the preceding year ($550 million from $600 million)—was the year of the downturn. Generally, the claim appears to have been borne out, to judge from figure 1.2, which shows shipment figures (in millions

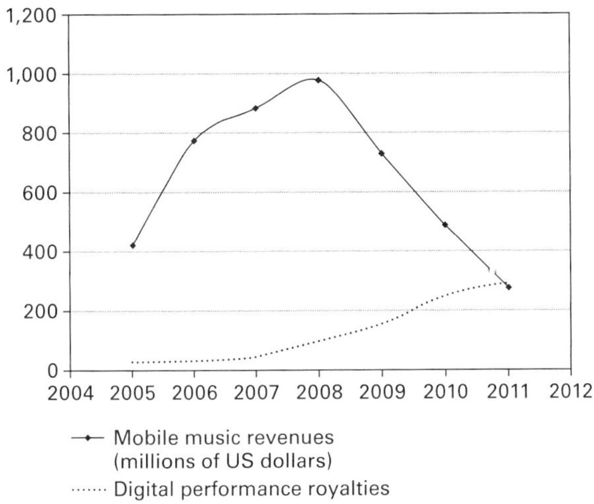

Figure 1.2
RIAA year-end shipment data for mobile music sales and digital performance royalties, 2005–2011.

of dollars) for mobile sales, which include "Master Ringtunes, Ringbacks, Music Videos, Full Length Downloads, and Other Mobile" products, and, for comparison, digital performance royalties. Accounting for the relatively small percentage of on-deck full-track downloads via mobile phones in the US market, we can see that the peak was actually in 2008 ($977.1 million), with the conjunction of that year's global economic crisis and the decreasing appeal of ringtones, ringbacks, and carrier-based downloads having a compounded effect.[248] Ringtones faced the brunt of the decline, facing a 34 percent decrease from 2008 to 2009, a 41 percent decrease from 2009 to 2010, and a 38 percent decrease from 2010 to 2011.[249] In contrast, digital performance royalties, including "payments to performers and copyright holders for webcasting, satellite radio, and other non-interactive digital music services" (including, for example, music listening via Internet radio), surpassed "mobile products" in 2011, reaching $292 million—while still remaining a fraction of peak ringtone revenues.[250]

Whatever the prospects for digital music revenue overall, the ringtone industry is by now undeniably redundant. But although that industry is terminally imperiled, its profits transformed into losses and the last of its firms folding or moving on to greener pastures, will the "traditional," for-profit ringtone ultimately disappear and quickly be forgotten as ephemera? This is certainly possible, as are one of two other outcomes, which are not mutually exclusive. In the first, the ringtone merely settles slowly into an effective holding pattern, or at least an asymptotically decreasing sales decline, in which some small subset of digital music consumers will continue to pay inflated costs for ringtones and some amount of profits, especially garnered through sales of exclusively available ringtones, will still be had by the music industry and other mobile-media firms. In the second, the price of the ringtone in the West continues to drop sharply, in an effort to remain competitive with other digital music sales. Here, Western firms would be following the trend in Japan. In all cases, though, the continued existence of the ringtone would, for the capitalist world-market, still amount to a death, in the sense that the ringtone would cease increasing in profitability and hence be consigned to the figurative digital dollar store, selling products that were once, but are no longer, leading-edge commodities. But unlike digital watches or digital alarm clocks or other once-leading edge products, the logic of production movement from core to semi-periphery to periphery may not be applicable, since the same digital information is merely being sold under two guises, as part of a repackaging tactic used by media monopolies to maintain their monopoly status within the world economy. The bad faith in the digital economy lies in its claim to evade or transcend materiality itself—conflicting material interests, technological constraints, and even the labor process itself, which is sometimes unhelpfully termed "immaterial."[251] Though the supersession of material reality is impossible, high-tech capital can eschew traditional patterns of accumulation in favor of quick fixes facilitated by the fungibility of digital information. For skeptics, however, some comfort can be taken in knowing that these fixes cannot evade, and are in the last instance undone by, the merciless logic of the dialectical reversal.[252]

II Ramifications of the Ringtone's Identity Crisis: The Social and Cultural Fallout of Technological Transformation

In an excellent treatment dating from 2005, Sasha Frere-Jones underscored the ephemeral and transitional nature of the polyphonic ringtone:

> Next time you hear your favorite song playing in full verisimilitude from someone's pants, give a moment's thought to the lowly, twinkling polyphonic. Transitional stages of technology often have their own imperfect charms, memorable in ways that no one could have predicted. Polyphonic-ringtone nostalgia is approximately six months away.[1]

The "imperfect charms" of the ringtone's earlier phases and potential nostalgia for its loss point to some of the ramifications of the ringtone's rapid history of technological development from monophonic to polyphonic to sound-file formats, as described in the previous chapter. As "attractions," such digital charms are the (degraded) utopian elements of the ringtone as novelty, but more is at stake in their efflorescence than entertainment.[2] Indeed, if we briefly consider the broader impact of mobile telephony, which some have characterized as a genuine Schumpeterian economic *innovation*—as opposed to a purely technical *invention*, the innovation's counterpart in the conservative Austrian economist's famous dichotomy—we might appreciate the degree to which its transformative effect has not only been profoundly productive, but also destructive.[3] Since the mobile phone might best be described as what economists call a General Purpose Technology (GPT)—a technology having widespread "disruptive effects" on the social fabric, such as automation-induced unemployment—we should appreciate that the mobile phone industry has certainly produced its share of GPTs, the same being true on a smaller scale of the ringtone industry as well.[4] If, then, Schumpeter's term "creative destruction" can be used appropriately here, eschewing both its originator's subjectivist understanding of capitalism and its corresponding appropriation by management gurus to lionize corporate downsizing and other predatory techniques of the "new economy," it should be framed within a broader understanding of the familiar discourse of planned obsolescence (and the fact that, contra Vance Packard, technological advance and stylistic obsolescence cannot be easily distinguished where technologies of cultural production are concerned) and of theories of social acceleration, whose complex determination must be understood, in the last instance, as a by-product of acceleration in the productive processes and systemic life cycle of capital accumulation theorized by Giovanni Arrighi.[5] The result of the modern capitalist economy's speed-up, Frere-Jones suggests, would seem to contract the temporal horizon for nostalgia, with the indexical origins of the aura, age value, or historical value of its objects becoming progressively telescoped in time along with the corresponding increase in the outmoding of commodities.[6]

But if all of the foregoing can be understood as an essentially unidirectional process related to the historical transformation of newness under capitalist modernity, it is not the whole story where the ringtone is concerned. For the problem of nostalgia identified by Frere-Jones should alert us to other forms of ringtone nostalgia, and as such, to a broader, entrenched dynamic of cyclical processes concerning the ringtone's historical referents. That dynamic concerns not only the more obvious forms of nostalgia facilitated by sound-file ringtones' recordings of old-fashioned phones or music conglomerates' exploitation of their lucrative back catalogs of groups such as ABBA and the Beatles for ringtone sales.[7] At a deeper level, nostalgia is built into the very structure of the popularity of older ringtone formats—at least for developed-world consumers in their twenties and thirties during the 2000s—on account of the reuse, with minor adaptations, of the sound production systems of earlier technologies, such as the Yamaha DX-7 frequency modulation synthesis chip for polyphonic ringtones, which can produce sounds that are not profoundly distant from the synthesizers on video game consoles such as the Nintendo Entertainment System (NES). These kinds of sonic affinities between historical synthesizer systems make polyphonic-ringtone arrangements of the popular Super Mario Bros. theme song by Koji Kondo reasonably comparable to their original source.[8] Bracketing, for the moment, the heavily youth-directed (especially teen and "tween") target market for ringtones, the cyclical process under consideration here is one that is most specifically relevant to that generational cohort for whom the ringtone functions as a form of implicit nostalgia. Indeed, the ringtone may have unwittingly participated in the 1980s fashion revival of the 2000s, following the 20-year-revival marketing meme that emerged out of the 1970s revival of 1950s fashion and culture—and which itself was produced by the historical forces of generational succession.[9] Perhaps, then, it is not surprising that the ringtone could be understood in its temporal dynamics as a substantive *fad*, occupying a median location between an innovation with long-lasting impact and a fleeting fashion.[10] Of course, reconsidering the younger cohort, who probably are the dominant consumers of ringtones,[11] we might appreciate the widespread popularity of the mid-1980s Super Mario Bros. theme song as being a revival that introduced younger consumers to that video game for the first time and thus as exemplifying what Fred Davis called "the geometry of intergenerational time."[12] Dovetailing with the planned obsolescence of the ringtone format (and thus cell phone models generally), we might well find a form of *planned revivification* operating at a different pseudo-cyclical rhythm.[13] In light of the above, one could undertake a similar exercise with slightly different generational coordinates regarding the monophonic ringtone, which had its single-oscillator correlates in digital watches and other devices from the late 1970s and the early 1980s.

The implicitly cyclical character of the ringtone thus contrasts with what appears to be the unidirectionality of its striking novelty. No doubt that novelty is real, and its real effects on various social and cultural domains (labor, websites, classical music, cultural production, public space) have had troubling consequences for modernity's denizens in everyday life. Particularly affected are the conditions of possibility of creative work for cultural laborers, who must account for yearly shifts in fashion and technologies, highly unstable employment, and small firms systematically succumbing to media and entertainment conglomerates and capital holdings firms. In the wake of these shifts, careers are made and lost (almost always without the benefits of union representation that appear in the broader wireless and entertainment sectors); innumerable websites, stored files, and electronic devices accumulate in cyberspace and in real space, leading to an agglomeration of digital junk within ballooning Internet archives and, more urgently, landfills and other spaces of hardware disposal; and real and symbolic cultural battles are waged and fought, with the outcomes now appearing largely to have been settled over the course of what one might tentatively describe as the "ringtone conjuncture," extending from the late 1990s to 2010. And yet, within the decade-long history of a fad industry's creation, persistence, and decline (if not complete disappearance), we might take notice of the "always already" character of its resultant effects, in which familiar stories of labor deskilling, the private enclosure of once-public information, the spread of Western tuning and tonality, and the appearance and disappearance of an artistic means of production and/or source of topical commentary all have at one time or another taken place, in variant forms and with slower periodicities, at previous moments within the history of global capitalist modernity.

2 Ringtones and the Deskilling of Mobile-Musical Labor: A Preliminary Investigation

On January 19, 2009, the cartoonist Garry Trudeau inaugurated a week-long foray into the world of ringtones. In a series of *Doonesbury* strips depicting a radio interview between a central character, the NPR DJ Mark Slackmeyer, and a minor character, the genre-switching rock star Jimmy Thudpucker, now working as a self-described "ringtone artist," Trudeau treats the scene as a pretext for humorous commentary on the ringtone phenomenon. As might be expected, the familiar marketing discourse of ringtones acting as a means of personalization looms large, with ringtones being "about self-expression." According to Jimmy, jerks are known to download Kanye West songs, and hedge-fund managers gravitate toward a trumpet-laden ringtone featuring the lyrics "It's *good* to be king! . . . It's *good* to be king!"[1] Also unsurprisingly, the short duration of the ringtone itself served in a number of gags, including Mark's befuddlement by the brevity of Jimmy's "latest release" and, more curiously, Jimmy's description of a "summer ring-tone festival called Baby Bells," in which an entire concert requires "about 80 acts."[2] More interesting, however, are the meditations on musical form and genre in these strips. Ringtones are understood as "haiku-like" forms into which one "can pack a whole world" and are highly variable in their staying power ("some of them are evergreens, others are *very* perishable"); the aesthetic advantages and limitations of composing new ringtones as opposed to using full-length songs also merit comment.[3]

Jimmy: . . . with the longer form, you no longer have control over when the user interrupts the song. If it's before the hook, his experience is compromised. But with a two-second song, no matter when he picks up, the user gets my whole message.

Mark: Which is?

Jimmy: Your phone is ringing.[4]

In his characteristically wry way, Trudeau draws attention to a wide range of issues surrounding the ringtone—for example, as can be gleaned above, his allusion to the ringtone's thinning of song content verges on the awareness of the ringtone as being nothing more than an advertisement for mobile telephony itself.[5] And yet, for all his perspicacity, Trudeau perhaps unwittingly equates ringtone production as being driven by musical *artists* and thereby ignores the complex division of labor found within the ringtone industry itself. Indeed, given the satirical form of *Doonesbury*, its general focus on members of the Baby Boomer generation, and the reality that most of its minor characters tend to remain "identifiable social types,"[6] Jimmy Thudpucker, the musically fickle but genial aging rock star, serves primarily as a figure of ridicule as a result of his involvement with ringtones—perhaps most explicitly demonstrated in his laughable participation in a vexed United Service Organizations entertainment tour to troops stationed in Bagram, Afghanistan. One might argue, then, that for all of the comic strip's capaciousness, certain possibilities—such as the depiction of relatively nameless ringtone workers, largely Generation X and Y members—would appear to fall outside its purview, perhaps even being unthinkable within its very form.

For a week, the largely invisible work of making ringtones received more public attention within US print media than it had before or since, and yet the picture on offer was tilted toward a caricature of those privileged enough to call themselves "artists" in the world of mobile entertainment.[7] In contrast, my purpose here is to offer a preliminary account of musical labor in the ringtone industry, including descriptions of the various types of labor found in that industry, discrepancies in payment and prestige between its numerous participants, and work/workplace arrangements, all produced by the technical transformations occurring over the brief and still ongoing history of what might be understood, perhaps paradoxically, as a global micro-industry. The point about technical transformations is of prime importance here, as the labor needs of companies hiring musicians to produce ringtones have varied substantially over time. Specifically, as the file format of the ringtone changed (see chapter 1), different forms of work and numbers of employees were required to create a sufficient volume of ringtones for purchase on a mass scale. These technical changes in the file format of ringtones have resulted in two thoroughgoing transformations in labor skill requirements in the ringtone industry. With the supersession of the monophonic ringtone by the polyphonic ringtone, a major skill increase in ringtone labor was required and effectively professionalized ringtone production, whereas the shift to sound-file ringtones involved a moment of *labor deskilling*—meaning that the skill requirements to do the work decreased, and precipitously at that.

Although on one level these changes are transparent, on the other the problem of defining musical skill in the ringtone industry is complex, rooted in both the variable and multiple abilities relevant to producing ringtones (encompassing various computer-technical skills, musical transcription and arranging, and competence in specific and often multiple musical genres), as well as the nature of skill acquisition—which can take place in and outside different institutions of musical learning and both on and off the job, in part as a result of the employment dynamics and economics of creative labor.[8] By wading into such murky terrain, I hope to initiate a broader conversation on the problem of musical skill, understood not, for example, from the perspective of music-educational imperatives and benchmarks, but rather from that of a highly unstable, ill-defined, and largely non-unionized *musical labor market* that nonetheless employs music workers at different levels of pay, task complexity, autonomy, and job security. Largely submerged at present, such an understanding of musical skill was more clearly codified, at least in the United States, during a period of lower public respect for the musical practices of subaltern groups (and, arguably, of lesser musical diversity in the country as a whole), of high union membership numbers for working musicians (especially in classical music and jazz), and of concomitant union-determined skills standards. In the present moment, then, an understanding of musical skill would have to account for skills acquired in numerous kinds of musical practices and the ways in which musicians routinely move between those practices within a fluid musical marketplace—making the task of defining skill much more difficult. Examining musical skill today in the context of the ringtone industry might make a partial contribution toward this effort.

Such an undertaking also makes it possible to engage with some of the broader debates within labor studies about the relationship between skill and employment, including the myriad investigations of and arguments about "the labor process," as inaugurated by the extraordinary work of Harry Braverman on the problem of deskilling.[9] The labor process debate, by and large, did not make its way into music scholarship. Instead, and with some notable exceptions, insofar as the loss of musical skill has been treated in any significant measure it is largely with respect to the skills of *consumers*, whose abilities to produce and reproduce music have, so the argument goes, been compromised through the rise of sound reproduction technologies (which include, in historical order, the piano, player-piano, and gramophone).[10] As recreation or entertainment drawing upon the social surplus, professional musical work has too often remained invisible to music scholars, and when it comes into view, it typically does so in terms of the trajectories of musical artists toward major professional success (high-selling recordings, a major performance/concert career) or historical canonization rather than in the context of the employment lives of mostly anonymous working musicians—many of whom never achieve such aspired-to successes and yet build lifelong if oftentimes precarious careers in music.[11]

Ringtone work often acts as just such a "second job" that pays bills for or supplements incomes of freelance performing musicians, whereas "musical labor" usually is understood as the labor that constitutes what we think musicians typically do (perform, record songs and albums)—with the paradox that much of this labor is unpaid or poorly paid. As such, skilled musical work like ringtone production often takes place on the margins of musical life itself, concealed behind the doors of small offices or home studios and rarely drawing attention to itself. When combined with the secretive nature of the ringtone industry (and mobile entertainment industries generally), this led my study to be less systematic than I'd hoped and more reliant on chance contacts and encounters. Nonetheless, the picture illustrated below hopefully begins to shed light on the varieties of musical labor and might provoke a more thorough (and much-needed) study of this subject.

The Origins of Ringtone Labor

Luke Holden, who in the mid 2000s held the title of Product/Marketing Manager at one of the major worldwide ringtone distributors, wrote a description in *Electronic Musician* of work expectations for ringtone programmers—i.e., the individuals actually making ringtone files—during the heyday of the polyphonic ringtone. In his words,

> it's rare for a full-time mobile-media programmer to make a living solely on ringtone programming. In most cases, film and TV composers, game-audio programmers, and audio editors will create ringtones to supplement their existing income. And keep in mind that mobile-content providers typically aren't interested in original music, so your composition skills will not be a big asset.[12]

Holden's description of ringtone labor, most accurately characterizing the life of and demands upon ringtone programmers in the United States at that time, emphasizes both the supplemental nature of their income and the lack of demand for original composition in the industry. But it was not always thus, and of course significant variation appears to be found in different parts of the world and especially at different points in the historical trajectory of the ringtone industry's development.

Returning to the moment of the industry's emergence, it seems that a number of functions—composer, programmer, entrepreneur—initially were more likely to be conjoined rather than separated. For example, it is worth recognizing that James Winsoar and many of the other early ringtone entrepreneurs were composers. Although Winsoar once described his main interest as "marketing," his early catalog of original monophonic ringtones at his "My Nokia" website grew to more than 100.[13] In another example, Vesa-Matti Paananen, often credited with being the first person to develop and market uploadable ringtones, worked with the Finnish wireless provider Radiolinja to develop the Jukeboksi (Jukebox) service in 1998. Here consumers and producers overlapped significantly, with the service creating a kind of user-generated content system well before the vaunted "Web 2.0" phenomena of YouTube, Facebook, etc. As Paananen described it,

> The original Jukeboksi service contained three main features:
>
> • TOP100 lists of the most popular ringtones (composed by end-users)
> • Harmonium composer [a ringtone creation program] and a private ringtone bank where tones were saved
> • A mechanism to publish ringtones from a private bank to public and pre-listen [to] them.[14]

In Paananen's system, collective creation was transformed into small-scale accumulation. Users received royalties from sales of their uploaded creations at percentages comparable to those in the ringtone industry as a whole before the advent of the sound-file ringtone. Paananen states

> I want to emphasize that everybody got their share and all ringtones were legal. If a ring tone cost 50 cents, then an artist got about 15 cents, the service provider/broker got about 10 cents and an operator got the rest[,] about 25 cents per download.[15]

Paananen's desire to demonstrate legality and scrupulousness are important in view of the semi-legal nature of the ringtone business in its first days. The majority of the earliest for-sale uploadable ringtones were unlicensed arrangements of popular songs, but legal reprisals by music publishers and the growing pressure for product differentiation in an increasingly competitive market had an impact on those companies seeking to circumvent the music publishing establishment—particularly in the well-regulated European, Japanese, and North American markets. For example, Anthony Stonefield's UK-based YourMobile, discussed in chapter

1, gave away free, unlicensed ringtones before facing legal threats from EMI; James Winsoar subsequently claimed that his competitor's ringtones, though "hugely popular," were "very poor" in "quality."[16] These differentiations in ringtone quality and legitimacy/legality contributed to a process of professionalization that diminished the role of consumer/user creation. In Paananen's words, "When professional ringtones started to get common, [end-user] composers disappeared and ringtone services started to be download only."[17]

Composition and Creation of Monophonic Ringtones

The increasing popularity of ringtones, particularly popular songs, and the resulting pressures toward professionalization also helped to create a division of labor within the ringtone industry, with firms beginning to hire ringtone composers and programmers with various aims in mind. Of course, it is not surprising that some musicians would find employment composing ringtones for firms even early on in the historical development of the ringtone. Among the most touted of such composers in the English-language press, Martin Plante took a circuitous but not atypical route to a part-time job as a ringtone composer. A classically trained pianist, a child of a violinist father and mother who was a professional operatic soprano, Plante first gained notoriety in the 1980s as a keyboardist in the now-defunct Montreal-based rock band Bundock. Working as a piano accompanist with Cirque du Soleil and with ballet companies, Plante was hired to compose original ringtones for a New York website, About.com, in early 2001. Giving the ringtones away for free, the website paid Plante based on how many visits the website received per month. Although the work at the time was a second job for him, Plante mentioned that "every month I get a sort of salary and at times the amount has been surprising, especially when the dollar was down, almost enough to live on."[18] Plante left About.com in 2008 and began producing free ringtones for another site, phonerated.com, according to his colleague Eric Bernatchez.[19]

Receiving less media attention than Plante but greater economic security, ringtone composers can be found among those workers who attend to numerous sound-related issues at mobile and electronics firms. It would seem that, often, these workers are the sole individuals involved in composing ringtones at their places of employment. Chia-Chun Chen, a Taiwanese music engineer and music teacher, worked for two years in the mid 2000s at the semiconductor manufacturing firm MediaTek, which produces microchips for mobile phones (among other products). Paid 50,000 New

Taiwan Dollars (about US$1,515) a month, and working 40 hours a week, she got the job when searching for an internship after having graduated from Florida International University with a master's degree in music technology. Her friend's boyfriend worked there and, as an interview, solicited a few compositions from her. She subsequently was hired to compose ringtones and any other functional sounds used by phones—probably preset sounds that would be hardwired into phones—and composed both monophonic tones and especially polyphonic ones, including as many as 64 voices. Working in this capacity, she seemed to have a reasonable degree of aesthetic freedom to make musical decisions. "I had to compose ringtones and any sounds a cell phone might use (phone on/off, message, game music, etc.)," she explained. "Sometimes I just arranged classical [music] or music that's out of copyright period. When I had ideas, I would compose some music. It's pretty interesting most of the time. But sometimes [when I wasn't getting any ideas], it would become hard."[20] In a similar but much more secretive vein, an unnamed worker at Samsung was interviewed for *PC Magazine* in February 2007 about his work composing ringtones (again, probably presets) for the Korean handset manufacturer. As the interviewer, Seth Porges, noted, "Because the competition between Samsung and crosstown rival LG is so intense, the company has asked us not to reveal his identity for fear of poaching."[21] The composer explained that he designs "the ringtones and function/menu key sounds that best suit the concept of that particular phone. Basically, I am responsible for all sound-related aspects of a phone." Composing on the Yamaha MA5 synthesizer chip, the composer drew inspiration from Ennio Morricone and Ryuichi Sakamoto. The latter "also designed ringtones at one time and gives me a lot of inspiration." The process at Samsung seems somewhat more collaborative than the one Chia-Chun described at MediaTek:

> First there is a theme for each cell-phone model. We compile a list of sounds that are suitable or go along with the concept of the phone, and make a demo of the possible sounds that go with the phone. Within our team, we then discuss it and edit the sounds and apply the result to the phone.[22]

The work of composing ringtones is not particularly glamorous: it is, in a sense, an updated form of writing jingles that advertise mobile telephony itself. In the case of the monophonic ringtone, it benefits from being a highly constrained art form. The composer has a single melody line of limited length (maximum 30 seconds) and range (up to four octaves) at her disposal, with little to no possibility of varying timbre or dynamic. The simplicity of such melodies, which must balance an aesthetic interest

Figure 2.1
"Scifi," a Nokia 5110 preset ringtone. Reproduced by permission of the Nokia Corporation.

(sounding "good," in the abstract) with a function (being an effective signal to the phone's user), seems most effective when they are short—perhaps only a few seconds. (Plante's monophonic melodies are under 10 seconds.) Although composing such a melody is by no means easy, it can be done by someone with a basic knowledge of Western music theory and the ability to encode a melodic idea in the relevant ringtone text language. Additionally, one must consider the reality that the signal function of the ringtone does not necessitate a melody and can be realized equally well—perhaps even better—with a sound effect (for example, a rising arpeggio or simple sequence of beeps). Moreover, with the possibility of relatively short note durations, rather dramatic sound effects can be created with a cascading series of beeps—tinkling, digital "sheets of sound," to borrow Ira Gitler's description of John Coltrane's rapid-fire, glissando-like saxophone playing circa 1957–58. Two such preset monophonic ringtones on the Nokia 5110 were called "Dizzy" and "Scifi," the latter being an accelerando chromatic wedge leading to a rapid trill. (See figure 2.1.)

Ringtone Programming and Polyphonic Ringtones

Relatively few musicians working in the ringtone industry were hired to create original compositions, and the percentage surely decreased with the growth in polyphonic ringtone sales in the early to mid 2000s. Keith Nowak, a spokesperson for Nokia, emphasized in 2004 that there were very few people who were employed to compose ringtones at the company and even fewer who did so as a primary occupation.[23] Hence, we must distinguish between the relatively rare ringtone composer and the much more common *ringtone programmer*, whose work typically involved not original creation but a process of musical transcription and arrangement. The era of peak employment for ringtone programmers was the mid 2000s, coinciding with the greatest sales volume in polyphonic ringtones. Whereas ringtone composers like those above often worked in mobile entertainment firms or handset (and other) manufacturing companies and typically also performed other tasks in addition to composing ringtones, most ringtone programmers worked for ringtone wholesalers or distributors—companies that sold ringtones to mobile entertainment firms. And, with the move from the monophonic to polyphonic ringtone, overlaps between ringtone composers and ringtone programmers probably decreased, particularly as the labor of ringtone creation was increasingly relocated to companies wholly devoted to the task of producing ringtones for mobile content sales firms.

How might one describe the employment arrangements for ringtone programmers, particularly during the crucial polyphonic period—which one person I interviewed described as a "golden age?"[24] There seem to be at least two kinds. The first and seemingly less common arrangement is one involving a large employment pool that is possibly given hourly payments and hired to produce ringtones for a large distribution firm. Although I have heard rumors of factory-like arrangements in East Asia involving a large number of ringtone programmers, each working at a workstation with a computer and keyboard controller, I have not found any concrete evidence of such a workplace to date. The closest example might be that of a Norwegian company called Soundonweb, which in the early to mid 2000s hired 25 musicians to construct ringtones for delivery to the SonyEricsson assembly line. Apparently the company had ties to Nokia as well, perhaps as a subcontracted ringtone provider for the Finnish handset manufacturer.[25] In a PDF of a corporate PowerPoint presentation, the company (now named Arphiola) boasts a catalog of more than 50,000 ringtones and, when accounting for different formats, up to 500,000 (mobile) content items—numbers that suggest the presence of a significant past and/or present labor pool.[26]

A more typical arrangement appears to be one involving a small team of ringtone programmers—perhaps from three to six of them—who are overseen by a manager. In an early example, Melodi Ltd., founded by Iain Kerr and Jason Shaw in Lichfield, Staffordshire in 2002, was a small ringtone company for the UK and Australian markets. A year later, the company boasted that it had hired four full-time musicians.[27] In the mid 2000s, three companies with which I had some contact followed this kind of arrangement. One of these was 3rd Ear, a major ringtone wholesale/distribution firm based in Sheffield that was not involved in the consumer-sales end of the ringtone business. In 2005, at what probably was the high point of its success, 3rd Ear had four or five ringtone programmers and supplied ringtones to T-Mobile, Orange, and some other wireless providers.[28] Another similar enterprise was The Source Mobile Channel, produced by Hudson Entertainment, an American subsidiary of the Japanese gaming firm Hudson Soft (known for producing the games *Lode Runner* and *Bomberman*). This spin-off venture from the hip-hop magazine *The Source* was an on-deck application that could be accessed through wireless carriers, selling polyphonic ringtones in exchange for pre-purchased tokens. According to the US-based project manager Arne Lucas, the company hired a team of four ringtone programmers based in Cleveland who would produce polyphonic hip-hop ringtones for the project.[29] A third project, very similar to the Source Mobile Channel, was an American-based project for a very large international ringtone distributor/vendor, led by Luke Holden. As in the previously mentioned case, Holden was the project manager, overseeing a group of five outsourced American ringtone programmers based in various parts of the United States.[30] "Nearly all ringtones are created by freelance programmers," Holden wrote in 2006. "Rarely will companies hire programmers as full-time employees."[31] To judge from the limited sample size I have encountered, this seems largely to be true—in contrast to the ringtone composers/mobile phone sound personnel described above—and only at 3rd Ear was the possibility of working on a salaried basis even mentioned.

The effect of this hiring method has at least two important corollaries. First, ringtone programmers typically work at home, in their home studios (if they have them) or at home computers with keyboard controllers, though some amount of training at a more centralized location may be involved. Although the work is mostly individualized and potentially isolating (or unencumbered by workplace interaction, depending on your view), at least in the case of The Source project the programmers there seemed to be in touch with one another regularly, helping each other out with difficult ringtone transcriptions as needed. Billy Dixon, one of the programmers at The Source who programmed much of the Def Jam catalog for the project, described it as follows.

> We all worked from home. . . . A lot of times we shuffled titles around to help each other. . . . This was rap, so it's, like, samples, ya know? It's tough to approximate samples with MIDI, or anything atonal with MIDI. So we would help each other out with 'em here and there.[32]

Second, ringtone programmers hired on a freelance basis would receive variable rather than consistent workloads and were typically paid on a piecework basis. An employment ad for a monophonic and polyphonic ringtone programmer posted at the Phat Tonez website in 2004 gives some sense of the work requirements and payments in that company at the time:

> **Freelance Ringtone Composer** Specialising in Hip Hop and Rap music you will produce Nokring RTTTL format and SP MIDI polyphonic ringtones (4 note polyphony) by listening to the original music and copying it by ear. You should be able to copy the rhythm, bass lines and melody with no wrong notes. You will be paid £10 per ringtone on a self-employed basis.[33]

(Note the discrepancy between describing the job description of "composer" and stating that the work involves "copying.") As is typical with musical freelancing in general, payment rates seemed to vary a great deal, and without a sufficient sample size and information at the level of hiring, one can only speculate as to the causes. The two UK-based rates I came across, at Phat Tonez and at 3rd Ear, both seemed to pay £10 per ringtone, which may be coincidental or could have been the going rate for a number of UK-based companies at the time. The US-based projects paid a great deal more. At The Source, programmers were paid $100 per ringtone at first (more precisely, $200 per tone, the work for which was divided among two programmers); the rate later dropped to $50 per polyphonic ringtone with the advent of sound-file ringtones, which were quickly making polyphonic tones obsolete.[34] The project led by Luke Holden paid even more—about $180–$250 per ringtone, depending on the format and nature of the specific musical and programming tasks involved.[35] The differences in these rates are striking, though it may simply be a matter of the company size and degree of ringtone quality sought in the projects involved. In the case of The Source Mobile Channel, the project was one of the later entries into the polyphonic ringtone market, having been preceded already by Def Jam's own ringtone vendor/channel, and thus the time-sensitive nature of the venture and need to compete by producing high-quality ringtones may have encouraged Hudson to pay a relatively lucrative rate. In the case of Holden's project, the large size of the company and the degree of quality control and quality assurance involved seem to have been characteristic of that company as a whole: Holden himself painstakingly checked the work of the programmers he contracted. Moreover, the difficulty of the job and the music-genre acculturation required for transcribing the specific ringtones involved may have inflated the rates in the American-based projects. In the case of The Source, the Japanese programmers at Hudson Soft appeared to have trouble programming hip-hop songs. A similar problem occurred with Holden's international employer, which often resulted in non-US-based ringtone programmers transcribing the rap lines as melody lines, creating monotonous and ineffective simulations of the source songs.[36]

Monophonic and Polyphonic Ringtone Formats and Labor

The work of producing the ringtones themselves has, of course, transformed drastically, with the degree of specialized knowledge required to do so peaking with the production of polyphonic ringtones (as well as hybrid tones—polyphonics that also included short audio samples). Programming a monophonic ringtone was a fairly simple affair, involving the production of a text string that encoded basic information about a melody (pitch classes, registers, note durations, rests, etc.), all within a 12-tone equal-temperament framework. One such format was that found on Nokia phones, the Ring Tone Text Transfer Language (RTTTL), which used a simple way of encoding pitches, a numerical code correlating directly to note duration (8 = eighth note, 4 = quarter note, 2 = half note, and so on), and another number referring to the octave (from 4 to 7, mirroring the phone's limited range—encompassing the four octaves above the A below middle C).[37] Additionally, to cut down on extra text (and thereby shorten the size of the ringtone file/message), the text string also included a header describing the ringtone's name and basic pulse, octave, and tempo of the ringtone. For example, the string

5thSymphony:d=16,o=5,b=100:g,g,g,4d#,4p,f,f,f,4d,4p,g,g,g,d#,g#,g#,g#,g,d#6,d#6,d#6,4c6,8p,g,g,g,d,g#,g#,g#,g,f6,f6,f6,4d6,8p,g6,g6,f6,4d#6,8p,g6,g6,f6,4d#6[38]

translates literally to the musical notation shown in figure 2.2, which can be seen as a crude monophonic approximation of the opening of Beethoven's Symphony No. 5 in C minor, Op. 67 (1807–08), realized at an octave higher and represented in staff notation with halved note values in comparison to the original. (Note, in addition to the generally defamiliarizing effect of the transcription and corresponding text language, the missing D6 sixteenth note following the penultimate D♯/E♭6 and the unusual pauses in this rendering, particularly the eighth rests, which exaggerate the separation of the archetypal sentential musical gestures in this passage.)

A more sophisticated (and complicated) encoding system, the iMelody format, was created by the Infrared Data Association (iRDA) and adopted for use in Ericsson, Motorola, and Siemens phones, hence "making it the first industry-standard cross-platform ringtone format."[39] This format incorporates sharps and flats (the latter expressed by "&"), a much greater amount of header data, a wider range of rhythmic subdivisions and octaves, and the ability to alter individual note volumes in the ringtone. A similar ringtone to that above, but now accounting for the ***ff*** and ***p*** dynamics at the outset and after the second quarter rest, respectively, might read as follows:

```
BEGIN: IMELODY
VERSION: 1.2
FORMAT: CLASS1.0
NAME: 5thSymphony
COMPOSER: Beethoven
BEAT: 100
STYLE: S1
VOLUME: V9
    MELODY:*5g4*5g4*5g4*5&e2r2*5f4*5f4*5f4*5d2r2V6*5g4*5g4*5g4*5&
    e4*6&a4*6&a4*6&a4*5g4*6&e4*6&e4*6&e4*6c2r4*5g4*5g4*5g4*5d4*6
    &a4*6&a4*6&a4*5g4*6f4*6f4*6f4*6d2r4*6g4*6g4*6f4*6&e2r4*6g4*6g4
    *6f4*6&e2
END: IMELODY
```

(Look for the "V6" code in the melody string, which decreases the volume from the higher level at the outset, V9.)

The mastery of different ringtone programming formats was already a problem during the monophonic period and was imposed by the fractured and competitive mobile handset industry and the invention of highly variable phone standards (a trend seen again more recently with proliferation of smartphone operating systems). In terms of labor, this often resulted in a number of possibilities: mastering the various formats, dividing up the labor between transcription and format conversion, and, perhaps most commonly, using software packages produced by the companies making the formats (Yamaha, Beatnik, Faith, etc.) that would automatically convert files according to the necessary specifications.[40] Negotiating disparate file formats became even more challenging with the advent of the polyphonic

Figure 2.2
RTTTL string ("5thSymphony") realized in musical notation.

ringtone, and it seems that freelance polyphonic ringtone programmers sometimes were even paid variable piecework rates based on the difficulty of programming in particular formats (accounting in part, for example, for the $180–$250 range provided by Holden).

In contrast to making monophonic ringtones, programming a polyphonic ringtone is a rather complex and time-intensive undertaking. It involves the creation of a multi-voiced MIDI synthesizer file, allowing ringtones to emulate the arrangements of more complex musical textures as opposed to reproducing melodies—with the effect of creating a kind of video-game music or elevator-music rendition of a song or a composition (or, with preset ringtones or other, rarer instances, original music of this sort). As the mobile phone MIDI synthesizer systems became more complex, with more parts and higher sonic fidelity, they increasingly began to approximate the audio files they emulated, particularly with the arrival of hybrid tones that could also incorporate short audio clips.

The task is essentially the same as producing any MIDI synthesizer file. Ringtone programmers would use a digital audio workstation software package (Cubase has been mentioned more than once in interviews), and typically they would input sounds through the use of a (piano-style) keyboard controller, aurally imitating individual parts as heard in the source audio clips to be transcribed (usually up to about 35 seconds in length).[41] Additionally, piano-roll-style editor interfaces would often be used to monitor and correct or alter the results from the keyboard controller. (As is the case with editor interfaces for digital image software, it is much easier to correct or alter given data rather than producing it from scratch.) When the transcriptions were finished, they would be saved as .mid files and then altered or adjusted using the necessary formatting and authoring tools for specific formats (Beatnik, Yamaha, etc.).

One of the critical issues to negotiate for a ringtone programmer during the height of polyphonic ringtone sales was the ever-changing nature of device formats, which resulted in a high degree of inconsistency in how many voices (or lines) could be used in these ringtones. To account for this problem, the predominant polyphonic ringtone format was known as Scalable-MIDI (SP-MIDI), which meant that the ringtone file would give priority to four main voices and add in other voices as were available on the phone according to a predetermined hierarchy of importance. SP-MIDI, commonly used on Nokia phones, was approved by the MIDI Manufacturers Association in May 2002 and appeared in phones later that year.[42] Of course, as phone formats increased the number of textural voices, ringtone programmers would get creative with the relative abundance of available channels and used them to create effects that were otherwise not possible on these relatively simple synthesizer systems. Holden mentioned that these systems lacked digital signal processing and that they

> had no reverb, delay or modulation effects to create what most would consider decent-sounding voices. So, as a workaround we might use 3 to 5 tracks as unison. . . . Maybe detune two tracks using pitch bend controllers at the start of the file (to create a more lush chorused-sound), then another track or two all decreasing levels set back an eighth or quarter note to add delay or "reverb" effects. You can bet pretty much every title was loaded full of 16 tracks—even if half were just "layer tracks" for delay/chorusing sound. We'd also sync the vibration mechanism of the handset to the kick drum and the light indicators to the melody's rhythm.[43]

Handsets would vibrate and light up like a battery-powered musical toy when receiving calls, transforming the polyphonic ringtone into a primitive multisensory spectacle.

One of the most crucial skills in creating ringtones derived from existing music, especially popular musics, is that of *hook finding*. During the monophonic and especially polyphonic moments of ringtone production, there was quite a bit of variability in who was responsible for finding hooks within the ringtone production company's division of labor: sometimes, the project manager (as in the case of Arne Lucas at the Source Mobile or at Holden's international employer) would select the hooks, whereas in other cases the programmer would do so. An example of the latter was found at 3rd Ear, where Andy Booth worked as a ringtone programmer for a few years. Andy shared a number of insights into polyphonic ringtone creation which both demonstrated the importance of hook finding as well as accounting for the limitations of polyphonic synthesizer formats with respect to particular musical styles:

> For transcribing, I get sent a recording of the song, sometimes just a little clip. I choose a memorable bit, usually the chorus or the hook, but it depends on what it is. For example, for Indian/Bollywood tunes (of which we do quite a few) it's usually a matter of reading the title and trying to hear the words of the title somewhere, which is usually the bit you need. With rap, it's a matter of hoping the rapper shuts up long enough to get a few melodic bars, though often these make nice tones, especially when the chorus divas chirp in. It all depends on the tune, and sometimes you just have to make the best out of a bad tune. . . .
>
> The biggest pain is usually the vocal melody line, soul and Indian singers being the worst. Sometimes you simply can NOT make a vocal line sound good, no matter how accurately you transcribe it. Bloody Jamiroquai for example. You can't do rap or "talking," which is a pain when selecting your clip sometimes. You just leave it out—nothing else to be done.[44]

Producing Sound-File Ringtones

The task of selecting the hook of a song (its catchiest material, often the chorus) in order to make a ringtone is, perhaps, one of the few elements of continuity between the earlier modes of ringtone production and the newer system put in place for making sound-file ringtones. Indeed, the entire structure of the industry, including its productive forces, was radically altered in a very short period of time. The original composition of ringtones was relegated to a marginal status within the industry, and ringtone-programming labor effectively disappeared. In its stead, a much smaller labor force, sometimes but not always composed of former ringtone programmers, began working as sound-file ringtone producers, which essentially involved the labor of *audio file editing.*[45] This form of excerpting labor requires a much lower degree of skill than polyphonic or even monophonic ringtone production, only requiring hook-finding and the use of a simple sound editing software package in which one would isolate the desired section of a recording, cut out the extraneous material and possibly add fade-ins and -outs, and bounce or export the final file in a compressed format used for the ringtone (often a 128-kilobytes-per-second MP3 or AAC file). Since such laborers are relatively few in number and are still active in the ringtone labor market, it is somewhat more difficult to find information about such work. However, the San Francisco-based media production, distribution, and marketing firm INgrooves has hired at least one person to produce sound-file ringtones since 2006 and was able to give me a picture of this worker's labor. Currently, at INgrooves, the work is given to someone who is paid an hourly rate to produce ringtones, which take 5–10 minutes each to make. The software platform of choice at the company is Audacity, an unsophisticated freeware package that is nonetheless adequate to the task—rather than more complex and expensive programs like ProTools or sound-file editors like Peak. The sound clips themselves can vary in length, being 15, 30, or even 45 seconds long, all according to the specifications of the buyer (the mobile entertainment firm or other entity selling the ringtones or using them as promotional materials, etc.). The sound file itself is usually an excerpt of a high-quality audio file (as in a WAV or AIFF format), which is then formatted as necessary and then converted into a relatively low-grade audio file. Sometimes, in addition to fades and excerpts, the file may be boosted in volume. Like assembly-line factory work, the entire process is labor-intensive, demanding a degree of concentration and musical attention, and a skilled musician is hired to do the job, probably in large part as a quality-control measure.[46]

Although I was not able to get a large sample of pay rates for this work, one can easily imagine that it is far lower than the highest pay rates associated with polyphonic ringtone production. Having transitioned out of polyphonic ringtone work and into audio tone production, Billy Dixon noted that he made $2–$10 per audio ringtone, "depending on destination carriers and file format (and package complexity)."[47]

Certainly, at times subtler editing interventions would be required of an audio file ringtone producer, above and beyond the basics of clipping files and formatting them to specifications. Billy Dixon's work in hip-hop, for example, required him to undertake intensive forms of sound-file editing on account of the prevalence of expletives in some hip-hop tracks, which are usually censored when appearing as ringtones. In his description of this censoring procedure, Dixon states: "The standard for this is and has always been 'reverse' editing, finding the cuss, and reversing it. Sometimes . . . I'd use subtle splicing especially in extremely explicit titles like Death Row, Snoop, Tupac, etc."[48] Dixon also noted that in doing audio editing work he used the shareware package Sound eXchange or SoX—a relatively powerful program that may have been of particular use in these delicate editing operations.

Such moments notwithstanding, the labor involved in making audio ringtones is routine and, on balance, relatively unchallenging. How one reacts to that routine labor—whether as drudgery or as easy work, for example—is surely a complex matter, depending on the varieties of labor subjectivity encountering that work. But perhaps the defining aspect of audio ringtone production is, as Dixon notes, its focus on "volume, as sometimes it takes 1 minute to make an audio tone."[49] In view of the time and care required to make a polyphonic ringtone, in contrast, ringtone programmers took a degree of pride in the accomplishment of transcription during the polyphonic period in a way that seems rather unlikely with audio ringtones. For example, Dixon noted that tracks by Gang Starr, DJ Premier, and the Wu Tang Clan were often difficult to transcribe ("although I did 'C.R.E.A.M.' and it wasn't difficult"), and Andy Booth of 3rd Ear said: "I'm not being big-headed, but tunes I've heard from other companies usually sound pretty crap compared to the versions I put out. (My 'Sex in the City' is peerless in my opinion; now how sad is THAT!)"[50] On the other hand, the greater skill and labor time required to make polyphonic ringtones could be also seen as especially demanding. Dixon in particular seems to have found the push to produce polyphonic ringtones quickly for The Source to be exhausting work, affecting his musical creativity:

It affected me heavily actually, during polyphonics, I made nearly zero music of my own. After, for a good while, I did audio tones, and during that phase, only really got into my own music again because I pushed myself. It was hard, taking what I do to create and express, and then doing the same thing without any creativity of my own to make money, for a job. It was confusing, for me anyway. I've only just in the last year made a little bit of my own music, so yeah, definitely a large dynamic phased itself through my experience.[51]

[illegible], who seemed to be working fewer hours in the industry, didn't seem particularly drained by the work: "If it's a song I like, I spend far too long figuring out stuff that nobody would ever hear, then listening to it and feeling all clever. I look upon my job as paid ear training that I *have* to do, otherwise I starve. I almost do, anyway."[52]

"Stars" in the Ringtone Industry

From the above, it should be apparent that most ringtone programmers and even most ringtone composers are anonymous workers, receiving little press or public attention and working until their sources of pay dry up. However, there is also a history of media and music industry celebrities, both erstwhile and not, who found their way into the ringtone industry. In the early days of the monophonic ringtone's rise to prominence, a number of former music industry stars or second-tier figures, if you will, were found to be associated with the business—which makes sense, since monophonic and polyphonic ringtones garnered the recording industry conglomerates and their stables of artists little to no profits. Boy George, who in the 1980s was the lead singer of Culture Club and who in the 2000s developed a second career as a successful electronica DJ, produced an early monophonic ringtone, called "Sonic Trigger Ringtone," that was distributed exclusively to Vodaphone subscribers in the United Kingdom.[53] In 2002, the Danish musician Lee Oskar signed with a Seattle-based mobile entertainment company called Vesaly Games to produce a series of exclusive, high-quality polyphonic ringtones. Oskar is perhaps most famous as a harmonica player and a founding member of the funk-rock band War. Able to construct catchy harmonica riffs for songs such as "Low Rider" and to innovatively combine his instrument with horn arrangements, Oskar decided to lend his melodic gift to the construction of ringtones, stating: "I'm a melody writer. Give me three notes, and I'll give you the best melody you've ever heard. You don't need a long stretch of time to create a melody. If you've got the right tone and tempo and a great, catchy riff, you've got the game."[54] Other musicians associated with once-famous rock bands, among them Deep Purple's guitarist Steve Morse, also signed deals with mobile entertainment firms to compose new ringtones.[55]

But with the appearance of the sound-file ringtone around 2002–03, the rules of the game changed quickly. Monophonic and polyphonic ringtones would be obsolete within a few years in most major markets. The new ringtone format allowed the emergence of the notion of a "mobile music" market, which essentially meant digital music distributed to mobile phones, and top-tier artists began to partner with wireless entertainment firms and even wireless carriers, particularly in the early to mid 2000s, in order to capitalize on this new market. For example, in late 2003, Jay-Z signed a deal with Nokia for an exclusive arrangement with the special edition Nokia 3300, which would include the "Black Album" preloaded onto the phone as well as exclusive wallpaper, ringtones, and monthly text and voice messages from the artist.[56] Similarly, 50 Cent and Snoop Dog signed distribution deals with the mobile entertainment company Zingy, leading one commentator to describe 50 Cent's 2004 arrangement in particular as a first: "I'm pretty sure this is a first! A worldknown artist signing an exclusive agreement with a mobile media company. Usually, it's the music companies signing up with media companies or wireless operators."[57] Artists and their labels surely viewed such opportunities as promotional bonanzas, using either preexisting content or advertising filler (not unlike artist-created radio spots, common in past decades) rather than focusing specifically on ringtones as an independent medium. For example, consider the presence of television advertisements on MTV1 and MTV2 in 2004 for ringtones created by the punk-revival band Green Day that themselves were advertisements for a new album, *American Idiot*, released in August of that year. Rather than merely reproducing singles from the album in sound-file ringtone format, the advertised ringtones presented spoken statements from Green Day band members like "Pick up the phone! It's your mother. I know. She's with me."[58] By the mid or the late 2000s, however, deals between record labels and wireless firms representing massive, multi-artist catalogs were more typical, completely supplanting the earlier single-artist arrangements and making ringtones a standard means of selling digital music for most mainstream music industry acts—so much so that when the hip-hop and R&B producer Timbaland released a promotional video in 2008 explaining his production of a new potential ringtone/mobile music hit, Keri Hilson's "Get It Girl," in a mobile recording studio built in tandem with Verizon, a YouTube user named "aahhhshaddap" tellingly posted the following: "wow timb. Producing your first ringtone. YAWN."[59] As such, the upper part of the two-tiered ringtone industry, with "stars" working at the top and anonymous workers at the bottom, seems not to have lasted, in that the aura of stardom associated with elite artists' signature ringtones dissolved once the ringtone became a standard means of selling recordings for all artists.[60] On the other hand, given the direct overlaps between the ringtone industry and the music industry more generally (i.e., ringtone workers often being gigging musicians; ringtone hits largely being the songs of music industry stars), such hierarchies effectively remain in place, if not within the specific labor categories pertaining to ringtone production.

Ringtones and Deskilling

However one evaluates it as a phenomenon, it seems incontrovertible that the end of polyphonic ringtone production, which Billy Dixon described as "a funny little industry" that came and went, was a critical moment of labor deskilling in the brief history of ringtone production.[61] Indeed, we might characterize the shift from the monophonic to polyphonic to sound-file ringtone as involving a moment of skilling and then subsequent deskilling, each produced by a significant technological change in the content of the ringtone commodity. In fact, one can also track productive skill requirements in inverse relation to the frequency of consumer-based ringtone creation, which (as mentioned above) was relatively common with the monophonic ringtone, then declined in the moment of the polyphonic ringtone (supplanted only by ringtone remix software programs, often found on phones themselves, which would give the consumer the semblance of productive control over the ringtone), and then increased somewhat with the sound-file ringtone (the labor of clipping and formatting being relatively accessible to consumers, as demonstrated by Xingtone and many other software packages).[62]

How do these processes of skilling and deskilling relate to standard accounts in the labor studies literature? Perhaps the most important point of comparison is still Harry Braverman's *Labor and Monopoly Capital* (1974), a masterpiece of Marxist labor analysis that brought renewed attention to problems of skill in labor under capitalism.[63] In reading Braverman, it quickly becomes evident that the skill changes occasioned by ringtone technology shifts are significantly unlike most of the cases in industrial and office work that he describes. For one, the change in ringtone technology appears not to have been primarily motivated by issues of controlling and fragmenting the structure of the labor process as a result of scientific management; rather, as I have argued elsewhere, the motivation was one of format upgrading in the interest of both technical advances by handset manufacturers and, especially, greater profits for the music industry as a result of increased copyright revenue through licensed recordings. And yet, such shifts were realized under the guise of a putatively increasing sonic fidelity, since, in being digital music files, ringtones would consistently be compared to digital sound files and, as a result of the ringtone's recapitulative historical development, would eventually converge with them.[64] Thus, observers would casually perceive in the ringtone not the degradation in the quality of the product on account of deskilled labor fragmentation and automation, but rather, that the ringtone's sonic quality improved greatly over time—this despite the fact that the actual craft of making ringtones (as synthesizer transcription) would decline in favor of a much simpler method of ringtone production. Insofar as ringtone production can be seen as *redundant labor* when viewed from the teleological endpoint of the digital sound file—polyphonic ringtone creation essentially being an attempt to copy another sound file using a MIDI synthesizer system—one might say that skilling and deskilling in the ringtone industry were *secondary effects* rather than primary ones, although shifts in skill requirements and the resulting changes in labor costs cannot have gone unnoticed by those interests advancing the cause of the audio file ringtone.

Braverman's work would then, at first glance, seem irrelevant to the preceding argument, but it proves resiliently pertinent in at least two ways. First, Braverman describes the emergence of what one might call a *short-term craft*, which seems particularly relevant to the ringtone's historical development. Braverman's example is the emergence, in the 1940s and the 1950s, of data processing, which "for a short time . . . displayed the characteristics of a craft. . . . The development of a data-processing craft was abortive, however, since along with the computer a new division of labor was introduced and the destruction of the craft greatly hastened."[65] Although the division of labor is not profoundly different in the cases of polyphonic and sound-file ringtone production (each involves some form of management and some form of employee labor, albeit often on a "flexible," freelance basis), it is clear that the craft of making polyphonic ringtones had a short-term existence, one that was not sustainable given the imperatives to sonic fidelity and the interests of the recording industry to sell audio ringtones. Dixon reflected poignantly upon his experience in the industry: "It was strange, being in my mid twenties, and watching this strange niche profession, polyphonic ringtone producer, become archaic, useless."[66] In the place of that niche profession, the (mostly) low-level audio editing work done by sound-file ringtone producers constituted a *de facto* deskilling, and perhaps, one might argue, contributes to the creation of a pool of laborers performing what one might call semi-skilled musical labor.[67]

Second, and perhaps more important, Braverman's observations provide a framework of comparison between the era of modern automation and a newer moment of intensified, micromaterialized digitization, as found in the Internet economy.[68] In contrast to the historical period Braverman examines (the early to mid twentieth century, mostly), the pace of contemporary industrial birth and decline, as well as broader accumulative processes, seems to be quicker—a product of an often-noted economic and social acceleration occurring at multiple time scales.[69] In a new digital business' more limited life cycle (and, typically, limited size), the amount of advance planning and managerial implementation required to automate on a mass scale is temporally and economically less feasible, according to the older Taylorist models involving the extensive scientific study of the labor process and the long-term implementation of automation through the creation of new machines.[70] Digital technologies, which obviate some of these previous necessities, appear at present to be an emergent dialectical sublation of the specific forms of deskilling within particular industries treated by Braverman into a *generalized form of deskilling* relevant to many industries at once, with generic software packages and computation methods covering basic business needs (accounting, stocking, distribution, shipping, sales, customer service, etc.) and production systems (including automation software) now being customizable for rather different industries and with ever greater forms of manual labor being replaced outright by digitally automated machinery.[71] But as a general-purpose technology, software nonetheless pushes in the direction of technical universalization that Braverman already perceived with more limited forms of computation-based automation that facilitated the greater separation of mental labor from manual labor, of conception from execution, in the mid twentieth century.[72]

At present, the promise of the digital still outpaces its realities. Companies often wrongly presume a guarantee of labor-time reduction through the adoption of upgraded software packages, but by and large upgrading does precipitate a far greater degree of deskilling and automation than was possible in the past. The digital, in this neo-Taylorist imaginary, thus offers the prospect of *deskilling as fantasy*, such that file format, software platform, and hardware upgrades are often (wishfully) understood to save labor relative to their earlier versions. The impulse to upgrade—which, at the consumer end, also drives the planned obsolescence-based commodity cycle—is thus seen as concordant with the interests and potentialities of labor-power reduction, perhaps not always consciously so but nonetheless acting such that the very mechanisms of digital upkeep now incorporate deskilling into their basic principle of operation, or at least seem to. The historical process of upgrading musical formats on mobile phones and within the ringtone industry may thus have seemed like a naturalized one—a part of what Martin Lister calls "upgrade culture"[73]—whose technical advantages and labor-saving features were presumed in both the business world and the broader consumer world, even if the labor deskilling it entailed was not necessarily envisioned with the utmost clarity by the industry itself. Built on the predicates of the "new capitalism," in which consumption transforms seamlessly into production and labor mimics leisure, the impulse to upgrade that binds consumers and capitalists together signals a presumption of perpetual technical progress that makes visible, with every subsequent purchase, the utopian dream of a world without increasingly compulsory "free" labor and the dystopian prospects for a world of surplus populations that needs no wage laborers.[74] We can be certain of one thing, however. Long located predominantly within the informal sector, musicians will continue to struggle there. The music will go on; seduced (and annoyed) by its strains, should we not pause to ask "At what cost?"[75]

3 Left Behind: Case Studies of Decline and Recapitulation in the Ringtone as Representation

In mid October 2002, a Welsh man and woman in their thirties who lived near the town of Aberdare waited at the Green Street Methodist Church with their respective families in tow. It was 3 p.m., and the organist had failed to turn up for the couple's wedding service. The groom's close friend, a 35-year-old roofer by the name of Darren Medd, had just purchased a new Samsung T300 mobile phone equipped with 16-voice polyphony and the ability to download ringtones; he knew the phone had a ringtone version of "Here Comes the Bride," probably as a preset already on the phone.[1] Before the service, Medd had played the ringtone for a couple of minutes, "with everyone laughing." When the organist failed to show, he held his mobile phone to the microphone at the altar and played it again. The bride, Tracy Muxworthy, marched in double time to the up-tempo rendition of the wedding staple, and giggled as she approached the groom, Ian Davies. The organist, it turns out, had been told that the wedding was at 4 p.m.; "he arrived in time for the hymns and to play the voluntary as the couple left the church."[2] The couple, a pair of social workers, were married in front of more than 80 guests, surely thankful for Medd's handy resource.

This brief tale is, on the one hand, clearly a product of the present-day digital world, the inhabitants of which walk around regularly with music-playing devices—their phones—and in which, one might add, few might actually be able to perform "Here Comes the Bride" on an organ. However, the moment is also marked by the fact that polyphonic ringtones were relatively new in the United Kingdom at that point, having first emerged only a year to two years earlier, and that they could produce a passable rendition of a piece of music for use in a wedding service (if eliciting a few laughs in the process). Perhaps less obviously, the music used in the service, a famous English adaptation of the bridal chorus ("Treulich geführt") from the opening of Act 3 of Richard Wagner's *Lohengrin* (1850), is not often heard as a polyphonic ringtone, despite its ubiquity in other contexts (and, indeed, as a monophonic ringtone). Although only a few years earlier preset monophonic ringtones often featured chestnuts of the Western classical repertory, by the early 2000s the ringtone industry was in full swing, and large numbers of phone owners were seeking to replace their monophonic and polyphonic preset ringtones with new ones of their choice—typically ringtone arrangements of popular songs. Although classical preset ringtones did persist into the polyphonic phase, the "moment" of Western classical music on ringtones was the monophonic period, which had by and large already passed. Moreover, the ease with which one could alter a ringtone and the means to do so changed in the polyphonic period. Gone were the old monophonic text-entry editors, through which one could input a ringtone of choice; in their stead, for-pay portals accessed via Wireless Application Protocol (WAP) on phones became the primary way of changing one's ringtone, at least in Europe and North America.

In observing such a moment as an historical event, we can see that it heralds the arrival of a new technological form and its incorporation into everyday social life. But, one may ask, in these moments, what happens to the older technological form that was supplanted? Following Gil Rodman, we might posit that the prejudice in favor of examining leading-edge technologies often obscures the fates of trailing-edge technologies, which quickly become unusable or unfashionable on account of the planned obsolescence of upgrading.[3] Disappearing from view while still often existing in a diminished capacity (or diffusing out from regions dominated by economic processes associated with the core zones to the semi-peripheral and peripheral ones), trailing-edge technologies and the social practices they engender are not only worthy of study in and of themselves; in addition, they also often help to place the new technologies and practices in historical context and provide opportunities for demonstrating continuities and possible points of origination.[4] In the case of the ringtone, the present chapter takes practices and qualities associated with the monophonic ringtone—the relative ease of manually inputting a ringtone of one's choice and the curious predominance of Western classical music in its repertoire—and seeks to ask what was lost as a result of technical advances seemingly universally acclaimed by all. Both require a certain kind and degree of musical literacy, but the way these two dimensions intersect is not simple. An attempt to untangle them appears in the chapter's conclusion.

Marketing and Distributing Ringtones Online (and in Print)

The monophonic ringtone and composer programs that appeared in the late 1990s and the early 2000s (discussed in chapter 2) gave rise to an even more curious phenomenon. At that time, one could easily observe the proliferation of websites and chat groups listing hundreds of song titles, each with instructions for inputting monophonic ringtones into different phone formats. Such sites were divided according to language, and they existed for ringtone enthusiasts interested in particular music genres or repertoires from all over the world—especially those from regions with high rates of mobile-phone use. Hence, one could find particular websites devoted to monophonic ringtones of Japanese anime music, Canto-pop melodies, Bollywood film songs, Hollywood film scores, Western classical tunes, and American TV shows. Some websites merely listed song after song in the relevant genre, often dividing them up according to handset manufacturers (which have different coding languages for ringtones). In other cases, a single song would be posted, with the melody translated into several different manufacturers' coding languages and listed in a column below the song title. On discussion boards, one might find participants asking each other how to code a particular song or how one would translate a particular set of instructions from one handset maker's coding language to another's.

Cryptic in appearance, the long lists of numbers and letters on such websites constituted a new kind of musical score or set of instructions for musical production, coded using a simplified version of Western musical notation consonant with many popular musical practices (note names, rhythms, registers, tempo in beats per minute). Understood in this way, these sites might be seen as part of a growing online archive of music information that also includes song lyrics, guitar tablature and chords, material on music groups and bands, reviews, and, of course, sound files (both pirated and original files, as in the case of remix websites[5]). Significantly, this haphazard archive is quite different from the institutionally sanctioned collections of scores, musical writings, journals, audio files, historical information, and reviews available online through universities or through websites such as Allmusic.com—unofficial sites are rarely thorough or comprehensive, often include mistakes in transcriptions of lyrics and musical materials, and are often sustained financially by highly irritating Internet advertising. But as part of a continuum of online musical materials, this desultory archive provides an important set of resources used by many musicians and listeners in everyday life and contains a sprawling repertory of music available online for performance use, fact checking, analysis, and composition.

The emergence of this online archive also represented an opportunity for book publishers to make quick profits. In a way not unlike the explosion of sound-file comedy and novelty ringtones several years later (see chapter 7), publishing companies seized upon the ringtone phenomenon and began to produce cheap books compiling the kind of information on the websites and chat logs described above. Indeed, a print-based precedent appeared in Japan as early as 1997, in all probability inspiring cost-free online variants: the multi-million-selling *Keitai Chaku Mero Doremi Book*.[6] Typically, ringtone books such as these begin by including a list of instructions explaining how to use them—which here means explaining how to input the encoded ringtones into one's phone, which may be a particular brand (especially the then-dominant global brand, Nokia) or may accommodate different phone models. They then go on to list a large number of melodies or effects culled from various songs and musical compositions, which are classified under various, often idiosyncratic rubrics. For example, the British book *Ringtone Mania! Platinum* (Wise, 2002) includes four main categories of musical sources ("Entertainment," "Chart Hits," "Decades," and "Classics"), with some fairly counterintuitive and sometimes revealing subheadings for a few of the categories.[7] As on the websites (see below and chapter 2), each song would be listed as having a note (or rest) duration, a note (pitch class and register), and a tempo. In some cases, as with the *Ringtone Mania!* series, copyrights for songs were secured—distinguishing these products from the unsanctioned online postings.[8] Such books were not exclusive to the dominant global economic centers, however, and could be found wherever a mobile culture was thriving during the height of monophonic ringtone popularity. Such books published in the Philippines, for example, included publications featuring Pinoy jokes and other text messages as well as collections of ringtones.[9] Unlike *Ringtone Mania! Platinum*, these seem to have been encoded in an "easy keypress mode" that didn't require knowledge of musical notation.[10]

The early monophonic code websites were similar in substance to the popular ringtone books. Since many of their "scores" required a basic knowledge of Western music theory, it is ironic that these sites seemed to flourish (and to some degree continue to persist) for East and South Asian repertories. One example from a Bollywood ringtones website will illustrate how they seem to work.

For Nokia mobile phone
Ankhein—Gustakhian
Tempo 125
4c1 8g1 8f1 4g1 8f1 8#d1 4f1 8#d1 8d1 4#d1 8d1 4.c1 8g1 8f1 8g1 8f1 8d1 8#d1 1f1 4c1 8g1 8f1 4g1 8f1 8#d1 4f1 8#d1 8d1 4#d1 8d1 4.c1 8g1 8f1 8g1 8f1 8d1 8#d1 1f1

For Ericsson cell phone
C g f G f #d F #d d #D d C g f g f d #d F C g f G f #d F #d d #D d C g f g f d #d F

For Motorola cell phone
4 C-4 G-4 F-4 G-4 F-4 D-#4 F-4 D-#4 D-4 D-#4 D-4 C-4 R2 G-4 F-4 G-4 F-4 D-4 D-#4 F-4 C-4 G-4 F-4 G-4 F-4 D-#4 F-4 D-#4 D-4 D-#4 D-4 C-4 R2 G-4 F-4 G-4 F-4 D-4 D-#4 F-4

For Siemens cell phone
C1(1/4) G1(1/8) F1(1/8) G1(1/4) F1(1/8) Dis1(1/8) F1(1/4) Dis1(1/8) D1(1/8) Dis1(1/4) D1(1/8) C1(1/4) G1(1/8) F1(1/8) G1(1/8) F1(1/8) D1(1/8) Dis1(1/8) F1(1/1) C1(1/4) G1(1/8) F1(1/8) G1(1/4) F1(1/8) Dis1(1/8) F1(1/4) Dis1(1/8) D1(1/8) Dis1(1/4) D1(1/8) C1(1/4) G1(1/8) F1(1/8) G1(1/8) F1(1/8) D1(1/8) Dis1(1/8) F1(1/1)

Samsung
1 5*** 4 5 4 2# 4 2# 2 2# 2 1 0*** 5 4 5 4 2 2# 4 1 5 4 5 4 2# 4 2# 2 2# 2 1 0*** 5 4 5 4 2 2# 4[11]

The song "Gustakhiyan," or "Gustakhiyan Hai," is from *Aankhen*, a 2002 movie directed by Vipul Amrutal Shah and starring Amitabh Bachchan as a fired bank worker who robs his former employer for revenge. If one follows only the Nokia instructions listed above (which are the most accurate), the ringtone melody should sound as illustrated in figure 3.1. The melody immediately strikes one as odd due to a simplification in the second measure.[12] But, having verified the melody with a website that sells monophonic ringtones, it seems to have been correct or at least sufficiently consistent with what was available on the market ca. 2004–5.

Figure 3.1
The "Gustakhiyan" ringtone

Listening to the song "Gustakhiyan" ("Transgression") itself helps to clarify how the ringtone was created. The song is essentially a Bollywoodized piece of dance music: a lavishly produced but highly generic version of an ethnic or tribal house track, on top of which is overlaid a male-female duet characteristic of Hindi film love songs.[13] The track begins with a moderately fast (126 beats per minute) "four on the floor" disco/house beat (0:00–0:08), then adds a sampled layer of a percussion instrument, perhaps South Asian in origin although it seemingly functions as a nonspecific membranophone signifying the "ethnic" (0:08–0:23).[14] The track continues by adding a syncopated synthesizer line on a reiterated tonic pitch (C3) that is processed using a filter sweep, or gradually varying from high frequencies to low frequencies and back up to high frequencies along the pitch's harmonic series (0:24–0:53), in the manner of much acid house—in fact, the synthesizer could be a Roland TB-303 bass synthesizer, the instrument that provides the characteristic sound of acid house music. At 0:54 the male singer enters, singing the melodic line in figure 3.2 in a highly stylized fashion (reminiscent both of earlier Bollywood singing styles and of English-language rock and dance-pop singers). The line is treated as a wordless refrain that returns throughout the course of the song, and it is not surprising that it would stand out to a ringtone arranger as the best melodic line to excerpt.[15] But it should be immediately apparent that this line does not include the simplification of the second measure and syncopation of C4 in the third measure of the Nokia version of the ringtone; this despite the fact that the melody is easy enough to hear for the purpose of transcription.

The commonsensical answer to this conundrum is that the original transcriber of the song made a mistake, which was then replicated in other contexts and ultimately normalized in the monophonic ringtone of "Gustakhiyan." Certainly this is plausible: the nature of online information is such that it propagates quickly, whether or not errors are present. Lacking professional editors who might catch such oversights, ringtone websites (like lyrics websites) often adopt preexisting transcriptions found online

Figure 3.2
The original vocal line of "Gustakhiyan."

rather than redoing the tedious labor of transcribing melodies. Nonetheless, the gap between the two versions poses a number of questions that by now probably are unanswerable. Who first transcribed the ringtone: a professional musician, or (more likely) an amateur? (In other words, did the error originate in a professional or amateur context before migrating outwards?) Was it transcribed from the song recording itself or from another ringtone? If copied from a professional (or semi-professional) effort, was it from a publication such as those described above or from listening to the ringtone itself? Was it done in a context in which permission was acquired to transcribe and sell the ringtone?[16] Despite the error, it is clear that the transcription is in some sense adequate. The transcription mistake occurs at the lowest point in the melodic line, where the singer slurs over the notes and the lower range is somewhat more covered over by the backing instrumental tracks—the melody thus remains fully recognizable and is still quantifiably different from the original.[17]

As the more lucrative polyphonic and especially sound-file ringtones and ringback tones became increasingly prominent on the subcontinent, do-it-yourself Bollywood monophonic ringtone websites drifted out of view, if not entirely—in part, on account of sales and use of secondhand phones in India and elsewhere in the world.[18] Nonetheless, as similar stores of accumulated information all over the world go unused, they float for some time in cyberspace like digital junk, demoted on the ever-changing (and now personalized) Google hierarchy before ultimately being deleted or removed.[19] As such, they correlate to the real junk of older mobile handsets with outdated features like monophonic ringtones. These phones are disposed of, and, if not resold, then gathered into mountains of electronic junk, and in some cases recycled, making room for newer handset models.[20]

Website domains, however, can be invaluable commodities in themselves, militating in some cases against the general amassing of information online. Monophonic ringtone code websites began to disappear in the mid 2000s, giving way to for-pay websites from which one could purchase polyphonic and then sound-file ringtones. As the ringtone boom took off, potentially lucrative domain names (e.g., ringtone.com, ringtones.com, and even ringtonesusa.com) were bought up and then ultimately transformed into portals for mobile entertainment firms, which reached customers via WAP access on phones and by WWW access on computers. Often, a user would be compelled to select a global region, then register a phone number to gain access to the site, and thereby be enrolled into unwanted ringtone subscription plans. (See chapter 5.) As a result some sites ended up pushing instead for single-payment approaches and advertising themselves on that basis.

The story of the ringtonesusa.com site is instructive in this regard. The site appears to have been purchased first by the Nokia Corporation in late 2001, advertising a link to the forthcoming usa-nokia.com site. Sometime in late 2002 the site appeared to be "under construction," and by March 2003 the site was bouncing directly to usa-nokia.com, which announced that its ringtones "are REALLY Free for a limited time," and although the site was "not yet fully open for business," it was promised to be entirely legal—unlike many of the earlier ringtones sales sites. (See chapter 1.) The site even claimed to be "working under the guidelines and policies set down by ASCAP, the US Recording industry watchdog." In April 2004, the site bounced to oingo.com, owned by Applied Solutions, a Santa Monica-based software-development company that had been purchased by Google in 2003.[21] In February 2005, the domain name ringtonesusa.com was up for sale, and seems to have remained so for much or most of that year. In a year's time (February 2006), a minimal interface registered the presence of the "Ringtone Download Warehouse," announced the launch of RingtonesUSA.com, and encouraged interested parties to sign up (with a mobile number and/or email address) to get a text message or an e-mail message on the site's opening day. By March, the site had launched and even received a little press, announcing that the company's strategy would be a "no-subscription model" in response to consumers' frustrations with subscription plans.[22] The completed site, featuring ringtones, wallpaper, and an apparent marketing relationship with the "Girls of Red Hot Dateline," evidently a phone sex or chat firm, seems to have been in operation at least into 2007. But by 2008 the domain name was up for sale again, and it is still available as of this writing. Clearly, the company was a flop.[23]

In none of the versions of the ringtonesusa.com site was there the kind of user-based, participatory information sharing (as well as piracy, or what Christopher Dodd has called digital "looting"[24]) found in the earlier monophonic ringtone websites mentioned above; likewise, this was typically not the case with other mobile entertainment firms' online platforms. Instead, the moment of the polyphonic ringtone signaled a certain closure of information, given that the files were much harder to transfer between phones and that the time-intensive coding of these files required specialized software and could not be easily mastered by amateurs, although Bluetooth-enabled phones allowed for ringtone sharing in less protected national markets. With the rise of sound-file ringtones, sharing did become much more possible, as sites like rapidshare.com or newer chat sites again offered platforms for freely distributing ringtones (including for political use). Still, something of the primitive culture of text strings was lost in the transition to more complex ringtone file formats, and in due course websites proclaiming the value of monophonic ringtones (such as monophonicringtones.net) began to pop up in nostalgic support of the old order.[25]

Ringtones and Western Classical Tonality

A Bach Companion (2000), an otherwise dutiful and informative EuroArts documentary on the life and music of J. S. Bach, has a bizarre opening sequence. A dusty wig sits on a mannequin head. Bach's famous "Air" from the Orchestral Suite No. 3 in D major (known more familiarly in its transposed version as "Air on the G String") plays. The wig is lifted by some anonymous hands from the mannequin and placed onto a human head lacerated with wounds resembling handwriting—perhaps the composer's own signature. With a cry like that of Frankenstein's monster, the reanimated corpse, which we take to be that of Bach himself, suddenly comes to life. The video then cuts to a close-up of a mobile phone, a Siemens M35, playing a monophonic ringtone of the "Badinerie" from Bach's Orchestral Suite No. 2 in B minor. The phone, seen in close-up, is held by a young man's right hand and lifted up next to his ear. As the video cuts to a more distant close-up of the man's face, he says, in German, "The whole world is celebrating the genius of Johann Sebastian Bach. For this reason we've decided to let our mobile phones ring only with his music."[26] The man then demonstrates, showing a mobile phone playing the famous opening of the Toccata and Fugue in D minor, then another phone playing the fugue subject of the same, and still another phone playing an undecipherable ringtone, also presumably by Bach; he smiles knowingly amid the cacophonous tinkling. Then the hands from the opening return. Just as they land on the wig-adorned head, near its ears, the ringtone music stops and Bach-Frankenstein screams again, perhaps in response to the ringtones themselves. The video then cuts immediately to a circular saw noisily cutting a piece of wood. Next, a construction worker carrying a boom box states that Bach was the "greatest composer" and that Bach's music helps him to relax when things get "stressful" at work. The scene ends showing the balletic movements of construction cranes, still accompanied by the D major "Air."

An example of the hackneyed expressionism found in much contemporary German-language media production, the scene is nonetheless fascinating. It suggests simultaneously that ringtones have reanimated the spirit—and even the dead body—of Bach the composer; that their jangling noise would be anathema to that composer, were he to hear it; and that Bach's music (and, perhaps, classical music by extension) has been adopted into mass culture in numerous ways, including via ringtones and even as workplace mood music by the German-speaking working class. The undead Bach, however, is not the only individual disturbed by these contradictory developments. In [illegible], echoing a sentiment that was popular during the earliest days of the monophonic ringtone's appearance, Angela Santo, a 23-year-old English civil servant, said "Those classical ring tones, the ones that are of classical music, I hate them with a passion."[27] Initially, the sound of the ringtone was inextricably linked to the greatest hits of older musics, especially Western classical music; indeed, the early monophonic ringtone—particularly the preset ringtone from the mid 1990s to early 2000s—was a veritable primer in some of the most overplayed melodies in the history of Western music. Who can forget the once-ubiquitous presence of truncated, cheap oscillator-tone renditions of classic pieces by Beethoven (Für Elise, Symphony No. 5 in C minor), Mozart (Symphony No. 40 in G minor, "Eine Kleine Nachtmusik"), Bach (numerous examples, including the A minor Two-Part Invention, the "Badinerie"), and Brahms (G minor Hungarian Dance, no. 5)?

However, lest one think that the early monophonic ringtone—the "classical" ringtone, as it were—is solely dominated by the great German masters, one only has to examine the inescapable Nokia Tune, the most classical of ringtones. The thirteen-note melody is an excerpt from the "Gran Vals" by nineteenth-century Spanish guitar composer Francisco Tárrega. The original piece is a multi-strain waltz in ABCDA form, with individual tunes moving symmetrically from A major to E major to B major to E major before returning to A major. The main tune consists of two parallel phrases, with the first (mm. 1–16) including a striking cadential passage (mm. 13–16) that ends the phrase. (See figure 3.3.)

The story of the Nokia Tune, which would come to be the "sonic logo" of the Nokia Corporation, begins in 1993. As part of a holistic branding strategy, Anssi Vanjoki (then Vice President, Sales of Nokia Mobile Phones; later Executive Vice President of the Markets unit and head of the "Mobile Solutions" division of the company) suggested to Lauri Kivinen (then Vice President of Communications; later Head of Corporate Affairs of Nokia Siemens Networks) that the "Gran Vals" be included as a ringtone on its 2100 series, since it was already being used in advertisements, and gave him the score of the "Gran Vals."[28] Kivinen, a musically literate choral

Figure 3.3
Francisco Tárrega, "Gran Vals" (1902), mm. 1–16.

Figure 3.4
The "Nokia Tune."

singer, then seems to have selected the cadential figure and added an extra tonic pitch at the end to provide a sense of finality or closure. (See figure 3.4.) According to one Finnish blogger, drawing on a Finnish news report, "The Nokia tune was [for the] first time played on a Compaq LTE computer. . . . The ringtone was so good that the notes were faxed directly to Oulu, Finland, where Nokia ringtones were developed."[29] In the early Nokia advertisements, much of the entire Tárrega piece was played on guitar; in contrast, the ringtone was a tiny snippet that could be usable as a flexible branding device, beyond the context of a formal advertising spot on TV or radio. Although the melody was included as one of the early monophonic preset ringtones in the 2100 series and after, simply and accurately titled "Gran Vals" (without providing any more information), Kivinen seemed to have greater designs for it, allegedly perceiving a kinship between the ringtone and Nokia's slogan, "Technology with a human touch."[30] Perhaps at Kivinen's instigation, between the release of the 2100 series in 1994 and 1998 the tune proved to be popular enough—perhaps in part among office employees, who were eager to acquire ringtone-playing phones themselves[31]—that the company renamed it the "Nokia Tune" (probably in 1998, but perhaps earlier) and made it the default ringtone on its phones.[32] However, it may be that its branding took place in Finland relatively early on but diffused more slowly out to the rest of the world.

There are several reasons why music of the Western classical tradition became prominent in early monophonic ringtones. According to a spokesperson for Nokia, after hitting upon the ringtone concept the company wanted to use public-domain melodies to avoid potential legal problems. That explanation has certainly been promulgated in the press as the most salient factor in the selection of default ringtones.[33] And in examining default ringtone selections, one does occasionally note the presence of Sousa marches, familiar Tin Pan Alley tunes (e.g., "Take Me Out to the Ball Game"), and other public-domain materials not generally conceived of as classical. Another reason may involve the residual distinction value of Western classical music as elite or sophisticated and associated with the upper class. From their inception in 1975 to as late as 1990 (and even later in the United States), cell phones were associated with elite and "yuppie" lifestyles, and this association persisted long afterwards. In one example of the "you are your ringtone" trope, a psychiatrist classified ringtone choices according to personality and class types, arguing that "traditional, conservative and cerebral types," especially those "belonging to an older age group who like the idea they are sophisticated," are likely to choose "classical music" ringtones.[34] But explanations mapping aspects of consumer identity (including actual class position) onto ringtone choice seem insufficient. For one thing, the preset melodies did not typically vary between countries or regional markets, except in minor ways.[35] Although such changes would inevitably raise manufacturing costs, they are eminently possible, leading one to wonder why manufacturers did not pursue such strategies to tailor their products to their target markets. Moreover, with the cell phone emerging as a mass-market item in the 1990s, the residual class distinctions of the handsets themselves seem to have given way to a recognition that, at least within the developed world and increasingly within the developing world, anyone can own a cell phone.[36]

Perhaps the answer lies in the fact that the earliest primary markets for mobile telephony—Europe and East Asia—are regions in which Western classical music is especially popular. Indeed, classical music may have been a bridge between these two earliest and still strongest global mobile telephone markets, as the ringtone concept and its classical repertoire spread quickly from Finland and Japan to Germany, South Korea, the United Kingdom, and beyond. In other words, the appearance of classical music on cell phones may have been a part of promoting their increasing popularity in Europe and East Asia, with intimations of upward mobility being evoked by the object itself. Although in the Asian context a cultural imperialist viewpoint would merely see this as a globalized stage of Western dominance, in which Western cultural forms are still viewed as hierarchically superior by post-colonial subjects, I would argue instead that the phenomenon must be understood in relation to the fact that Western classical music is now, and has been for some time, a central part of global mass culture.[37] That this mass culture is inseparable from the mass consumption of cell phone handsets thus seems all the more fitting.

A second explanation for the disproportionate presence of classical music on mobile phones in the 1990s and early 2000s lies in the structure of monophonic ringtone sound production itself. The earliest means of producing sound on cell phones were quite limited. A phone could produce a single melody with an unflattering digital timbre that was tuned in a twelve-note equally tempered chromatic scale. The melody was of relatively short duration (maximum 30 seconds), was played within a small range (three or four octaves), and allowed for rhythmic subdivisions into multiples of 2 only.[38] Such limitations encoded a number of musical assumptions into the monophonic ringtone. They prevented both standardized and non-standardized tuning systems of the Arab tone system (in 24-tone equal temperament), the variable Indonesian temperaments that produce *pelog* and *slendro* scales, and African tuning practices and scales.[39] They encouraged the coding of rhythmically simple melodies, making unusual rhythmic subdivisions difficult to create. And they fixed music into a single melodic line with set pitches and simple timbres, making even simple unpitched music difficult to create. Of course, owing to the technological limitations of the cell phone in the late 1990s and the widespread global presence of Westernized tunings and musical styles (which had been promulgated in a somewhat analogous way in the course of the twentieth century through audio recordings of Western musics, fixed-tuning instruments such as fretted guitars, and, eventually, the digital keyboard synthesizer), such a sound-production system probably seemed both cost-efficient and eminently suited to the purpose. The situation was roughly analogous for polyphonic ringtones, which theoretically could produce non-equal tempered tunings but which seem to have been used little in this capacity, in part because the global popular styles that would make desirable ringtones are very often already in or easily approximated by 12-tone equal temperament.[40] (Indeed, the only detuned MIDI examples I have encountered in ringtone production have been for effects, and none of the ringtone workers I have interviewed ever detuned MIDI to reproduce nonstandard tuning systems in their ringtones.)

The inaccuracies and reproductive limitations inherent to the early forms of the ringtone suggest that its original purpose was not to accurately reproduce music but to *represent* it, offering amusement while simultaneously fulfilling its functional role. As Shawn Conahan of Moviso has aptly noted, "we're not selling music—we're selling a representation of a song someone has heard."[41] The monophonic ringtone's structural constraints produce an abstract form of a particular musical selection by excerpting its main melodic line and playing it back with an indistinctive timbre and a mechanical rhythmic articulation; in the process, music is arranged and adapted to fit its new guise. The ringtone is therefore not unlike other forms of musical representation and instrumentalization; as David Suisman notes, piano arrangements, piano rolls, and phonograph records all "*represented* music . . . in ways that could likewise be considered forms of adaptation."[42] Although this would be true for any repertory converted into ringtones, I would like to advance the argument that the Western classical music that appeared in early ringtones was eminently representable and abstractable music in at least three ways. First, such music (including virtually all of the public-domain music that appeared in default ringtones) stems from an era in which sheet music and musical scores were the primary vehicles for composing, understanding, interpreting, and commodifying music in the West. Indeed, this is the first inescapable legacy of Western musical technology: it produced a musical practice whose audible result was inseparable from its written form. From this perspective, a ringtone essentially functions as a means of abstraction not dissimilar to a musical score—and even more like reference pages or books that notate the first or main theme of a particular piece of music. A second, related aspect of this music that provides a means of abstraction is its formal focus on important melodic themes, tunes, or motives, which are highlighted for the listener as being particularly important within the long-range flow of the music. Variously connected to ideologies of romantic organicism (especially in nineteenth-century classical music) or a catchy melody line (as in popular songs and advertising jingles in the era of mass culture), the music provides particular melodic fragments that one can easily recall, a feature that can be marshaled for use in ringtones.[43]

Third, the music in question all derives from the period of Western music history in which a relatively consistent set of compositional practices predominated. Chief among these practices is what music theorists characterize as "common practice tonality," a term used to describe the period of Western musical history from about 1600 to 1900. As Brian Hyer elegantly explains,

> Perhaps the most common use of the term ["tonality"], then, in either its noun or adjective forms, is to designate the arrangement of musical phenomena around a referential tonic in European music from about 1600 to about 1910. However this arrangement is conceptualized, musicians agree that there are two basic modal genera, major and minor, with different but analogous musical and expressive properties. It gives rise, moreover, to abstract relations that control melodic motion and harmonic succession over long expanses of musical time. In its power to form musical goals and regulate the progress of the music towards these moments of arrival, tonality has become, in Western culture, the principal musical means with which to manage expectation and structure desire. It is thus understood to be essential to modern Western music: it determines the coordination of harmony with melody, metre with phrasing, and texture with register, thus encompassing—within its historical domain—the whole of music.[44]

One of the fundamental aspects of tonal music, as the music theorist Heinrich Schenker observed, was the way in which melodic lines seemed interpenetrated with harmonies or chords and vice-versa, mutually reinforcing tonal structure.[45] Arguing a similar point from the perspective of information theory, the musicologist Leonard Meyer argued that tonal music includes various forms of redundancy that allow the listener to comprehend quite complex musical forms and styles.[46] For a monophonic ringtone, then, the isolated tonal melody is able to encapsulate and communicate—or represent—the whole of an excerpt's musical structure without including all of its components. Even in the case of the Nokia Tune, many listeners are able to implicitly hear the chords that underlie the melody (which are found, almost superfluously, in the polyphonic and sound file versions of that ringtone)—the E6-D6-F♯5-G♯5 and C♯6-B5-D5-E5 parts of the melody include mostly pitches associated with an E-dominant seventh chord (or a V^7 chord in A major), whereas the B5-A5-C♯5-E5-A5 mostly outlines an A major triad (or I chord in A major). In contrast, many of the American and American-influenced popular musics that draw on blues modality demonstrate what David Temperley describes as a "melodic-harmonic divorce," such that blues-based melodies don't necessarily "line up" with the chords that support them and therefore require a multi-part arrangement to produce the characteristic harmonic structure of the music.[47] Thus, although monophonic classical music ringtones offer a pale imitation or representation of their source music, monophonic-ringtone versions of popular songs often seem that much more inadequate and laughable. As one ringtone programmer quipped, "You can't make Brazilian funk on a monophonic ringtone."[48] This deficiency was certainly convenient for advocates of ringtones' technological upgrading.[49]

The profitability of popular music was the primary motivation driving the transformations in the ringtone's reproduction technology: polyphonic and especially sound-file ringtones were able to reproduce the distinctive timbres and textures of popular musics in ways that were simply not possible with monophonic ringtones. Here one must consider the importance of *sound*, as opposed to abstract pitches, in contemporary popular music—epitomized by the production of the "track," or the most important compositional layer besides songwriting and arrangement/instrumentation/orchestration[50]—and its dialectical relationship to the rise of audio recording, the third fundamental technological contribution of the West to musical experience and practice after mensural notation and the mechanization of instrument production.[51] The shift toward a sound-production-based aesthetic has had huge consequences for the music industry, placing much more economic and political power in the hands of recording companies and less in the hands of sheet music/song/composition publishers and transforming the very aesthetic premises of what music should be, which is something heard as recorded sound *first* (the hegemony of which is attested to by the current boom in headphone listening occasioned by the iPod). As a consequence, Western and Western-derived popular musics define and sell themselves not through abstract melodies or harmonies but through distinctive, even brandable timbres.[52]

The ringtone's shift from notated, public-domain, mostly common-practice-era classical music to copyrighted and ultimately recorded popular music had numerous consequences, some of which are explored in chapters 1 and 9, including the acceleration and ultimate demise of the for-profit ringtone industry itself. But that shift also had consequences in more modest quarters—for example, in the new preset ringtones that accompanied newer, sound-file-capable phones. Although one might point to Motorola's "Hello Moto" ringtone as an important exemplar (because it incorporated a speech sample as part of a hybrid polyphonic ringtone) in the English-language world and especially in North America, the most important of the new preset ringtones are those on Apple's earliest iPhone of 2007. The most famous of these tones is "Marimba" (figure 3.5), a riff about 4 seconds in duration (including the subsequent pause).[53]

From a melodic perspective, "Marimba" is far less conventionally tuneful than the Nokia Tune. Instead, it features two strikingly different qualities. First, it is built on the rhythmic-textural principles of hocketing and syncopation, with the upper-part attacks sounding whenever the lower parts are

Figure 5.3
Andy Neesley's transcription of the iPhone "Marimba" ringtone.

held or silent and vice versa and with the attacks themselves often sounding off the beat; this textural differentiation helps to accentuate a motivic echo effect in which the lower-voice third B4-G4 (m. 1) is recalled in quasi-augmentation (or slowed down) in the upper parts (G5-B5, B5-[G5]-G5, mm. 1–2). Second, the ringtone is, true to its name, performed on a marimba or some other mallet instrument that produces a rounded and woody but percussive timbre.[54] The combined effect is somewhat akin to various kinds of African and neo-African musics. One might think of balafon music, but the clear and cheerily diatonic, major-triad-plus-added-sixth or pentatonic tonality and marimba timbre (now naturalized in the West) evokes a sonic style closer to home: musical minimalism, particularly in a certain African-influenced vein recalling Steve Reich's music. (Indeed, several Blackberry alert tones of the late 2000s had a marimba timbre, and one of them appears to quote Reich's 1971 composition *Drumming*.[55]) The use of this kind of music for making preset branded ringtones is overdetermined, at least three coordinates being relevant. First, minimalism has long been associated with advertising and with a relatively "new" corporate ethos emerging from the tech industry and beyond in the 1980s and the 1990s, the connections between advertising and minimalism being literalized in numerous examples and perhaps sharing a homological kinship.[56] Second, keyed mallet instruments have long had powerful associations in the realm of sonic branding: perhaps the most famous of sonic logos is Walter Werzowa's for Intel (also see chapter 8), which features in its last four notes a combination of "marimba, xylophone, bells and a 'secret recipe' of instruments."[57] Secret or not, the marimba/xylophone timbre stands out quite prominently. Third, marimba effects of this sort have also adorned the soundtracks of prominent feature films. For example, Thomas Newman's soundtrack to the 1999 film *American Beauty* featured a Dorian-mode marimba riff (in a tune called "Dead Already") similar to the iPhone "Marimba" (and, in fact, extremely similar in mode and general character to the "Xylophone" preset iPhone ringtone, though with a significantly faster tempo).[58] In all these cases, the historical locus is quite clear: whereas the Nokia Tune's origins in a nineteenth-century musical landscape are apparent enough, the iPhone is rooted in the musics of the late twentieth century and the early twenty-first century—a period shaped by a consumer "ecosphere" saturated with multimedia and with articulations with advertising of numerous sorts. The ringtone's numerous points of determination help to fashion a cosmopolitan, up-to-date, corporate, and vaguely New-Agey sonic sensibility suitable for mobile subjects in the era of neoliberal globalization. Perhaps the mobile phone's catching up to the present, then, is less remarkable than its dalliance in musical and historical formats of an earlier era.

Conclusion: Historical Recapitulations and Post-Literacy?

A comparison between the two cases given above—between the decline of new, quasi-scores for monophonic ringtones and the decline of Western classical music, whose revival by the ringtone was surely unexpected—seems in order. As a certain kind of historical window or laboratory, the pressures upon which include the limits of technical reproducibility and control over copyright permissions, the ringtone effectively illustrated a shift from one phase in Western modernity (a phase dominated by the logics of notation and notated music) to a later phase (sometimes termed postmodernity) in which orality and increased proprietary and corporate control, reigned supreme. Of course, insofar as polyphonic ringtones and (especially) sound-file ringtones could provide adequate representations or reproductions of a much wider array of musics, the decline of Western hegemony in favor of a postmodern pluralism decentering the West also became evident in the ringtone, even if certain Western materials (e.g., the ambient, world-music-tinged minimalism of "Marimba") and Western-derived cultural products still retain an outsized presence in the ringtone market and global culture markets more generally, arguably reinscribing (or even sublating) the musical hegemony of the West historically produced by tonality, equal temperament, AABA song form, and rock-based genres.[59] The marked decrease in musical literacy of a traditional sort marks the long history of modernity in Western art music. According to Richard Taruskin, "the literate tradition of Western music is coherent at least insofar as it has a completed shape. Its beginnings are known and explicable, and its end is now foreseeable (and also explicable)."[60] That endpoint, for Taruskin, is augured by a twentieth-century "orality" lubricated by digital controllers.

In his view, "it is now not only possible, but increasingly common, to create, 'perform,' and preserve music that is recognizably within the traditions of classical music without ever using notation. Music, without any necessary loss of conceptual complexity or novelty, can now take leave of the eye. The lineaments of a postliterate age are clearly discernible."[61]

Taruskin's narrative anticipates a postliterate moment by noting that a "postliterate age will have arrived" when "a majority of composers" fix their works "digitally and reproduce them via synthesizer or via MIDI" while other composers will still continue to use traditional scores. Taruskin seems to propose a dichotomy between digital and traditional representation, but reality has already blurred the two. Digital notation packages have become standard for even the most traditional of composers—facilitating the outsourcing of editorial labor to composers themselves—and present-day compositional practice often exhibits fluid movement among notated, MIDI, and sound-file representations of music, often coordinated by automated, multimedia performance systems like Max/MSP. The monophonic ringtone's generation of a temporary notation system—and, thus, a new kind of music literacy, built on the "language" of machine code—suggests yet another aspect of this more fluid and complicated reality: the possibility that a multiplicity of musical representation systems, many of which retain a literate character, will pervade music making of the future. One might think not only of ringtones, but also of the sight-reading system of musical video games like *Guitar Hero* and *Rock Band*, as discussed by Kiri Miller.[62] Moreover, these systems will have wildly varying shelf lives, which will make preservation and storage much more problematic in view of the inadequacies of MIDI (readily apparent within the world of ringtones, let alone more sophisticated forms of musical production). A single, convergent standard doesn't seem to be on the horizon, and preservation will require multiple forms of expertise, including Western mensural/staff notation. Concomitant with this proliferation of representational modes is the proliferation of platforms—already evident in the keyboard synthesizer, the digital workstation, the programmable video game sound chip, sound embedded in websites, the mobile phone's sound system, and elsewhere—new exemplars of which present the possibility of, yet again, recapitulating and reusing familiar technologies, cultural materials, and compressed historical trajectories, in one version of what Jay David Bolter and Richard Grusin call "remediation."[63]

The narratives described above, even if merely recapitulatory, suggest a teleology of progress whose utopian horizon might initially be presumed to be equivalent to the goal of technical progress itself. Yet if we consider Jameson's flawed yet insightful solution to the problem of mass culture (that it always discloses moments of reification and utopia), we may suspect that this thinking is merely the mystification of utopia by the cult of technological advance—Adams' "dynamo," if you will—and seek to identify utopian moments under the purview of the monophonic ringtone.[64] The most obvious of these is its facilitation of consumer-based creativity, discussed in chapter 2. But another may have to do with its peculiar form of abstraction. In 2001, while at a news conference in Jerusalem, the conductor Daniel Barenboim heard the unmistakable sound of Wagner's "Ride of the Valkyries" as a monophonic cell phone ringtone. Recognizing that Israeli citizens were already listening to music by Wagner (which, owing to the composer's anti-Semitism and later appropriation by Nazi culture, was implicitly banned from concert performance), Barenboim decided to break the ban and to attempt to push Israeli society beyond the politically stifling reification of the Holocaust.[65] Although the story would have been theoretically possible on any ringtone format, might it be the case that the very *diminutive* effect of the "Ride" as a monophonic ringtone in one sense served to render it sonically harmless, thereby creating an opening for the performance of Wagner's music through a recognition of the relative *lack* of social power that music has? Such an image of a potentially "dangerous" music would, in this case, be the exact opposite of the use of the very same piece in the famous helicopter attack scene in the 1979 film *Apocalypse Now*, in which its terrifying power amplifies the power of the American military combatants within the diegetic world and thus presents a sickening and ironic representation of music's danger when placed in the service of Cold War-era American military imperialism.[66]

This diminishing of music's symbolic power through its abstraction and saturation in everyday life, as demonstrated in the Barenboim example, results in the erasure of historical memory as a form of cultural amnesia. Even as a technologically and historically specific moment in (and not a general feature of) the commodification of music, the reifying force of that amnesia, which reduces music to an effect of all-pervasive and abstract commodity exchange, should be self-evident.[67] But affirming in certain contexts the utopian aspect of *forgetting* may in the end amount to identifying the utopian dimension of capitalism itself—a position that any dialectical analysis must both affirm and contest.

4 The Ringtone and Its Aesthetic Subgenres in Contemporary Classical Music and Media Performance/ Installation Art

Two privileged sites for witnessing the diminishing relevance of the ringtone in the 2000s are those of present-day classical music and the global art market—spheres that include many interrelated, semi-autonomous, and complex subdivisions, such as those between the modern orchestra and "new music" (itself fractured along aesthetic lines) or those among museums, "new media" art exhibitions, and for-profit galleries. An important distinction within these spaces is that between cultural producers whose artistic media are either extremely isolated from new technologies like cell phones (typically the case with the modern orchestra) or uncomfortably close to them (such as "new media" art scenes). On both sides of that divide, we might first attend to the reality of artistic domains' need for producing continual relevance to prospective consumers in increasingly crowded culture (arts and leisure) markets; the ringtone fad, when still current, offered relatively immediate access to media attention—and, in some instances, corporate capital—and upon its decline suffered a loss in appeal.[1] But this only tells one part of the story: as the ringtone transformed into a sound file, it converged with other audio playback technologies and therefore became far less distinctive as a sonic and aesthetic entity, and thus less desirable for artistic appropriation. While mobile phones, as portable computers and media devices, continue to serve as objects of fascination for artists and musicians of various stripes, the ringtone per se—with a few exceptions—no longer seems to be on the agenda of contemporary, "high art" cultural production.

The primary goal of this chapter is to track the emergence of new genres of contemporary music and art enabled by the monophonic and polyphonic ringtone—the medley composition and Nokia Tune piece in contemporary art music, the ringtone sound installation, and ringtone sound/music performance in new media art—and to chart their decline as the sound file became the dominant ringtone format. But a secondary theme, not always explicitly articulated below, concerns the contradiction between art's parasitic absorption of a fad, demonstrating a bid for attention and relevance, and the ringtone's role as an object of social critique, distancing art from that fad. As the fad became less novel, the ringtone itself became less interesting sonically. As aestheticized treatments of it became more widely known, the tension between interest and disinterest structuring the contradiction of ringtone art slackened. The techniques of aesthetic engagement were correspondingly (further) banalized, and ringtone artwork, already a peripheral phenomenon, was ultimately reduced to the gag that motivated it from the outset. The two most important and compelling representatives of ringtone art discussed below—Salvatore Sciarrino's *Archaeologia del telefono* (2005) and Golan Levin's *Dialtones: A Telesymphony* (2001–02)—best embody the contradiction of fad and critique. Both surfaced before the ringtone had become passé and fundamentally indistinguishable from other ambient music.

Ringtones in Contemporary Classical Music

In the broad domain of Western classical music, the ringtone functioned largely as a novelty phenomenon. Ringtones, ringtone melodies, and even mobile phones were used to draw attention to themselves as marked, even invasive presences within broader musical and performative flows. Employed in works and arrangements by composers, performing musicians, and orchestral conductors to make Western concert music more current and socially relevant or to ironically distance it from contemporary social phenomena, the ringtone and related paraphernalia surfaced primarily in two forms. In the first, ringtones participated in a number of medley-style compositions or performance works, typically in which familiar melodies were performed on mobile phones (as ringtones) in tandem with larger musical ensembles. For example, in October 2001 the percussionist and conductor Bernd Kremling of the Drumming Hands Ensemble in Würzburg, Germany wrote and led the premiere performance of a composition for 30 mobile phones and 30 percussion instruments titled *Von der Hand zum Handy* (From the Hand to the Cell Phone). The piece juxtaposes African-influenced percussion music with the quotations of works by classical composers (including the C major prelude from Book I of Bach's *Well-Tempered Clavier*, Bach's cantata chorale movement known in English as "Jesu, Joy of Man's Desiring," the Alla Turca from Mozart's A major sonata K. 331, and Liszt's Hungarian Rhapsody No. 2), each first played by a ringtone and then continued by a marimba, and tunes like

"Old MacDonald Had a Farm," which is accompanied by a two-step (oompah) drumbeat.[2] Only a few months earlier, a 10-minute medley titled *Spring Cellphony* was performed in Jerusalem by a symphony orchestra, as part of a science exhibit at the Bloomfield Science Museum. Amplified ringtone-ized snippets of familiar tunes like Rossini's *William Tell Overture* and works by Bach and Mozart were then continued by the orchestral versions of those same works, offering audiences a direct comparison between ringtones and their sources. Toward the end of the performance, a cell phone rang out audibly from the audience. According to English-language press reports, "the symphony's conductor stopped, pivoted and shot a dirty look at the audience." Of course, this was intentional, a joke commenting on the pervasiveness of mobile phones in Israeli life and, perhaps, on the phenomenon of "cell phone rage," in which errant ringtones interrupting performances received vitriolic responses by audiences and writers.[3] In 2003, in a comical inversion of this phenomenon, the American musicologist Peter Schickele, as a performer of music by his composer alter ego P. D. Q. Bach, performed his *Variations on an Unusually Simple-Minded Theme*, a concerto for piano and orchestra based on an opera tune and, in some performances, requiring orchestra members to stand and sway like players in a swing band; in at least one performance, the soloist's cell phone went off just as the solo part was to make its "grand entrance"—and the soloist took the call.[4]

The second category might be described as "Nokia Tune" compositions or improvisations: music that makes prominent use of the famed ringtone melody discussed in chapter 3. Since the familiarity of the melody makes its recognition as a ringtone not dependent on mobile phone playback, works and performances of this sort range widely; one can note at least two different approaches. In the first approach, the Nokia Tune begins the composition and typically features centrally throughout most of the entirety of the music in question. So, for example, in the French-Canadian pianist Marc-André Hamelin's *Valse Irritation d'après Nokia*, which was published in 2008 (though it was composed at least as early as 2006), the work immediately announces the theme within a kind of jazzed-up nineteenth-century salon waltz texture—one might think of Bill Evans embellishing slightly upon Chopin and Brahms. (See figure 4.1.) Rather cleverly, the end of the composition uses mostly dissonant intervals in the piano's upper register to imitate the sound of touch-tone dialing, suggesting that, after the dizzying reverie induced by the phone's ringtone (triggered by the call of a prospective lover?), the phone owner dutifully and perhaps a bit sheepishly returns the call. Or, since Hamelin tends to perform the work whenever a

Figure 4.1
Marc-André Hamelin, *The Ringtone Waltz* (2008), mm. 1–4, 40–43. Reproduced by permission of Theodore Presser.

ringtone interrupts one of his concerts, it may be that the music figures a harmonious blending of ringtone and virtuoso piano performance—an imaginary solution to the real social contradiction of the rogue ringtone heard in concerts—before the caller conspicuously dials his or her phone. (However it is taken, the joke is not lost on audiences, who laugh audibly at the work's ending on one recording.) In another example, in 2008 the young Canadian composer and pianist Vincent Lo composed a three-part Bach-style "Nokia Fugue" featuring the standard fugue exposition, episodes, *stretti*, and a closing section with a pedal point on the dominant. (See figure 4.2.) Garnering more than 500,000 YouTube hits between the different versions of the fugue he posted online as of mid 2012, Lo's exercise ranks favorably among a large swath of improvisations and compositions based on the Nokia Tune available on the website/media platform.[5]

The second, less common approach to employing the Nokia Tune in compositions is to position it in the middle of a work or a movement, producing it most clearly only at the end and thereby maximizing the aural effect of hearing the tune by treating it as a culminating or coda-like commentary on the preceding music. A good example of this technique can be found in Jocelyn Pook's *Mobile* (2002), a six-part contemporary madrigal written for the King's Singers, setting a text by the poet Andrew Motion.

Figure 4.2
Vincent Lo, "Nokia Fugue" (2008), mm. 1–15. Reproduced by permission of the composer.

The work juxtaposes historical topoi, contrasting a slowly paced neo-modal music characteristic of early British folk-music inspired composition—as in the work of Ralph Vaughan Williams—with a distinctly pop-oriented texture featuring rapid, repeated-note accompaniment figures in some of the voices and a somewhat syncopated melody line in the others. Punctuating these historical juxtapositions, which set texts contrasting past modes of communication and existence with the present, are the appearances of the Nokia Tune, which in concert is heard twice as a monophonic ringtone (played either by a cell phone or as a sound recording) at the end of one of the sections of slower or faster music.[6] Toward the end of the composition, after the second airing of the ringtone, the first countertenor imitates the ringtone with a schoolyard-like taunt, "na-na na na, na-na na na, na-na na na na"—representing either a mocking reaction to the ringtone (surely articulating the audience's repressed juvenile reaction to the ringtone and mobile phones generally) or hypostasizing it into a social force that mocks the ringtone-and-mobile-phone-saturated audience. (See figure 4.3.) As this imitation ringtone repeats, the other voices produce a wordless ii-I progression in A major/Ionian (with an additional voice moving up to the tonic, sounding like a variant on the Renaissance-era Burgundian cadence); the music progressively increases the attack speed

Figure 4.3
Jocelyn Pook, *Mobile* (2002), mm. 65–70. Reproduced by permission of Chester Music.

while maintaining the harmonic rhythm (creating a minimalist effect that recalls Philip Glass) and adding in repetitions of the word "talking." The very end returns to the slower, neo-modal music, setting the words "In the here and now of after Babel, the sounds all twist in a single braid / So the words we chose are those we share, and the talk we talk the lives we make." Although it clearly pokes fun at the frenetic pace of contemporary life, even viewing it somewhat critically, *Mobile* ultimately offers an optimistic treatment of the power of mobile communication to overcome divisions in the world and facilitate the discourse that makes life meaningful. But the piece also seeks to show the similarity between past and present moments in modernity, which consistently mobilizes the universal of human communication, and to illustrate capitalist industrialization's process of social acceleration, which sutures together contemporaneity with bygone eras.[7] Indeed, the ending's return to the slow music suggests that the present, seen as glacially slow from a technologically advanced future, will one day be understood as history.

The categories described above—ringtone medley compositions and Nokia Tune compositions—can be said to have crystallized, if incompletely, by the early 2000s. A particularly ambitious example of each was composed in the middle of the decade, with each shedding further light on the priorities and impulses of the categories more generally. The most developed composition based in the medley style is David Baker's *Concertino for Cellular Phones and Symphony Orchestra*, which was premiered by the Chicago Sinfonietta and commissioned by its director, Paul Freeman, on 1 October 2006. The work received a great deal of attention in the press, including a prominent treatment in the *New York Times* calling it "a perverse commentary on the scourge of modern concert halls."[8] The composition features several sound sources: a symphony orchestra, a cell phone soloist, and a few additional musicians playing ringtones on phones along with that orchestra, and the audience, which is divided up into different groups based on spatial location and given signals to indicate whether or not to play their phones' ringtones (these being stoplight-like green and red lights in the premiere performance). Structurally, the *Concertino* divides into several sections, with the overarching logic grouping them into five parts. First, an original motive (A-B♭-F-E) dominates the music up to rehearsal letter E (the motive is also turned into a ringtone on the soloist's phone); another original theme extends until rehearsal letter F. In a long second part of the piece, a jazz-funk riff appears in the low brass and woodwinds, and is accompanied by a number of different motives and themes, many of which are quotations from the standard repertoire of orchestral literature. A third, multi-part section (rehearsal letters I–O) includes the most explicitly medley-oriented material, presenting ringtones of the last movement of Brahms' Symphony no. 4, Rossini's *William Tell Overture*, and Rimsky-Korsakov's *Sheherazade*, each of which is then followed by an orchestral version of the same music (although comparatively, the three orchestral versions are progressively more distant from their original sources—by the third example, *Sheherazade* has been set in an orchestral mambo/salsa style). The fourth section (rehearsal letters R–Y) revisits the second, returning to the jazz-funk riff (and a variant) with orchestral quotations, and the final section (rehearsal letters AA–end) revisits the opening four-note motive as part of a coda/conclusion to the piece. Roughly following an arch-form design, the *Concertino* seems to build up to its central medley section—which appears to draw the most laughs from audiences—and thereby situates the obviousness of its quotations within a musical framework saturated with less readily apparent references to other works.[9] The use of cell phone ringtones in the work ranges between mass-cacophonous forms by audience members—often as introductory or punctuation moments (the latter typically in tandem with *tutti* orchestral section endings)—to more targeted uses presented by the cell phone soloist and orchestra phones, as in the medley section described above. Thus, whereas the former tends toward the kinds of collage-oriented, Ives- and Cage-inspired uses cited by the composer in interviews, the latter lends itself to more didactic and humorous utilizations of ringtones.[10]

A clearly structured work that brings audiences into a participatory relationship with concert music, Baker's *Concertino*, then, is much less "perverse" than one might suppose. In fact, I'd like to suggest that it can be interpreted in at least two ways, based on two of its apparent ambitions. First, it arguably represents an attempt by Baker, a prolific composer, to incorporate himself symbolically—if not in actuality—into the orchestral canon.[11] Crucial to this reading is the effect of the jazz-funk riffs mentioned above; they seem to serve as the work's motor for generating the welter of quotations. Since they probably are quotations from other works by Baker and/or are characteristic of his work, they place his music (and, given their stylistic roots, perhaps "black music" by extension) in a central position within the history of Western classical music. (Dvořák's Negro-spiritual-influenced Symphony No. 9 is prominently cited in the piece, too.) The arch-form effect seems to provide somewhat serious, even ominous music as the outer frame for the work (with imitation siren effects depicting the realities and even politics of urban life), and the sections with the jazz-funk riffs seem to present Baker (or a subject like him) circulating amid numerous Western art music references. The innermost section, in contrast, reveals a humorous, perhaps even trickster-like play with the source material of ringtones, depicting the ironies of sonic existence in the contemporary classical concert hall and perhaps demonstrating the gradual impact of persons of color on the broad domain of classical music more generally. As such, the narrative moves Baker gradually within and then back out of the world of (largely white) concert music, perhaps presenting a perspective on the composer's career, which began in the world of jazz and progressively moved, via avant-gardist experimentalism, into more traditional concert composition in the classical mold (while never abandoning jazz). The *Concertino* arguably reveals the difficulties facing contemporary black composers, jazz-based composers, and American composers in cracking the American orchestral repertory (something that only a few post-1945 composers seem to have done with any consistency).[12]

Despite what might be understood as an autobiographical narrative, Baker's larger ambitions for the work seem to be much more self-effacing: the *Concertino*'s primary task is that of drawing new audiences to the concert hall and educating them about classical music. It is, in other words, a significantly *pedagogical* project, which reflects Baker's longstanding work as a teacher and figure of musical pedagogy (especially but not exclusively in jazz, composition, and orchestration); moreover, it seems to be characteristic of the medley style of ringtone compositions in general, in contrast to the more pointedly social-critical Nokia Tune works.[13] Thus, although the participatory aspect of the piece provides fun and may give people a sense that they are "making history,"[14] it also serves as bait for increasing audiences for orchestral concerts and then getting them to listen to classical music. Such a draw seems characteristic of the commissioning orchestra, the Chicago Sinfonietta, whose ambitious programs have made it a rare success story in an American orchestral landscape facing major crises of viability.[15] Inseparable from the financial troubles that American orchestras have faced and continue to face as a result, most recently, of the 2008 recession, Baker's composition signals the fact that "new music" in the United States simply cannot be taken for granted as it can in other parts of the world, and that new compositions in the classical tradition are always something of a fraught proposition in an aesthetically pluralist environment prejudiced against state-based funding for the (most established) arts.[16]

The state of contemporary classical music in Europe is a rather different matter, and even though state funding of the arts has long been imperiled in the era of neoliberalism, the institutions of concert music there are relatively financially secure when compared to the United States. Generally better represented in Europe, unsurprisingly, are the works of the aging postwar European neo-avant-garde, whose compositions have not been immune to the seductions of the Nokia Tune—which, owing to its and its trademarking company's European origins, is arguably even more familiar within Europe than in the United States. Even as recent a work as the celebrated Hungarian composer-conductor Péter Eötvös' 2010 opera *Die Tragödie des Teufels* (The Tragedy of the Devil)—his seventh full-length opera—for the Bavarian State Opera featured a character (the devil, no less) with an old-fashioned (mobile) phone whose ringtone, heard two minutes into the opera's first part and played by an accordion, turns out to be the familiar 13-note melody.[17]

But perhaps the best-known Nokia Tune work emerging from this orbit is the Italian composer Salvatore Sciarrino's unusual chamber-ensemble composition of 2005, *Archeologia del telefono*. The leading Italian composer of his generation, Sciarrino (born in 1947) has developed a fascinating and much imitated compositional practice based predominantly on sonorities usually located at the borderlines of normative art music—"extended techniques" such as bow noises, vocal breaths, harmonics, multiphonics, and other effects.[18] These usually marginal sounds help contribute to a perception in the composer's music of "silence," a term often invoked when discussing it, and not without merit—much of the music hovers in a very quiet dynamic range, with relatively few sonic events taking place.[19] Yet if we follow David Metzer's argument, silence (or, better, quietude) is not meant to convey a profound interiority (of stillness in nature or of traumatic loss), as might be the case with Webern or Nono, for example; rather, in Sciarrino, it provokes expectation and creates a charged atmosphere of waiting—as in his operas like *Lohengrin* (1982–84, based on Jules Laforgue's reinterpretation of the myth and Wagner's opera), *Luci mie traditrici* (1998, on the subject of Gesualdo's murders), or the Kafka-inspired *La porta della legge* (2009).[20] Another crucial component of the composer's sonic material is what might be termed his *transcriptive* imaginary: Sciarrino often draws on and imitates "natural" sounds in his scores, realizing them on traditional musical instruments, either singly or in combination; these sounds include the seemingly hyper-precise imitation of chirping crickets, wind, thunder, birdsong, and even the flapping of a bird's wings—his distinctive and often-used flute technique.[21] This naturalism is quintessentially modern, being dependent on the close and repeated listening practices enabled by the microphone and sound recording. But Sciarrino's putatively transcribed sounds are not applied as if part of an audio-photorealistic transcription of a natural soundscape. Instead, they are displaced from their original contexts and rearranged into strange auditory vistas reminiscent of electroacoustic music, crossing into the threshold of dream states and the limits of perceptible reality itself, often for dramatic effect—terms like psychological realism, "surreal landscape," or, in the composer's own terms, a "disquieting realism" aptly describe the effect.[22]

These techniques and aesthetic prerogatives are employed in *Archeologia*, which is unusual in Sciarrino's output for several reasons. For one, the technological per se isn't usually the focus of Sciarrino's music, even if its transcriptive imaginary is predicated on audio recording technology. Very few of his pieces thematize it explicitly, such as *Efebo con radio* (Boy with Radio), from 1981, and even this work features benign invocations of an earlier phase of popular culture, especially American popular song—in which Sciarrino is well versed. By contrast, the intended effect in *Archeologia* is different: the mobile phone (and its ringtones) is a form of popular culture that Sciarrino despises, and in his program notes to the piece he pathologizes the mobile phone, comparing it to smoking. Sciarrino's verbose program note is worth quoting at length for its meditations on mobile phones and ringtones, including a trenchant, if implicitly elitist critique of personalization:

> I do not know who honestly wishes to investigate whether [wireless] repeaters are harmful. In the case of the mobile phone it was known, beginning with the earliest models (which are now about to be declared antiques), that they damaged the brain. But we were intoxicated by the first waves, did not worry about them, and instead outlawed smoking, whose harm has some history behind it. As a result of the global dissemination of mobile phones, we now have the opportunity to talk constantly, in an illusory kind of exclusivity that takes the place of proximity; at the moment there is a great deal of confusion between, or rather an annoying mixture of, marketing and private life. We are constantly awaiting a call that is supposed to pull us out of our indifferent isolation. A "personal" ring, selected from among the obvious nuances of the little joke. Too many people believe that they become more personal by playing with the same kind of banalities. Meows, birdcalls, nostalgic phone rings, an endlessly rich shop window full of brief tonal clues, ambitious and fleeting, exposed individually and alone, in order to impress one's neighbors. In my imagination I follow these devices to the point where they will one day lie, sad and serious, on the display shelves of a museum.[23]

It is apparent from the above that Sciarrino's reference points were the audio files permitting the wide variety of ringtones he mentions. As such, the temporal horizon of this composition is self-evidently the present or very recent past (ca. 2005) and is therefore unlike much of his work—a great deal of which draws on mythological, medieval, and early modern reference points (though often filtered through modern reinterpretations, as in the work of the Franco-Uruguayan symbolist poet Jules Laforgue). The political valences of these historically oriented endeavors cannot be easily reduced or simplified; Gavin Thomas views Sciarrino as a kind of nostalgically inclined and apolitical counterpart to the German composer Helmut Lachenmann, whereas Nicholas Till positions him within a not apolitical practice of postwar opera critique in formerly fascist countries through anti-operatic composition.[24] Nonetheless, apparently in part on account of the political conservatism of Italy in the mid 2000s, Sciarrino seems at that time to have turned his attentions increasingly toward contemporary problems, as in the Kafka opera (which is intended as a critical, if "quiet" commentary on the present, issued "without shouting"[25]) or *Superflumina* (2006–2008), an opera built around the character of a homeless person, drawing on the Bible, and inspired by Elizabeth Smart's 1945 poetic novel *By Grand Central Station I Sat Down and Wept* (itself alluding to Psalm 137), again with possible reference to contemporaneous economic conditions in Italy; the worsening political and "social situation" in Italy in the late

2000s and its effect on cultural policy seem to have prompted this turn to the present.[26] From the composer's perspective, the invasive mobile phone would appear to be a product of the neoliberal, Berlusconi-led deregulatory regime that has threatened state funding for the "high" arts in Italy, and hence it makes some sense, from this vantage point, however incomplete it may be, to link classical music and the mobile phone in an inverse relationship.[27]

Archeologia begins, after a nearly inaudible bow-hair scratching on the body of the violin, with a low grumbling motive in the double bass (mm. 1–2) that includes a held tone followed by a rising glissando only a half step in intervallic distance from G♯2 to A2.[28] When combined with the crescendo *dal niente* that accompanies it, as well as a quick G2-A2 figure that follows and retraces (and expands) the previous gesture's path, the sound seems on the one hand quite like one of Sciarrino's characteristically minimal motivic figures while also managing to sound like a mobile phone buzzing on "vibrate" mode. A low bass drum rumble—thunder, or the movement of large vehicles heard at a distance, perhaps—rounds out the source material in the work's first nineteen measures; in the twentieth, upper string harmonics in tremolo are sounded; then a low piano dyad, combined with silently depressed piano keys producing sustained tones, is heard as part of a larger sonic complex (m. 23). New forms of material gradually accrete, with more florid variants of the double bass gesture in the other string parts (m. 24) and then slowly descending string harmonic glissandi (mm. 34–35), the automotive likeness of which is palpable (although the Doppler effect isn't rendered exactly). A signal event is heard in m. 39: a bell-ringing sound not unlike an old-fashioned phone ringer, produced by a sonic complex featuring a prominent tremolo figure in the crotales. Another is heard in m. 67, with bassoon multiphonics and subsequent pizzicato strings on the bridge helping to produce the effect of a louder signal and off-click sound (perhaps a more apparent vibrate mode, or intimations of a busy signal). Gradually, organically, Sciarrino builds up his material, in the process producing what seems to be a panoply of telephonic sounds—occupied/engaged signals, disconnected signals, old-fashioned ringer signals and other odd ringtones. Building to a pitch of greatest intensity toward the end of the work (mm. 157–172), the music becomes replete with the most obnoxious busy or ringback signals (again with bassoon multiphonics). A softer, quickly repeating disconnect (or busy) signal then sustains itself for several bars (mm. 172–179) before a variant of the obnoxious wind multiphonic signal repetitions returns, only to include a slightly buried version of the Nokia Tune, performed on the piano—clearly in its polyphonic or perhaps sound-file version. (See figure 4.4.) It would seem that the composer's description of the future historical museum of mobile phones has come to life within his score; many reviewers interpret the piece this way, emphasizing its comic qualities. But it also depicts a world seemingly without human subjects—perhaps as a critique of the naturalization of mobile telephone sounds but more probably as a pessimistic view of a machinic future without humans. Interpreted as an aural depiction of Marxian reification and personification, it not only portrays how thing relations transmogrify into social relations; it also portrays how they do so via the evacuation of the social itself within a dystopian imaginary.[29]

And yet more is going on in this strange work. As was alluded to earlier, some of the sounds appear to depict spaces and movement outside the indoor environs of a putative mobile phone museum, such as the bass drum rumbles and descending glissandi. Most strikingly, the woodwind (usually bassoon) multiphonics are cut off just as one or more of the strings present aggressive pizzicati (often below the bridge), creating a combined gesture that could sound like the harsh sound of a signal disconnection but is much more physical. To me, the effect is of a slammed door, with the wind multiphonics sounding like car horns—and when repeated in periodic fashion, as in the climactic section, we almost seem to be hearing a car alarm going off. However one interprets the sound, it is prudent to suggest that the mobile telephone is *not* the exclusive determinant or provenance of the sonorities heard in the work and that the phone is, like Baker's *Concertino*, situated acoustically within its urban environs, creating an elusively social sound world of discord at odds with the synthetic natural soundscapes for which Sciarrino is renowned. Blending public and private space, the composer illustrates an encroachment on "private life" mentioned in the program note and, in the process, memorializes a despised technology in his imaginary museum and the technological modernity on which his naturalism is based. Perhaps because of this exposed contradiction, *Archeologia*'s criticism of contemporary modernity and its unruly noises, if not exactly paranoid, is unnerving and strident. Betraying the composer's characteristic aesthetic and historical distance and perpetuating the role of the Nokia Tune as an occasion for social critique, Sciarrino's music here verges uncomfortably on an alarmist outcry. If the composition manages to avoid shouting its dismay to audiences through an exaggeration of ostensibly telephonic sound, it nonetheless audibly raises its voice.[30]

Figure 4.4

Salvatore Sciarrino, *Archeologia del telefono* (2005), mm. 179–81. Reproduced by permission of Edizioni Musicali Rai Trade. "Nokia Tune" indicated by author.

Ringtones in Media and Sound Art

Operating within somewhat different artistic frameworks from those of contemporary classical musicians, new media artists have found ways of exploiting the technologies of cell phones and ringtones more directly through installations and interactive works. In all cases, these works use mobile handsets as compound objects, often requiring corporate funding and sponsorship by mobile phone manufacturers and wireless providers. These works for the most part serve to promote mobile telephony and even particular companies, often explicitly at corporate and technological expositions. Among such works include an installation by Nokia Finland at the Hanover Exposition in 2000, in which Nokia phones were affixed to a wall with a painted tree backdrop and were themselves painted like birds, periodically sounding ringtones as birdsongs. Perhaps this was an attempt to naturalize the handsets and the sounds they produce, although it also may have been reference to the often-reported story that birds frequently imitated monophonic ringtones.[31] Other cell phone works and installations in such venues include a one-minute piece for seven ringing cell phones at the opening of the International Funkaustellung in Berlin in August 2001 and a Virgin Mobile promotional event in September 2001 in Leicester Square in London, at which several hundred people gathered together to attempt to set a world record for the greatest number of mobile phones ringing at the same time.[32]

Many of these new works are to some degree built on a reified principle of interactivity, in which audience members' activities help in real ways to constitute or construct the work itself.[33] The intended effect is to give the gallery visitor or concert listener a sense of agency in the process of experiencing an installation or performance, and interacting audiences are often absolutely necessary to the existence and success of these works. Yet it is often the case that an individual's contribution to these works is perfunctory or superficial, perhaps creating a false sense of agency on the part of attendees, whose efforts merely serve to glorify the creator of the work in question. A representative example of such an interactive work is the 2001 installation *Handywolke* (Cell Phone Cloud) by Peter Hrubesch and Dirk Scherkowski, which was installed at the Berlin Communication Museum and partially funded by Siemens Mobile. A cloud-like arrangement of 1,200 Siemens mobile phones is suspended under the main glass dome of the museum. Museumgoers' movements and dialing of phones below in the installation room help to determine the overall sound produced by the installation, and outsiders can also phone in and talk directly to the museum visitors.[34] Such artworks are reminiscent of displays at science and technology museums, in which children are meant to interact with the display (and perhaps each other) by pressing a few buttons. Marveling at the technological wonders of the feat installed before their eyes and ears, museum visitors are reduced to the status of children at a high-tech jungle-gym and, despite the stated intentions of the artists and museum, are not particularly intended to reflect upon "the contradictory aspects of wireless communication."[35]

Among the best publicized of such works is *Dialtones: A Telesymphony* (2001–02), a collaborative effort, conceived and organized by the media artist Golan Levin, to combine with relative precision the ringers and ringtones of 200 cell phones to produce an extended multimedia work that could "make a sacred space for the enjoyment of these phones."[36] Levin worked in collaboration with several musicians and programmers, most notably the electronic musician Scott Gibbons and the sound sculptor Gregory Shakar. Their composition has three sections: one in which audience members' own cell phones are used as an "orchestra," a solo section for cell phones played by Gibbons, and a section in which "soloist" and "orchestra" play together. The technical requirements of the multimedia work were extraordinarily complicated, working on three levels. First, audience members' phones were pre-registered in a database and were either assigned new ringtones or were allowed to play already existing ringtones. Second, Levin and Shakar manipulated graphical user interfaces live onstage. The interfaces allowed each person to activate groups of cell phones in the audience, who were seated in a large grid, by touching an image of a grid on a computer screen. Third, the team of technicians led by Yasmin Sohrawardy worked out a complicated switchboard mechanism. With the help of access to the high-speed mobile switching center of MobileCom Austria, Levin and Shakar were able to ring as many as sixty phones simultaneously.[37] In addition to the sonic aspects of the performance, each member of the grid-arranged audience was given a small red LED light that illuminated when a cell phone rang within a meter; similarly, audience members whose phones rang were spotlighted from above. The audience and the light effects were reflected by a huge (12 meters × 6 meters) angled mirror, which was hung toward the front of the stage.

Dialtones wasn't cheap: the initial two performances in Linz, Austria, were estimated to have cost about $100,000, and the subsequent seventeen concerts in Switzerland to have cost about $200,000–$350,000.[38] According to Levin, the entire operation turned out to be at least as much of an exercise in fund raising as a creative activity, with nine months devoted to the former and less than three to the latter.[39] Some of the major funding sources for the Linz performances included jet2web Mobilkom Austria, Aculab Germany, Cell phoneart.net, Design Machine, Ericsson, the Daniel Langlois Foundation for Art, Science, and Technology, Nokia Austria, and Siemens.[40] In contrast, the Swiss performances were solely funded by Swisscom Mobile. Levin explained the difference as follows:

> The right way to understand this is that [no commercial sponsors] had any faith in the concept in 2001, and so we were only able to raise very small amounts of support with great difficulty (e.g. the sole contributions of Siemens, Nokia Austria and Ericsson consisted of 10 loaned phones apiece). By 2002, the concept was proven and major companies lined up to use the event as an advertising prop. The truth is that the premiere was mostly self funded by me, and I consented to present it "commercially" the following summer in order to help pay off my credit card debt. Many new-media artworks go through this life-cycle.[41]

Dialtones is a remarkably well documented and commented-upon project—a fact best confirmed in the extensive "Final Report" produced by Levin and his colleagues. In addition to providing patient and accessible descriptions of the technical dimensions of the project, the report includes two components that merit consideration: an "Artists' Statement" and musical instructions for performance. In the former, Levin et al. advocate for the "Cagean" possibilities implicit in listening to cell phones and ask "What might we learn about our interconnected selves, in their high, pure tones?"[42] The reference to Cage can be a misleading bit of artspeak; although numerous unpredictable elements were present during the performances, including those based in the limitations of the communications network, audience members' phones, and control over audience behavior, ultimately, as Levin points out in editorial remarks on an interview with a chance/indeterminacy-preoccupied interlocutor, "the Dialtones project was trying to wrangle an extremely complex and inherently unpredictable communications system into some semblance of order. We were trying to negotiate chance, rather than seek it out."[43] What, then, was the project attempting to illustrate, beyond presenting an aestheticized, "deranged beauty" drawn from its unlikely source objects?[44] Toward the end of the statement, the ambition becomes clearer when articulated as presenting the "musical reification" of the "sprawling and enveloping omnipresence" of the "global communications network." And, later, "In *Dialtones*, the phones, and not their owners, speak to one another."[45] As with the Sciarrino composition, the notion of "reification" is crucial here and, when paired with the complementary term "personification," suggests a situation in which social relations take on the character of relations between things (reification, or thingification) and thing relations take on the character of social relations (personification).[46] The fact that audience members for the most part passively observe their phones communicating with the artists on stage—and with themselves and one another by means of the spotlighting and the LEDs—problematizes Levin's enthusiasm for interactivity and participatory art.[47] Discussing this dilemma, and in the process implicitly describing the strange kinds of reifications and personifications that result from a consumer's identification with a phone (or other commodity), Levin said:

> I felt kind of awkward about Dialtones when I made it, because I'm so concerned with interactivity, and it is concerned with ownership. But the strange thing actually is that people feel surprisingly close to their cell phones; they're actually a little extension of their body. The piece is more interactive than I'd expected, because all they had to do was own the phone. But when they were sitting in the performance space and suddenly their phone rang, something happened in the mind. Probably, something like, "Someone's calling me, I have to answer my phone." A mental event transpires there. And to catch yourself and realize that it's a concert is something that was played out nearly 8,000 times in a half-hour performance. People kept getting caught in this weird digital event. What's important about that is that it's the difference between being in a car and driving and getting rear-ended by someone, and thinking, "Hey, someone hit my car," and "Hey, someone hit me." You actually think the latter, though they actually hit your car. I think that the concert made that clear with the mobile phone.[48]

Another extensively documented aspect of *Dialtones* is its explanation and reproduction of performance instructions, in two forms. First, the "Final Report" offers musical transcriptions of 84 separate ringtones, in very small print, selected from the more than 100 composed for the work by Scott Gibbons and Gregory Shakar. "The ringtone melodies," Levin et al. note, "were collected into approximately twenty thematic and textural groupings, with colorful nicknames like 'drones', 'bugs', 'twinkly', etc."[49] Gazing upon this catalog of ringtones is somewhat dizzying, but even from a cursory inspection one can ascertain a few points. In line with the statement above, the ringtones can be grouped into clear categories, some using slow whole notes as drones, some featuring rapid ornamental and arpeggiated figures serving as effects, and still others producing a rhythmicized beat or groove of various types; similarly, the metronome marking or BPM given for each ringtone can vary widely, ranging from markings around 20–30 to more than 250. Underlying these differences are the ubiquity of 4/4 meter, suggesting a broader rhythmic regularity in the music—the implications of which become clearer below. Second, the report also includes a performance score for the first and third movements of the piece, presenting unexplained (but mostly readily interpretable) categories like "regular rings," "pulse," "bugs," "chordy," "drones," "casio," etc., and presenting cues for categories (e.g., Q, W, Z, E and 9, 2, 5) on a time chart with visual representations of the textural shifts and general flow of the piece. The image roughly confirms the claims, made elsewhere in the report, that "within each of the three movements, the composition is structured as a sequence of sound-textures" and "there are about fifteen sound-textures in all, each approximately two minutes long."[50]

Listening to the performance of *Dialtones* on the Staalplaat recording confirms and clarifies some of the impressions and descriptions given above. It begins with the sound of a single cell phone ringer, drawing laughs from the audience. With the commencement of the first section, we hear the growth of phone ringer and ringtone textures that are reminiscent of forest soundscapes, replete with imitations of birds and insects. Around 1:23, the textures shift as drones appear, simultaneously recalling organs, synthesizers, and cicadas; within the texture, the Nokia Tune is sometimes present. Around 3:19, the music shifts into a kind of minimalist, pseudo-Baroque arpeggio texture, which intensifies with the addition of more loops for some time. The music grows more dissonant as the section continues, and a new ostinato pattern ultimately appears (5:30), then is accompanied by a looping, arpeggiated figure (7:10) that continues to the end of the section. Section II also begins with a gag, with "soloist" Gibbons clearly dialing a number, evoking laughs. After a series of repeated note figures (0:13–0:47) and arpeggiated triads (0:47–1:35), a bird like-melody anticipates an atonal, mostly octatonic ostinato (beginning in proper at 1:56), on top of which numerous bird-like ringtones and others are superimposed. Eventually drumming sounds (around 4:00), created by the phone's tapping against an amplified pad on a table, providing an electronic-sounding but also vaguely African dance-beat to accompany the proceedings. A new ostinato appears (5:16), this one tonal, with birdsong melodies and fragments of Bach ringtone references superimposed on the ostinato. Cicada/organ-like drones appear (6:33–7:12) to transition the music into section III. Ten minutes in length, section III builds on the drones, with low humming drones appearing as accompanying material. New Philip Glass-like ostinati appear periodically, accompanying increasingly growling, industrial low sounds—these too draw laughs, initially sounding elephant-like, but quickly come to resemble jackhammers and other construction equipment. The music intensifies to a significant pitch around 7:00–8:00, with the music then receding into a final set of superimposed ostinati recalling earlier ones used in the piece. The music decays for the next two minutes, ending with a single, non-melodic phone ringer—and, hence, returning to the sound that inaugurated the entire affair.

We can interpret the visual, sonic, and technical references of *Dialtones* in three ways.

First, the music in combination with the visual effects seems to evoke a kind of atomized connectedness associated with global digital communications. The lights which cause areas with ringing phones to light up and the mirror reflecting the lights activate the otherwise inactive grid, resembling both some kind of illuminated microchip and a time-lapsed apartment building, in which lights go on and off during the night according to a semi-random logic. At this level, the music seems to be about *connectedness*, communicating the notion that we as participants are part of a bigger global phenomenon around us—and the participants holding their cell phones up like lighters at rock concerts only underscores this sense.[51] Musically, the dominance of grooves in a largely quadratic syntax—which make the arrhythmic sections seem like effects or unsettled moments that click into place once the groove enters—also provides a sense of rhythmic organization and entrainment to a series of bodies largely encouraged to stay still and keep quiet, a sonic sense of unison reinforced by the piece's tonal center of A-880.[52] Although moments of recognition and communication between audience and artist did take place, by and large audience members held their phones up for performance by someone else. The frequent description of the audience as an "orchestra"[53] is, in this context, misleading—the kind of push-and-pull between individual and collective agencies was largely absent here, or rather, the agencies-in-fusion were those of the artists and the phones themselves; instead of instruments performed by individual musicians, audience members held their phones as puppets whose invisible strings would be pulled, magically, at a distance. Hence, one might suggest that the connectedness produced by the work is strikingly alienated, if also potentially mystified.

Second, the sonic references seem to evoke a narrative pattern that switches between the "natural" sounds of birds and insects; "human" sounds of ringers, phone dialing, drumming, and heavy machinery; and the "reflective" sounds of minimalist, minor-mode or modal ostinatos that provide an air of pseudo-profundity to the music. From this view, the music seems to be a high-frequency, video-game-like tone painting of the history of human development and human society's evolving relationship with the natural world—but evoked as if starting from the daydream of a ringing or dialing phone. The work proceeds by portraying the intensifying conflicts between nature and human society, the potential catastrophes of which are hinted at, but never represented, toward the end of the piece. Such a message recalls that of another minimalist masterpiece, Godfrey Reggio and Philip Glass' 1982 film *Koyaanisqatsi*. Borrowing its title from a Hopi word meaning "life out of control," the film's time-lapse images of nature and human society, taken and edited by Reggio, provide testimony of rampant global development and its untold ecological damage. The film image is perfectly juxtaposed with Glass' repetitive score, which communicates profound stasis and monumentality at times and buzzing activity at others. Indeed, *Dialtones* might be viewed as a contemporary re-composition or re-imagination of that earlier work—but in this case, Glass' medieval and Wagnerian minimalist grooves are replaced by loops of beeping timbres produced by cell phone ringtones and dialing tones, evoking video-game tunes and electronic dance musics (such as minimal techno). And for the later work's historical moment—which was that of the monophonic ringtone—such a revisiting of the Glass-Reggio collaboration seems to have been particularly apt, given the way that cell phone ring-signals and monophonic ringtones have become almost "naturalized" in modernity. In writing corporate research reports overemphasizing the natural qualities of cell phones, cultural studies scholars like Sadie Plant have helped to make the cell phone seem like a natural phenomenon, arguing that ringing cell phones are like a form of "electronic birdsong."[54] And with the previously mentioned reports that birds have learned to imitate ringtones and that birdsongs are popular as ringtones, the actual conflicts between nature and capitalist technology are yet further mystified.[55]

Third, the technical setup of the piece hints at a particularly ominous aspect of cellular technology. The fact that every person's phone is preregistered in a database, with many but not all phones being reprogrammed with new ringtones, paints an image of a society constantly under surveillance, with each individual citizen's behavior being dictated by a hegemonic force—and the apparent freedom of some to retain their message is contrasted with the fact that not all participants maintain the "integrity" of their original voices. With profitability and policing going hand in hand in mobile telephony, the piece's intensification toward the end signals both the ever-greater expansion of a society of surveillance and the paranoia that the awareness of such a society engenders—itself exacerbated by the tinny, if shimmering timbres of the massed cell phones, in a kind of hallucinated social tinnitus.[56] Yet the overall effect is contemplative, retreating away from the precipice of chaos and back to the daydream occasioned by the single ringing cell phone. When combined with the (mostly) comfortable musical materials that animate its musical narrative, it remains unclear whether it ought to be read as a social critique of mobile telephony and the ringtone, or if it is meant to simply aesthetically "ratify" its existence in a constructed "sacred" space—with the ambiguity perhaps being the work's primary strength.[57]

Although the messages of *Dialtones* might, in some ways, seem profound and compellingly contradictory, its means also might be seen as inflated, gimmicky, and highly contingent upon corporate and arts-foundation sponsorship.[58] More modest cell phone and ringtone works have been created that are often able to project equally powerful ideas and aesthetic experiences while still remaining relatively independent of the advertising role that big budget works must often play. Special mention must be made of Allison Craighead and Jon Thomson, whose independent online store produced some of the earliest atonal ringtones and perhaps the first silent ringtone—a conceptualist gesture that also anticipated a now commonplace method of avoiding specific callers.[59] One work by Craighead and Thomson elegantly investigates the ideologies of the cell phone in ways similar to *Dialtones* and other interactive ringtone works. Visitors to their 2000 installation *Telephony* are encouraged to physically dial the numbers of Nokia cell phones arranged in a grid on the wall, or to call these phones on the wall from their own mobile phones.[60] As the phones are dialed, they are programmed to dial each other in turn, and a massed texture built from the familiar Nokia Tune results. In addition, elevator-style ambient music based on the Nokia Tune accompanies the phones, adding a strikingly New Age touch to the experience. Finally, a real-audio link to the US Naval Observatory's Master Clock periodically reads off the precise hour and minute of Eastern Standard Time. The result, I would argue, communicates both the sense of the enclosed space created by the cell phone, in which phone users become almost completely autonomous from their surroundings, and, ironically, a military-ready awareness of information that each bubble inhabitant seems to need. Emphasizing the (false) experience of autonomy over the message of alienated connectedness in *Dialtones*, *Telephony* thus communicates the militarized paranoia of the inhabitant of a cell phone bubble with a detachment that elegantly plays with some of the same ideas found in Levin's piece.

Of course, the differential effects of *Dialtones* and *Telephony* are due in part to their differences at the basic level of artistic form. *Dialtones* is a musical performance piece, whereas *Telephony* is a sound-art installation. Levin himself has pointed out this basic distinction in cell-phone-based artworks. For present purposes, we might subdivide his distinction into ringtone-performance works versus ringtone installations.[61] The former might seem to overlap somewhat with the ringtone compositions using cell phones (especially the "medley" approach) described above—*Dialtones* is in many ways composed and scored rather traditionally and is, after all, self-characterized as a "tele*symphony*."[62] Indeed, if *Handywolke* and the Nokia tree for the Hanover Exposition can be seen as more typical ringtone installations, the concert-like qualities of ringtone-performance works would lend themselves to descriptions as symphonies. For example, Simon Turner and Marcus Moore produced their corny, pun-laden *SIM-phone-ya (New Ring Cycle)* for the 2002 Cheltenham Music Festival. An "interactive symphony written for the ring tones of 30 mobile phones," it was apparently composed in four movements, in one of which "all of the audience members were invited to turn on their mobile phones," thus both recalling *Dialtones* and anticipating Baker's *Concertino*.[63] Nonetheless, works that overlap between performances and installations still use the label "symphony," perhaps as a marker of aesthetic legitimacy, ambition, or both. Specifically, the Danish singer Tobias Trier created his *MobilSymfoni* in 2001 for the Aarhus Festival, which involved "twenty mobile phones . . . suspended from a ceiling," which "were caused to ring by a live performer, who dialed them up using another four phones below."[64]

A similar hybrid work that avoids the "symphony" label, *Wählt die Signale! Ein Radiokonzert für 144 Handys* (Dial the Signals! A Radio Concert for 144 Mobile Phones), was an installation at the Hamburg Kunsthalle in 2003, accessible for 12-hours only and billed as a "concert." Ligna, a "free radio group . . . of media theorists and radio artists Ole Frahm, Michael Hüners and Torsten Michelsen," began working in 1995 and produced the work as one of two separate entries for the award competition of *trans-mediale*, a yearly media art festival based at the Berlin Academy of Arts. A complex undertaking involving a radio station, radios, and radio listeners to Freies Sender Kombinat (FSK), in addition to the mobile phones, *Wählt die Signale!* involved the composition of new individual ringtones by Röhm, who based them on the spectral content or overtone series of the ambient noise (a 52-hertz = G♯ hum) in the Kunsthalle passageway in which the phones were located, arranged on a platform in a 12 × 12 square. Each phone had its own number, which was given at the FSK website. Callers could call during the 12-hour active window for the installation (or "performance"), during which calls to the mobiles generated the entirety of the music. The video and audio of the ringing mobiles were broadcast out via Internet and the audio alone by radio, as well as directly to special locations in Hamburg, Karlsruhe, and Maastricht; the online representation also presented a grid like display showing which phone was ringing at what time (the grid here being reminiscent of *Dialtones*). As Frauke Behrendt notes in response to a CD of five-minute snippets of the piece released by Ligna/Röhm, the effect "is best described as a minimalistic carpet of sound. The composed ringtones are not reminiscent of the ringing of telephones, they aren't shrill or 'annoying.' The combination of 144 tones produces consistently new textures, which are reminiscent of ambient music."[65]

Be they installations, concerts, or hybrids of the two, the works described above remained comfortably ensconced in the worlds of gallery, festival, and institutionally located media art. But in view of the mobile phone's very ubiquity, accumulating a critical mass of mobile phones to produce comparable effects is not dependent upon this kind of economic support. Indeed, a number of mass demonstrations of simultaneous playback of mobile phone ringtones have been organized for various purposes, including as part of participatory corporate sponsorship stunts (as in the Virgin event mentioned above), although such practices have also migrated into political demonstrations. But, it is also the case that relatively unofficial performance events adopted the practice as well—suggesting the ease with which a form can migrate between corporate, political, and aesthetic contexts. Drawing on the practice of "flash mobs" (impromptu and typically absurdist actions of loosely coordinated individuals, often linked by text messages, social networking sites, or other "new" communicative modalities), the New York-based performance group (or "prank collective") Improv Everywhere—whose numerous stunts have involved masses of people performing spontaneous musicals in public places or riding on the subway without pants—appropriated the concept of the ringtone-performance work on February 18, 2006.[66] Even calling it their "Cell Phone Symphony," and contrasting it with *Dialtones* (deemed "A more sanctioned Cell Phone Symphony"), the group gathered 120 participants on a cold winter night near the Strand Bookstore, grouped into pairs, and then divided into two coordinated halves. Sixty participants entered the store and checked their phones (in backpacks or bags, etc.) in the store's storage cubicles.[67] Once those sixty participants were comfortably inside the store, the other sixty participants began to dial the phones, which had been divided up in advance according to phone brands and set to default ringtones—the Nokia phones would play the Nokia tune, the Motorola phones would play the "Hello Moto" ringtone, and so on. "Agent Kula" served as the "conductor." He led the performance, activating individual phone brands as sections, "like a symphony. Instead of the 'string section,' we would have the 'Samsung section.'" Cycling through the different sections and ringtones—Nokia Tune, "Hello Moto," "Snaggle," Samsung, etc., before all of the phones were called—the performance seems to have lasted a few minutes at most, the employees and patrons of the store largely finding the whole event perplexing and amusing (although the store's manager was extremely displeased and initially took to banning identified participants from the store). Drawing on the sonic familiarity of branded ringtones, the creation of textures based upon these turns the logic of symphonic "movements" and orchestral "sections" on its head, one-upping the cool irony of *Telephony*'s Nokia Tune wash. Although primitive in technique, this work may be the most viewed and heard ringtone performance on the planet, at least so far—the video documentation of the event has been viewed more than 3 million times on YouTube alone.[68]

Conclusion

The mass popularization of the cell phone symphony by Improv Everywhere and the fact that by 2006 sound-file ringtones had become mainstream seem, in many ways, to signal the last hurrah of the ringtone as aesthetic source material. The reason for this is relatively obvious: as the ringtone progressed from its monophonic to polyphonic and sound-file guises, it declined in sonic distinctiveness. The greater sonic fidelity of the sound-file ringtone relegates the sonic aspect of the cell phone to the same bin as other portable sound-file playing devices, such as the iPod. In contrast, the tinny, often harsh sound of the monophonic ringtone both generated a series of characteristic sounds, produced by individual phone models, *and* a number of branded melodies, sonic logos, and microcompositions that would ultimately outlive their earlier incarnations as monophonic or polyphonic ringtones (the Nokia Tune being an excellent example). Indeed, almost all of the works described above seem to be dominated by the aesthetics of monophonic ringtones—using the Nokia Tune or the specific timbres of monophonic ringtones as their sonic basis. In the case of the exceptions, such as the Improv Everywhere "Cell Phone Symphony," an aesthetic tradeoff was implicitly and perhaps unintentionally made in favor of formal simplicity, with copies of the same polyphonic and polyphonic-derived sound-file ringtones producing out-of-phase textural blankets of sound entirely dependent upon these ringtones' recognition in a public space. As Golan Levin put it rather compellingly, "presenting an 'orchestra' of phones became a less interesting proposition once any phone could sound like an orchestra."[69] In fact, one could claim that there were three phases of ringtone art corresponding to ringtone technologies. The first draws on the tinny sonic timbres and melodies of the monophonic ringtone, permitting mass playback and generating distinctive effects as well as being the butt of musical jokes. The second draws on, and even more frequently ridicules, the cheesy MIDI aesthetic of the polyphonic ringtone—now less likely to function in collective playback contexts except very simple ones (consider the hokey Ringtone Dancer, who dances in a ridiculous Mexican-wrestler-like outfit to a polyphonic ringtone dance-beat version of the main tune from Tchaikovsky's *Swan Lake*, as indicative, if not necessarily paradigmatic[70]). The third uses sound-file ringtones, now transformed into portable playback devices, with the media capabilities of mobile devices coming to the fore and the ringtone as joke or object of critique concomitantly receding.

What, then, has happened to the ringtone in the worlds of media art and contemporary classical composition? By and large, the second decade of the new millennium has witnessed the supersession of the ringtone, with two primary tendencies emerging. For media and sound artists, as well as composers gravitating to such work, the ringtone was, in retrospect, nothing but a way station on the road toward a more sophisticated portable media device capable of many new, converged possibilities, including locative media projects and new digital instruments (which nonetheless typically imitate other, more familiar instruments). For example, Ge Wang's Mobile Phone Orchestra and the iPhone and Smartphone apps produced by his company Smule (including the musical instrument Ocarina, the music game Leaf Trombone, the Autotune app I Am T-Pain, and the auto-harmonizer Glee) have emerged from the world of academic computer music and, through clever pop-cultural references, have become mass sellers in a new app market. Additionally, the Mobile Music Workshop group, which has held numerous workshops on mobile media devices and art in Europe and North America, has promoted new mobile sound art projects, "challenging the common understanding of mobile music as 'ringtones and i-pods.'"[71] In contrast, contemporary classical composers have largely drifted away from the ringtone and, in a few cases, toward the remaining sonically distinctive aspects of mobile communications. Nathan Davis' delicately minimal and well-received[72] composition *Bells* (2011) features audience members' mobile phones in tandem with a professional musical ensemble, but the audience members dial a pre-given number and press touch-tone codes to call up different sounds, including live processing of the performed musical instruments and archival recordings of "VLF (very low frequency) radio recordings of space, satellites, morse code, spy number stations, telephone sounds (sine tone pad combinations, ring back, etc.). And bells of course." The role of bells as historical, one-way communication devices was mimicked in the work via the communications network, and the sonic degradation induced by that network in the dialed-up audio played back on audience members' phones was central to the piece. In Davis' words, "All these sounds work well in the 300–3000 Hz range of phone lines, and with the compression that is usually introduced. The distortion and filtering are part of the sound, and different phone speakers highlight or distort different frequencies differently."[73] Likewise, another composition from 2011, Steve Reich's *WTC 9/11*, a string quartet premiered by the Kronos Quartet and based on the composer's personal experience of the 2001 World Trade Center attacks, involved a tense phone call to his son, who was living very close to the World Trade Center. Reich's composition begins with the analog sound of a disconnect signal—still heard on today's phones, including mobile phones—the E/F half-step interval of which is then imitated and harmonized by the ensemble in a characteristically Reichian way, involving musical transcription and musicalization of "non-musical" sounds, and combined with documentary recordings of voices in the manner of *Different Trains* (1988) and *Three Tales* (2002). The speed of the disconnect tone's repetitions additionally sets the tempo for the entirety of the piece.

To the extent that ringtones will persist in classical music and performance art, they will do so as a minority tendency. On the one hand, sound files have the capacity of reproducing older ringtone sonorities and so will make possible the continued existence of ringtone compositions time-stamped by a now-obsolete technology—indeed, recent performances of Baker's *Concertino* have done this. Even a relatively modest local theater work such as Alan Berks' *#Ringtone* (2009–10), in the Minneapolis-St. Paul theater scene, reproduces the sound of the monophonic Nokia Tune in its sound design, thereby preserving monophonic sonorities by audio file playback.[74] Much rarer are efforts such as Teresa Foley's Locally Toned community art project, which treats the ringtone as an opportunity for sonic documentation and portraiture and which makes its growing collection of sound-file ringtones freely and publicly available by website and

MMS. Foley's incessant field recording of individuals' or small groups' self-designed ringtones seems to use the residual short duration of the ringtone itself as its primary aesthetic constraint, although repetitive and sometimes irritating sonorities that may function as effective signals are prevalent.[75] Here, the ringtone performance or installation has transmuted—in part, I would argue, out of formal necessity, due to the [illegible] of the [illegible] file format—into an example of original ringtone composition (see chapter 7).

III The Ringtone's Dialectical Reversals

In the magisterial final chapter of *Marxism and Form*, Fredric Jameson offers an insightful summation of the workings of dialectical thought, one aspect of which is germane to the present discussion:

> The basic story which the dialectic has to tell is no doubt that of the dialectical *reversal*, that paradoxical turning around of a phenomenon into its opposite of which the transformation of quantity into quality is only one of the better known manifestations. It can be described as a kind of leap-frogging affair in time, in which the drawbacks of a given historical situation turn out in reality to be its secret advantages, in which what looked like built-in superiorities suddenly prove to set the most ironclad limits on its future development. It is a matter, indeed, of the reversal of limits, of the transformation from negative to positive and from positive to negative; and is basically a diachronic process.[1]

Jameson continues by giving a remarkable capsule history of the Cold War competition between the United States and the Soviet Union in airborne military technologies. In his schema, the United States' superior atomic weapons led the Soviet Union to devise powerful rockets that could carry their bulkier, inferior nuclear warheads; that invention, in turn, led US scientists to miniaturize their payloads and, in the process, to develop lightweight transistor-circuit devices that proved capable of transmitting more information.

The means of narration employed in Jameson's description is a compelling heuristic, and I would like to propose here that we view the development of ringtone technologies similarly. In our case, however, the two "antagonists" are not rival states but rather the interrelated firms of the mobile media industry, on the one hand, and the vast consuming public in possession of cell phones (more than 5 billion people worldwide in 2010, according to the International Telecommunications Union), on the other.[2] The story here concerns the familiar progression from monophonic and polyphonic ringtones to the sound-file ringtone, and can be related to the chapters in part II of this book as follows: Whereas those three chapters mostly emphasized processes of negation in the shift to the sound-file ringtone, with the increasing automation of labor, proprietary control of ringtone website sales, the disappearance of a new form of mass music literacy, and a decreasing aesthetic distinctiveness all contributing to an autumnal depiction of the ringtone's decline, what ensues is a meditation on the possibilities that developed as a result of the ringtone's later (and ultimately final) form.[3]

The sequence of dialectical reversals between industry and consumer, then, appears to work as follows: First the industry develops a novelty technology, which, unpredictably, consumers find fascinating. Then the industry grows in response to demand and begins advocating for increasingly advanced ringtone playback systems in a convergence of interests in the ringtone industry's two halves, device manufacturing and content production/distribution.[4] In particular, the drive toward greater sound fidelity in ringtones sustains further research investment by handset manufacturers in new technologies *and* guarantees increased monopoly profits by the biggest content providers (music/media conglomerates), who partly seek to recoup losses in other sectors and partly out of a typical capitalist interest in greater and more secure accumulation. The music industry in particular pushes hard for the sound-file ringtone, as it could then sell music as profitable samples *and* decrease contracted labor costs, while destroying, neutralizing, or absorbing independent mobile media firms. But consumers respond to the new sound-file technology by freely creating and exchanging ringtones, using established file-sharing and file-conversion methods, or even by becoming less interested in the music ringtone alone—which is now no longer a musical representation invoking a memory of preferred music ostensibly in the service of identity confirmation and display but instead an excerpted reproduction of that music. Consumers gravitate toward sound effects, human voices, nostalgia items (old telephones or train whistles), animal sounds, other "non-musical" (in the sense that they are not owned by the music industry) ringtones, and musical selections not (yet) licensed for sale by mobile media aggregators, not to speak of other mobile entertainment commodities (wallpaper, audiovisual media, video games, apps). These developments push phones further toward greater multifunctionality; they also lead to greater internal competition over ringtone sales—as Hollywood begins franchising its own ringtones, for example—and competition over the means of ringtone distribution and production, ranging from Xingtone's much-imitated do-it-yourself model to the increasing prevalence of music phones and Apple's iPhone, which allied easy and inexpensive ringtone production to the company's massive iTunes library.[5] And so on. (The unfolding of this historical process was treated in chapter 1.)

We can interpret the results of the shifts in ringtone consumption as being part of a broader opening of possibilities afforded by the ringtone's rebirth as a sound file. Merging into a multimedia continuum of new digital devices, software, and cultural forms, it facilitated a welter of mass creativity whose expressions have been found on YouTube, on various music/sound/podcasting sites (e.g., Odeo, SoundCloud, and MySpace), on weblogs, and on cell phones themselves, with text messages, pictures, videos, and other content increasingly transferable from phone to phone.

This part of the book consists of three extended case studies of sound-file ringtone creation and use, characterized by individual, non-professional, and even anonymous cultural production and in some cases ballooning into mass popular phenomena. The subject of the first case study is Crazy Frog, a global ringtone fad whose historical significance rests in the politics of the pre-recessionary conjuncture in the European Union. The second study involves the dramatic successes and false starts in social movements' political utilization of the ringtone. The third study addresses the aesthetics of original sound-file ringtone composition, focusing primarily on recordings on the Ringtone Records label.

5 The Annoying Thing: Crazy Frog and the Strange Career of a Sample

In late December 2010, Rupert Murdoch's media conglomerate, the News Corporation, ended its involvement in mobile entertainment and sold off the Fox Mobile Group (FMG) to the Jesta Group for an amount not disclosed to the public. The new owner of FMG would command the brands Jamba, Jamster, Mobizzo, and Bitbop. The president of the Jesta Group, Jason Aintabi, claimed, in the face of all existing evidence, to see "a very bright future for the company and its brands."[1] One suspects that the sale price was far less than the hundreds of millions that News Corp. spent in acquiring the division's main label Jamster/Jamba, which was during the mid 2000s one of the most successful mobile entertainment firms in the world on account of its flagship ringtone product, Crazy Frog. Although the dramatic decrease in ringtone sales by the time of the FMG sale led to a probable devaluing of the company, the aggressive (and in many ways illegal) marketing of this product led to dizzying initial successes and subsequent, spectacular failures, which affected not only the fortunes of Jamster, but also the ringtone industry as a whole. Nonetheless, the particular method of the ringtone-product's creation, the back story of which spans nearly ten years, demonstrated the mass and global marketability of a single ringtone, as well as the way that amateur-produced Internet detritus could find its way into the production of popular music.

This chapter examines the history of Crazy Frog, interpreting its unprecedented ringtone-derived aesthetic form and its popularity as a kind of Freudian family romance involving newly empowered child consumers. Moreover, the phenomenon itself is viewed as a cipher for the racial fantasies, animus, and melancholia that followed the decline of the European empires (especially the British Empire). As will be discussed, however, Crazy Frog also had profound economic consequences, not only for its parent company but also for the ringtone fad as a whole. Though it was a trivial novelty (complementing the ringtone as novelty), it had wide-ranging economic and cultural effects. Thus, Crazy Frog and its dramatic overreach and decline must be seen in light of the pre-recessionary conjuncture within Europe (and particularly in the United Kingdom, where it was especially prevalent).

The Origins of Crazy Frog

First we must ask the most important question: What *is* Crazy Frog?[2] Its tale begins in 1997, when a recording of a Swedish teenager (Daniel Malmedahl) imitating the sound of a two-stroke moped engine was posted online. The recording, an MP3 file, floated around on the Internet for some time unbeknownst to its creator, appearing online in conjunction with various images—including, in 2001, a static image of a race car that is known as the "Insanity Test," meant to test how long one could look and listen without laughing.[3] The sound file was eventually found in 2002 or 2003 by Erik Wernquist, a young Swedish theater actor/director and animation artist. Wernquist created an animation sequence, called "The Annoying Thing," that featured a strange-looking frog and was accompanied by the sounds of Daniel Malmedahl. He posted "The Annoying Thing" online, initially without knowing who had produced the sounds. After encountering the animation and incorrect assertions on online message boards regarding the sound file's provenance, Malmedahl contacted Wernquist and demonstrated over the phone that he was indeed the voice behind Crazy Frog.[4] The two partnered, and by 2001 they had licensed their joint creation to Ringtone Europe, Jamster België, and other mobile media companies for the purpose of advertising ringtone sales. In 2004 Jamba!/Jamster licensed the sound for use as a ringtone itself and included the animation as part of a quickly expanding marketing campaign promoting that ringtone worldwide. The ringtone soon became a high-sales item and became quite familiar to the European public, particularly in the United Kingdom. A transformative moment took place in early 2005 when Henning Reith and Reinhard "DJ Voodoo" Raith, members of a German production group known as Bass Bumpers, produced a mash-up arrangement of the Crazy Frog voice with a simple synthesizer version of Harold Faltermeyer's hit tune "Axel F," originally part of the soundtrack to the 1984 film *Beverly Hills Cop*. By May 23, 2005, the single had quickly earned £10 million in sales (as of late 2006, it had made at least £14 million), outselling Coldplay and Oasis. Wernquist created a popular video for the single, featuring the character Crazy Frog being chased by a bounty hunter seeking to destroy it. The single eventually topped the charts in the United Kingdom, France,

Spain, Sweden, Switzerland, Ukraine, Norway, Denmark, Belgium, Ireland, Australia, and New Zealand, reaching the top two or three in many more countries. The Crazy Frog version of "Axel F" failed to achieve comparable popularity in the United States and Japan.[5]

A second Crazy Frog single by the same production team soon followed, this one lacking Malmedahl's "Crazy Frog sound" and based on Gershon Kingsley's 1969 synthesizer-pop tune "Popcorn," which formerly had been covered to great success by Hot Butter in 1971. The simultaneous chart presence of both Crazy Frog hits throughout Europe, in Australia, in Canada, and elsewhere was part of a broader effort by the Jamba group to saturate multiple markets with Crazy Frog recordings and products. The two songs were also included on a 2005 CD titled *Crazy Frog Presents Crazy Hits*, which has charted well in many parts of the world (including the United States). The success of the early ringtones and songs led to a mass of Crazy Frog-related paraphernalia flooding the market, peaking around 2006: CDs, toys and dolls, video games, a board game, apparel items, a key ring, a backpack, a lunchbox, an air freshener, stationary, automobile decorations, and even a portable karaoke microphone. The profitability of this venture led to multiple efforts to invest in Jamba and connected entities. On 12 September 2006, the News Corporation, in tandem with the e-banking and online billing firm VeriSign, bought a majority stake in the ownership of Jamba—and with it Crazy Frog, by far the company's most successful product—for £100 million (US$187.5 million). By 2008 News Corp. would buy it in its entirety from VeriSign for another $200 million, spending a total of $388 million on the company.[6] Paradoxically, this massive sale might well be understood, in retrospect, as the final triumph in the saga of Crazy Frog, which despite a trickling of continued sales and products has not reversed its inexorable decline into distant memory.

What was the set of structuring conditions that led a Swedish teenage boy's vocal imitation of a moped engine to transmogrify, beyond recognition, into a global cultural phenomenon and a multi-million-dollar enterprise, itself a component of the larger global ringtone industry? I would like to suggest here that the forces at play were economic and media-structural on the one hand, and the cultural manifestation of a particular conjuncture in the European imaginary on the other—and here I would like to recall Marx's famous distinction between the *forces of production* and the *relations of production*, although I mean the application of these concepts to technology and society, respectively, only in a loose sense. The story under consideration, however, is not merely one of fortuitous convergence. The spectacular success of Crazy Frog has led, during its rise and fall, to various forms of reaction against the product, its marketing methods, and what it might appear to represent to those not predisposed toward it.

In the journalistic literature on Crazy Frog, commentators often noted with glee that the first Crazy Frog hit single was the first demonstration of a truly authentic *ringtone hit*. That is to say, the Crazy Frog version of "Axel F" emerged from the ringtone of the Crazy Frog sound file and was dependent on that sound file for its chart success rather than the other way round, as most ringtones parasitically reproduce well-known hit songs. We should not, however, subscribe to a simple mechanical determinism. There was nothing inevitable about Crazy Frog's success. It was the product of a fortuitous sequence of events in which the universalizing dictates of global capital productively encountered a set of cultural forms and images in a context of sheer contingency. Nonetheless, the economic and technological conditions of possibility for the ascent of Crazy Frog set the stage for its appearance in early 2005 in Europe and the broader world. The first of these conditions was the development of the ringtone itself: its growth from a marginal fad in the mid to late 1990s into a multi-billion-dollar global business by the early 2000s. The second was the particular transformation within the ringtone industry toward a model of ringtone production based on sound files rather than synthesizer files. Third, we must account for what Manuel Castells calls "the global hypertext"—the coexistence of intertextually linked media working together in a fluid continuum that shape the everyday lives of modern consumers: bulletin boards, mobile media websites, pop-up advertisements, and e-mail messages conveyed through the Internet; text messages and telemarketing calls dispatched to mobile handsets; the ubiquitous television and radio transmissions that still dominate ambient visual and sonic space; billboards, magazines, and yet older media forms that persist in the public sphere.[7] Making clever use of this media continuum, in part by focusing strongly but not solely on its *digital* parts, Jamster embarked upon a massive marketing campaign, in 2005 spending approximately £45 million (about US$85 million) on advertising for its products, the most lucrative of which were Crazy Frog wares.[8] In this sense, although Crazy Frog seems to have gone viral in the manner of an Internet meme, in many ways it was exactly *not* this on account of the impact of advertisements—even though early memes, including the earliest animal-oriented ones from 4chan (such as "Caturday/lolcats" in 2005), may have helped to create a more hospitable marketplace for it.[9] Fourth, the success pattern for Crazy Frog required the existence over the preceding 15 years of animation styles based on AutoCAD-style software packages and not on older styles based on drawings or scale models. Made commonplace in films and in the video games of the 1990s, many of which were produced first in Japan, these animation forms gave a particularly bionic and "hyper-real" look to the simulacra of animation.[10] Fifth, the pervasive global presence of cell phones effectively meant flooding the world with cheap, low-grade sound systems. Since expectations for sound fidelity on the cell phone medium were quite low at the time in most parts of the world, music and sound clips used in such contexts did not have to be particularly subtle sonically, even if ringtone creators and composers themselves had to take the limitations of the medium into account.

The Consumer Demographics of Crazy Frog

If the economic and technological forces at play are relatively self-evident, it is slightly trickier to identify the cultural imaginaries and contexts that collide with those forces to provide what the anthropologist Anna Tsing has characterized as the "friction" of capitalist production.[11] Here, one of the most important questions to ask is "Who is buying Crazy Frog?" In lieu of precise consumer information regarding Crazy Frog purchases, we might draw conclusions based on commonplace assertions about it and more general studies of recent retail purchasing trends in Europe. For starters, the producer of the Crazy Frog "Axel F" single, Guy Holmes, characterized the consuming demographic as "children and 'big kids.'"[12] Indeed, many reports on Crazy Frog include offhand comments on children and youngsters being responsible for its success. More extensive reports on consumption in the United Kingdom confirm that pre-teenagers and children are growing as a powerful marketing demographic, both as part of well-known, longer-term trends of corporate marketing to children and more recently because of declining birth rates and delayed childbearing in Western and Central Europe. In following these recent trends, parents have fewer children and wait to have children until they have greater financial security; hence, parents spend more money per child, raising children's disposable incomes significantly.[13] With high-tech gadgets such as cell phones, computers, and iPods becoming increasingly a part of children's lifeworlds, and with cell phones often being pushed on children by safety-conscious parents, the newest trends identify *independent* consumption by children as increasingly common. In some cases, this has led to rather unusual circumstances. For example, children in Germany—the largest consuming market for Crazy Frog products, in part due to the Jamba group's German origins—have increasingly acquired extensive debt by rapidly incurring cell phone charges that they cannot afford, which is doubly problematic for

German consumers less familiar with the credit cultures associated with the US and much of Western Europe.[14]

And yet, the success of Crazy Frog is not explained by the whims of child-consumers alone. As is well known, children are ultimately dependent on their parents to pay for the things they buy. This leads marketers to target intergenerational family relations, implicitly and sometimes explicitly encouraging children to manipulate their parents into buying desired goods.[15] One of the most interesting features of the Crazy Frog phenomenon was that, in spite of its so-called viral marketing via ringtone and digital media, 95 percent of the Crazy Frog "Axel F" singles were purchased in stores rather than online, which suggests that children were buying the CD recording of the song with their parents in tow.[16] Indeed, the role of parents in the story of the recent boom in child consumption is crucial, in part because parents appear to be determining consumer choice in many cases—even if peer pressure and advertising are still the dominant factors determining purchases by children. Stephen Robertson, the marketing director of the retail firm Woolworth in the United Kingdom, described the impact of parents' nostalgia on the toys currently available in stores.

> The toy market has benefited from nostalgia in recent years with the reinvention of great toys from the past, such as Strawberry Shortcake, Cabbage Patch Dolls, Weebles, Rubik's Cube and Noddy. . . . It's the adults who are disposed to wanting to buy them for their kids. The best toys really do have an enduring appeal.[17]

Parents, both as proxy consumers for their children and as designers of children's commodities, are skewing the range of choices available in children's markets in an effort, as one marketer put it, to "give children the childhood they had."[18] Perhaps, then, parental nostalgia explains something about the curious trend in the Crazy Frog singles production method: a simple formula in which a classic tune from the 1970s or the 1980s, such as Queen's "We Are the Champions" or Technotronic's "Pump Up the Jam," is given a generic "Eurotrash" Dance/Club-remix treatment and quickly released on the market. I would argue that it is not merely the "enduring appeal" of these former hit songs that creates Crazy Frog's hits, but also a reflection of an appeal to parents by repurposing their past favorites, which are in turn updated in a contemporary, post-subcultural mass-popular style accessible to children.

Perhaps more interestingly, the parent-child dynamic and demographic consumption trends are written into the cultural representations of Crazy Frog itself. Sonically speaking, we must first recall the strange voice of Daniel Malmedahl, then 17 years old, as found on the original recording.

Its strangeness lies, in part, in its sounding altered and sped up (although it may not be). It is reminiscent of a voice on helium, or of Ross Bagdasarian's Chipmunks recordings of the late 1950s. Here is a possible transcription of its "text":

A ding ding ding ding dididing ding bing bng
Pscht(-ng)
Dom, bom bom bedom bey bom
Bedum bum bum b-b-bom bom boe—m bom bom
Ra ma me ma-b'ra
Ba bla blarorrrrrim brram drrrrrrran drrrrrrrrrrra
Ba ba ba ba beh
Wheeeee![19]

Though we might perfectly understand the sounds to be an imitation of an internal-combustion engine, they also sound like a spastic child babbling or otherwise carrying on incomprehensibly and giddily. If the frog, then, figures the child or adolescent in the abstract—one that is nonetheless implicitly *male*—in the context of the mash-ups with familiar songs, a listener might have the distinct impression of a child running amok within the song itself—a sense the producers have achieved by filling in spaces in the texture with "ding, ding" and other Crazy Frog sounds extracted from the original recording, hence creating a call-response dynamic between the music and the frog. (See figure 5.1.) For example, Faltermeyer's relatively subtle synth-pop instrumental "Axel F" is built on the principle of cumulative layering and (preset) instrument changes, perhaps evoking the shifting landscapes of Los Angeles through which Eddie Murphy's character Axel Foley often finds himself driving. The Crazy Frog version is much blunter and more staccato, leaving longer gaps within the texture that are often filled by the Crazy Frog voice sample; as such, it emphasizes an opposition between the frog voice and the dance-music version of the original song, which has now been transposed down a minor third from F to D, sped up from 117 BPM to a more frantic 138 BPM, and performed with a straightforward dance beat as rhythmic backing and a simpler palette of synthesized sounds—a Eurotrash novelty song, not so distant from the world of Eurovision.[20] The "utopian" moment of the Crazy Frog version, such as it is, appears around 1:15–1:23, where the backing music stops temporarily while the frog does his thing (the lengthiest uncut playback of Malmedahl's original recording in the entire song), eventually lifting off (on "Wheeeee!") into the ether through an ascending vocal glissando saturated with reverb, the moment of aesthetic

Figure 5.1
Above: Harold Faltermeyer's "Axel F" (1985), 0:02–0:10. Below: Bass Bumpers' Crazy Frog cover of "Axel F" (2005), 0:06–0:13.

transcendence appearing with the return of the backing music at the original song's "B" section or bridge material. The sound-image of adolescent freedom, with a motorcycle ride into the distance almost figured as superhuman flight or vertical takeoff, allegorically releases the youngster temporarily from the binds of the rhythmically rigid and doubly antiquated song material.

The visual representation of Crazy Frog (figure 5.2) further serves as an allegory for pre-adolescent consciousness, though in some rather curious ways. First, it does so through a relatively standard tactic of representing children as animals—a common feature of many child cultures, and in this case a mass-cultural capitalization of child animism.[21] Presumably, animals are deployed in an effort to allow the child, deemed unable to relate to adults, to develop an imaginative investment in Others who are, at the same time, represented as being rather similar to them. Hence, Crazy Frog is merely another addition to the massive global menagerie of children's products, in which animals act as vehicles through which children understand the world of human social relations. Indeed, Crazy Frog is part of a de facto online bestiary that includes a number of online and ringtone characters (among them Sweety the Chick, Nessie the Dragon, Schnappi the Alligator, and Lampi the Bat). Many of these characters are imitations of the products of Japanese and Korean *kawaii* or "cute" culture (as in Hello Kitty or Pokémon). Second, although Crazy Frog appears to be male, resembling a young boy, there is a sexually underdeveloped quality to it, which is underscored by his ambiguous though probably masculine genital organ (which might be a penis, as one writer argued, or rather epiglottal, as another writer suggested, or testicular).[22] The "lack" or "inadequacy" of the phallus that characterizes Crazy Frog's physical appearance is in turn represented in most of the videos depicting him

Figure 5.2
Crazy Frog.

riding an imaginary vehicle, an invisible or fantasized motorcycle. Interestingly enough, in the videos (as in the beginning of the "Axel F" video or the original "The Annoying Thing" video) the lack of the phallus often signifies a state of fullness or completeness. The frog's complete ease with his condition and the world generally might be said to advocate for the fundamental normalcy and adequacy of the pre-adult life stage. In the one moment in which Crazy Frog acquires a "true" phallus, toward the end of the "Axel F" video, it is as a missile intended to kill him. After jumping up, *Matrix*-style, into the air, Crazy Frog lands on top of the missile, now straddling it as it takes its course (2:18–2:26, 2:30–2:33).[23] The result is the destruction of the original missile sender, the bounty hunter, figuring the squelching adult authority. In all examples, Crazy Frog escapes, always somewhat blissfully ignorant of the genuine danger he may have faced, therefore maintaining a degree of childlike innocence in an adult-contrived fantasy of permanent childhood. Hence, despite in most ways being male, particularly through his flight helmet and garb if not his genitals, the "adventurer" or "daredevil" persona is juvenile and not as fully masculine in the way an adult male is figured.[24]

As one might glean from some of the foregoing comments, the adolescent-parent dynamic in Crazy Frog is paramount but cannot account for all of the character's strange qualities. And here it may be appropriate to emphasize that despite the power of the youth/adolescent market in propelling Crazy Frog to stardom, many accounts of Crazy Frog emphasized its popularity with students and office workers.[25] Crucial for this adult demographic is the ironic appeal, on multiple levels, of owning and publicly displaying some Crazy Frog product—the ringtone, the "Axel F" song, a ringtone of the song. Amusing themselves by owning an annoying novelty product intended for children, young adults imaginatively participate in the Crazy Frog phenomenon in a different way, I would argue, in the process perhaps perceiving some of the more unusual representative features that go beyond the novelty of animalia and even his gender troubles. One aspect of this participation may involve the reclamation of an annoying tune that functions like a persistent, unforgettable advertising jingle; another concerns its relationship to the 1980s revival in the 2000s. (See the introduction to part II.)

Crazy Frog's facial features might be read in various ways, sometimes appearing "black" and sometimes appearing "East Asian" in a stereotypical sense. Regarding the latter, one might ponder the look-alike contest sponsored on a Taiwanese television show, in which the winner was deemed to look "90 percent" like Crazy Frog, after which he proceeded to demonstrate his break-dancing talents for the show's audiences and hosts—even if this incident says less about Crazy Frog's generic "Asianness" and more about his eccentric physiognomy.[26] But trying to identify the ambiguous physical raciality of the frog may be something of a conceptual dead end; we might do better to consider the racialized contexts in which the character appears. In this light, let us reconsider some aspects of the "Axel F" video. Toward the video's end, after the frog has straddled the missile, the video cuts to a skyline shot of the modern cityscape in which the action takes place. As the frog and the missile zoom toward the bounty hunter, the video cuts to an explosion near some buildings in the background, positioned in such a way as to obliquely allude to the September 11 attacks.[27] Indeed, in view of the frog's role in the explosion, it would appear that he has become an inadvertent suicide bomber of sorts. At the last instant, however, the frog appears out of nowhere, climbing up a building in the immediate foreground, ready to deliver the final "bam bam" of the song, and we see the frog's face covered in soot, ostensibly caused by the explosion (2:46). Any hints of Crazy Frog as "blackening up" here are tenuous (the moment is redolent of a post-explosion cartoon convention), but when added to the other layers of racialization (as black, as Asian, as Arab terrorist) the notion of Crazy Frog here as being in blackface or appearing like a camouflaged guerrilla militant is perhaps not absurdly far-fetched, particularly when we factor in the frog's association with Eddie Murphy in *Beverly Hills Cop* via the "Axel F" song and, from another angle, the frog voice's musical role as a kind of adolescent African "trickster" character vis-à-vis the musical selections with which it is combined.

Perhaps even more striking is the novelty dancehall hit "Ring Ding Ding" by L.O.C., which appeared in the summer of 2005 and reached the number 1 spot on the BBC 1Xtra dancehall charts. The song itself has quite a complex back story, rooted in the Jamaican production practices of "versioning"—that is, creating entirely different songs using the same backing track. The backing track or "riddim" used in "Ring Ding Ding," created by the highly regarded producer Robin "Snowcone" Fuller and titled "Applause," is a minimalist, four-bar pattern that is extended ingeniously through the variegated use of syncopated percussion and synth layers and through "stop-time" effects (a rudimentary version of which is built into the pattern itself, as can be seen in the fourth bar of the riddim's basic structure. (See figure 5.3 for the basic, unsyncopated layers of the riddim.[28])

By the end of 2005, Fuller had produced 35 different songs based on "Applause," with its stripped-down hardcore dancehall beat devoid of any fashionably exotic instrumentation. Many of the songs based on that beat are highly characteristic of hardcore dancehall music and include references to gangster violence and shootouts evoking Jamaican street cultures in Kingston, London, and New York. Perfectly characteristic, for example, are Sizzla's "Run Out Pon Dem" or Elephant Man's "war tune" titled "Hurr," replete with the typical gunshot sounds, although the subject matter of the various songs range more widely to include the

Figure 5.3
The basic pattern of Robin "Snowcone" Fuller's "Applause" version riddim (2005).

anti-gun violence message song "Wait a Bit" by Capleton and Rik Rok's pleading lover's tune "Tell Me Why."[29] L.O.C.'s song is a straightforward party number that promotes a new dance based on the physical movements of Crazy Frog ("We made a dance and buss de scene / And dis is how we really play it"), which is supposed to appeal to women ("De girl dem like it when dem see it / And a whi dat dance gon hit unique").[30] The two-part chorus includes a fractured, falsetto melody set to these words that adds chromatic and even whole-tone inflections to a Dorian-modal pitch content (on B♭) and floats above the riddim in a tauntingly childlike sort of way, with the musicians perhaps imitating the sound of a female chorus.[31] (See figure 5.4 for an approximate transcription.) The second part of the chorus then launches into a rhythmicized imitation of the Crazy Frog voice that accentuates its engine-like character (a possible transcription of the lyrics being "Ring ding ding ding di-ding ding ding / Ring ding ding ding di-ding ding ding / Daaawww ding ding ding daaawww").[32] The effect creates a caesura that acts as a musical analog to motorcyclists revving their engines at stoplights or in parking lots. In the video, apparently based in a barbershop in the Tottenham area of London (an ethnically diverse and significantly Jamaican area), the musicians demonstrate the dance during the chorus, which is based on the wrist movements of Crazy Frog's revving of his imaginary motorcycle's engine and a loping, semi-squatted dance step that might be seen simultaneously as somewhat frog-like and yet clearly evoking a sitting position on a motorcycle.[33]

The playful tone of the video is maintained throughout, with the phallic quality of the imaginary motorcycle now realized in a more explicitly sexual fashion, as in some cases women mount the backs of men on the dance floor, the latter now serving as "motorcycles." At other times, the musicians are sitting on actual motorcycles, with women clutching them from behind. As one commentator has noted, the lyrics of the song are intoned in the accents and "slang of the London streets"—and it would make sense that this song's geography is circumscribed to London, in view of Crazy Frog's extraordinary popularity in the UK and particularly in London. I would argue that, in addition to being a novelty dance-craze number, the song is also a humorous commentary on the phenomenon of Crazy Frog itself, with an apt analysis of the production technique of Crazy Frog's music expressed in the line "Any song any riddim Crazy Frog fit it." Moreover, L.O.C.'s appropriation of the mainstream and arguably white-coded Crazy Frog effectively inserts it into the irreverent, multi-racial popular culture of present-day Britain, which Paul Gilroy has recently

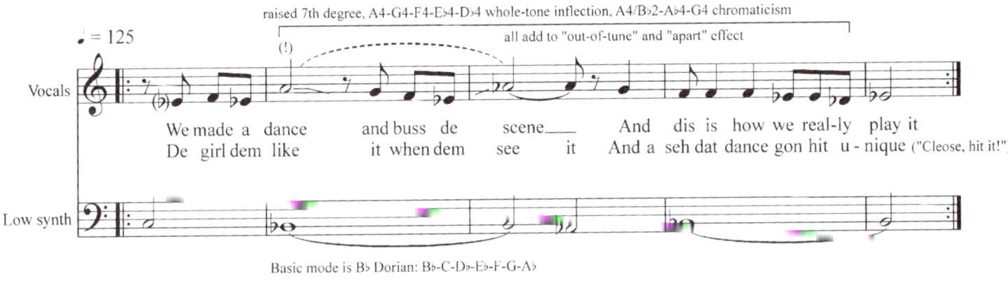

Figure 5.4
L.O.C.'s "Ring Ding Ding" (2005): first part of chorus, with falsetto, out-of-tune, and dissonant vocals.

characterized as a "convivial culture" and which, for him, is powerfully characterized by the music of the cockney rapper The Streets and the racial pranks and "chav" parody of Ali G.[34]

The Backlash against Crazy Frog

Crazy Frog attracted a great deal of negative attention, some of it wanted but much more of it unwanted. Eliciting annoyance from anyone within earshot of the frog's babbled vocals seemed to be, paradoxically, part of the strategy for its success. Indeed, Wernquist's original name for the frog is "The Annoying Thing," which most commentators find highly appropriate.[35] It would seem that part of the weirdly compelling quality of the frog's voice was in its irritating sound, as discussed above. But Crazy Frog's annoying quality, initially an effective marketing feature, was greatly exacerbated by the unprecedented and mind-numbing intensity of the marketing campaign that promoted the frog to what many television viewers found an alarming degree. According to one study, in May 2005 alone, Jamster purchased 73,716 advertising spots across all British TV channels, resulting in some 87 percent of the population seeing the advertisements 26 times—nearly once every day—and some 10 percent seeing it more than 60 times (twice a day). The dual impact of this campaign was predictable. On the one hand, it aided immensely in sales of the ringtone and single; on the other, it led to exasperation on the part of audiences with the advertisements and the voice and image of Crazy Frog.

And yet, in some cases, the negative reaction to the frog's ubiquity in the United Kingdom and Europe ballooned on a scale far out of proportion to its ubiquitous presence in European auditory culture. That is to say, a new, quasi-medical condition—Crazy Frog rage—seemed to overtake large segments of the global population. This rage found expression on weblogs in which sadistic fantasies of destroying the frog flooded the Internet highway. On 13 March 2005, an ode to its destruction was posted:

> i think we should stick that damn frog down the toliet
> then boil it mash it and stick it in a stew
> the flush the remains down a toliet
> or cut its head off and stuff it with pooo hehe
> *evil laugh* squeash it until it pops yes until it pops[36]

Other posters even proffered fantasized torture scenes. Example:

> i surgest buying it a proper moter bike, make the brakes faulty so it has a crash. So it will have to be in a wheelchair. then tie it up and keep poking it so it stays awake. so it can't sleep, then finally when done toturing, kill it by cutting every bit of it's limps one by one. . . .[37]

Many homemade video clips and even online video games were produced in which the frog was shown as maimed, beaten, bruised, cut, decapitated, and murdered, providing cathartic satisfaction for those libidinally invested in hating the frog. Consider the online game "Crazy Frog Baseball," in which the user directs the cursor to a baseball with which to beat and ultimately kill the frog; as of November 15, 2006, the frog had been killed 23,473,198 times. Opportunists took advantage of Crazy Frog rage—for example, a Crazy Frog virus, identified in March 2005, used the promise of an animation depicting the destruction of the frog to deceive unsuspecting spam-mail recipients into contracting and spreading it.[38]

Surely Crazy Frog did not merit this much attention—or did it? If teenagers and adults were motivated to put this much effort into expressing their rage, was the rage entirely about the frog and its advertising campaign, or was its target someone or something else? Might the focus of this rage have also encompassed the consumers purchasing the ringtone and singles or, even more ominously, what the frog may have appeared to symbolize? My supposition is that, in their rage against the consumers (who are mostly children), Frog haters participated in an expanded family romance of sibling rivalry or parent-child conflict and were actually annoyed by pre-adolescent consumers flush with disposable cash flooding the air with frogwaves—with an element of class *ressentiment* at play. But if Crazy Frog is hated as the Other in the European imaginary, we might be alerted to the tendency opposing the convivial culture in contemporary Britain, which exhibits what Paul Gilroy has deemed "postcolonial melancholia." In this case, the failure of British social democracy has rendered young people bereft of social provisions and has left them with only the fantasies of past imperial glory, couched in racial ideology, as found in the white football-hooligan culture or, for young blacks, the false promise of Islamic fundamentalism. For those libidinally invested in the ideology of nationhood, fantasies of torturing the state enemy—who are often blacks and Muslims labeled as terrorists—provide one way of participating in the body politic. Indeed, with the revelations of the Abu Ghraib torture program in early 2004, including the well-known photographs that flooded the world media but were surprising only to those previously unwilling to admit to themselves the sick cruelty of American imperial power, media-violence-saturated young people were practically primed to react to figures of Otherness in such a way. Re-injecting immigrant paranoia directed at second- and third-generation non-white Europeans, in part through the American-imposed war on terror, the politics of national and supranational belonging in the mid-decade conjuncture in Europe seems somehow linked to the fate of Crazy Frog, whose consumers would appear to participate in a prefabricated, and thus easily discredited, form of consumer cosmopolitanism.[39]

And yet, this interpretation of the racial-political unconscious of Crazy Frog must be understood as speculation that needs to be situated within explicit conflicts concerning the phenomenon—two of which unfolded in a short period of time after its appearance in the United Kingdom. One situation of prime importance was the moral panic over the previously discussed genitals appearing on the body of the frog. During the first moments of the Crazy Frog advertising campaign in early 2005, parents complained to various authorities and public news outlets that Crazy Frog's genitals made it inappropriate for children. In the best-known case, sixty people complained to the United Kingdom's Advertising Standards Authority (ASA), asking that the image not be shown or that the offending genitalia be censored from the image. Although on 2 February 2005 the ASA ruled that there was no impropriety in the depiction of the frog, the Jamba/Jamster group soon began voluntarily censoring the image—particularly in the US market, where such concerns, had Crazy Frog taken off there, may have caused a much greater uproar.[40] At first, the image included a censoring black box covering the frog's privates, but soon a genital-less version of Crazy Frog became the more common representational solution.[41]

Much more significant in the tale of the frog, however, was the scandal associated with the Jamba group's advertising practices, which came under heavy scrutiny in mid 2005. The essential problem was that the television advertisements for Crazy Frog lured consumers into paying for a subscription service for mobile entertainment rather than just the original Crazy Frog or the "Axel F" ringtone. Unsuspecting customers, believing they had made a one-time purchase, would soon find text messages or even new ringtones being sent to their phones, with bills far in excess of what consumers expected and weekly subscription fees as high as £5 (US$9).[43] Although the television advertisements included an explanation of the product being advertised, it was usually presented in small print, and a large number of children unfamiliar with such marketing practices unwittingly purchased the service after watching the ads. A lengthy legal and corporate-regulatory process ensued, involving successful efforts to have the misleading advertisements removed from the air and consumers refunded the service fees. For example, on 21 September 2005, in one important ruling by the ASA, the Crazy Frog advertisements were required to be shown after 9 p.m. in order to limit children's access to them. The effective endpoint of the saga was the ruling in late December 2005 that mBlox, the billing service company for Jamba, would be fined £40,000 (US$72,000) and told to reimburse customers seeking refunds for the undesired service. The ruling was made by the Independent Committee for the Supervision of Standards of the Telephone Information Services, an industry-funded regulatory body in the United Kingdom that "regulates premium-rate calls." ICSTIS, however, was legally unable to fine Jamba itself, since the company technically is not a "service provider."[43]

The greatest significance of the story, however, was not the negative financial impact on mBlox or even the entire Crazy Frog enterprise. Rather, the misdeeds of Crazy Frog were felt on a much larger scale: the Crazy Frog subscription scam had actually eaten into ringtone sales themselves, which began plummeting in late 2005—well before the 2008 recession curtailed ringtone sales in the US and elsewhere. Financial news reports admonishing Jamba for ruining the ringtone gold rush became increasingly common, but soon reports began to adopt a wider view of the broader decline in ringtone sales. As Rob Wells, the director of the new media division at Universal Music UK, said, "you can put it down to price, piracy, and the Crazy Frog effect."[44] In fact, in its very ineptitude the Crazy Frog scam appeared to burst the inflated ringtone bubble and unmask the scam of the ringtone itself, an overpriced service that, particularly with the advent of the sound-file ringtone, is nothing more or less than a simple editing and reformatting of a sound file. But if the first authentic ringtone hit also sounded the death knell of the ringtone, this may serve as a judgment on the lack of long-term viability of the ringtone medium, or, even more importantly, as a figure of the formation and development of much new media in late capitalism, in which the maturation process of a new medium is so rapid that its birth is nearly indistinguishable from its death.[45]

Conclusion

Its manifestation in the United Kingdom notwithstanding, how should we determine the geographical location of Crazy Frog? On the one hand, Crazy Frog can be located primarily in the crises over generation, class, and race in the pre-recessionary conjuncture in the United Kingdom and in the European Union generally. From this perspective, it is clear that the frog's manifest, determinate content is that of the European pre-adolescent, and by extension, the politics of (white) childbirth in the EU. But the unconscious, latent content of the frog might be its racialization, which significantly is also the hidden heart motivating white European paranoia about declining childbirth rates.[46] And forming the backdrop of this scene, moreover, is the European Union's subordination to US imperialism and the European public's rejection of this role, from the right and the left, which becomes bound up in both conservative white European fears of losing a coherent identity and leftist fears of the disintegration of social democracy and labor syndicalism—realities masked by narcissistic boosterism for the EU project.[47] And yet, on the other hand, at various levels of mediation the specter of Asia looms in Crazy Frog's background. The frog's animation style, the character's facial features, its oblique relationship to East Asian "cute" culture, and even its kinship with Asian video games such as Konami's Frogger all mark an undercurrent of some synthesis of different East Asian influences. But much more important is the overdetermined presence of Asian and Asian-influenced technical apparatuses in the production of Crazy Frog—and here we might consider the significance of digital electronic keyboards (in Crazy Frog and even the original "Axel F"), video games, mobile handsets, animation software, and the ringtone itself. One would not want to remove Europe from the equation, however, in view of the centrality of the European consumer in the success of Crazy Frog and the importance of Scandinavian contributions to ringtone culture generally—from the Finnish firm Nokia's early development of ringtones to the Swedish origins of Crazy Frog. Perhaps, then, the most significant way in which

Crazy Frog is Asian is that it is a global phenomenon that is overwhelmingly *not American*. As exemplified in the global ringtone industry and in the texts and subtexts of Crazy Frog, the strategic interconnection of the two other globally dominant economic zones, Europe and East Asia, signals the decline of American economic and geopolitical power and the much speculated-upon rise of East Asia as the future center of capital accumulation in the world system.

6 The Voice of the Politician and the Geographic Dispersion of the Political Ringtone

In November 2004, a new sound was heard from mobile phones in the Indian state of Gujarat. A ringtone featuring a clip from a speech by the popular right-wing chief minister of the state, Narendra Modi, had been produced by an anonymous professor of information technology at Bhavnagar University and was quickly gaining in popularity. The speech sample originated in Modi's reply to a question by the former chief minister of Gujarat and opposition party leader Amarsinh Chaudhary during the 2003 state parliamentary session. The sample's text is as follows: "Gujarat's future should be considered above political calculations. Concern should be for tomorrow's Gujarat. Betterment of Gujarat cannot be achieved by keeping an eye on the chair."[1] This seemingly anodyne statement—which may implicitly or explicitly signal Modi's Hindu-nationalist politics and problematic relationship to the infamous 2002 riots, which aided his reelection that year[2]—also marks an important step in the evolution of political ringtones—that is, ringtones used to index and/or support political movements.

The notion of a "political ringtone" might well strike some readers as odd. But in fact, the nature of "personalization"—a central part of the marketing ideology of the ringtone—was bound to transform the ringtone into a kind of aural equivalent of the bumper sticker or the political t-shirt (the American consumer's principal means of casually making political statements).[3] Early in the historical development of the ringtone, pre-political and nationalist uses were frequently found in monophonic forms all over the world. Examples include the prevalence of national anthems and Russian gangsters' use of the *Godfather* theme on their phones.[4] But, as the Modi example illustrates, political ringtones have become more refined since the rise of the sound-file format, which has allowed texts and voices to appear with ease. In what follows, I trace the development of a new kind of political ringtone, based on the sound-file format, that has been used effectively for political mobilization: a ringtone that draws on the recorded voices of politicians caught speaking in marked, often unscripted, and indiscreet ways. This new kind of ringtone treats the unprecedented sonic documentation of political figures as material for a new opportunity for commentary on contemporary politics, the very spectacularity of which privileges moments of authenticity (including covert conversations, verbal blunders, and outbursts) that elucidate the inner workings of the state. Moreover, a significant number of such ringtones remix these speech recordings with backing musical tracks to produce a kind of "mash-up" of the two sonic sources, the content of the musical tracks affecting the popularity of the ringtones themselves.[5] Although political ringtones have appeared worldwide, in the process becoming a *global form*, I will focus on interrelated examples in the Philippines, the United States, and the Spanish-language world, tracing a chronological sequence in 2005–2007.

Manila

The arrival of the ringtone as a political weapon first registered in global mass consciousness in the summer of 2005, in the context of a scandal in the Philippines. The story begins on 6 June with the release of tape recordings of President Gloria Macapagal-Arroyo speaking to Virgilio Garcillano, an elections commission official with close ties to the president. The press secretary, Ignacio Bunye, played a series of wiretapped recordings in a press conference, some segments of which were soon spliced and used in anonymously composed ringtones. The now-infamous spoken text, seemingly referring to illegal tampering with the results of the 2004 presidential elections in the Philippines, is as follows: "Hello Garci [Garcillano]? Will I still lead by more than one M [million votes]?" During the course of the scandal, the ringtones became immensely popular in the Philippines, apparently being downloaded more than a million times. These ringtones were, in fact, inseparable from the subsequent groundswell of protest that nearly brought down the political administration of the country. As the mobile-media activist Anthony Cruz put it, "in this protest, the ring tone is the medium and the message."[6] What were the conditions of possibility for such a remarkable phenomenon?[7]

The use of the mobile telephone in Filipino politics extends back several years, most famously in the People Power II revolution of 2001. In an essay on cell phone politics in the Philippines, Vicente Rafael describes the centrality of the role of text messaging within Filipino culture since the late 1990s and in particular within People Power II, in which President Joseph Estrada, a populist figure and a former film star, was deposed in a civilian-led coup that led to his succession by Arroyo. Rafael's main argument is that this revolution was framed in the media as a product of the Filipino middle classes, as demonstrated by a fantasy of instantaneous communication provided by text messaging and the respectably reformist politics espoused by members of "Generation Txt," a tech-savvy younger generation of middle-class Filipinos with access to mobile phones.[8] Part of a longer history of a technocratic middle class angling to gain control within what Benedict Anderson has described as a "cacique democracy," Generation Txt has repeatedly demonstrated the political utility of cell phones in dramatic fashion, this sub-cohort's mouthpieces having at times attributed an almost mystical power to the cell phone itself.[9]

The feel-good quality of People Power II—which was first spoiled by the violent protests against Estrada's arrest in April and May of 2001[10]—did not last particularly long, however. Arroyo's presidency was beset by a number of scandals and controversies, including an armed rebellion by hundreds of military officers and soldiers, a scandal involving her husband's alleged infidelity and campaign fund embezzlement, and the about-face in her decision to run for election in 2004 after initially forswearing a reelection bid. But the most significant problems arose in regard to the potential rigging of the 2004 elections themselves. The constituent events of the 2005 scandal unfolded as follows. Three days after the 6 June press conference, Arroyo denied having cheated in the elections but refused to make mention of the tapes themselves. On 10 June, Samuel Ong, a whistleblower and former deputy director of the National Bureau of Investigation, confirmed the voices of Arroyo and Garcillano on the tapes and claimed to have a master tape recording that could incriminate the president; mass protests began swelling throughout the country, calling for the president's impeachment. Ong, fearful for his life, ensconced himself at a seminary that was soon surrounded by supporters. On 11 June, Arroyo's office formally declared her innocence. That day, authorities began preventing protesters from gathering without permits. More tapes continued to surface, and on 12 June, the country's independence day, Arroyo denounced the entire phenomenon, while increasing numbers of tapes, CDs, and transcripts began circulating throughout the country. The government issued a statement that, because of the illegality of wiretapping, TV news stations were forbidden to play the tapes or air the transcripts. Within a few days, the recordings and transcripts began to appear on the Internet as MP3 and Word Document files. Meanwhile, Garcillano himself seemed to have disappeared, though he was apparently seen preparing to go abroad on 19 June, and two days later the Congress began investigating the wiretapping of the president's phone. At this point, the furor and protests had subsided temporarily, although the crisis was still active. On 27 June, three weeks after the scandal first broke, Arroyo admitted that it was her voice on the tapes and publicly apologized while refusing to resign. On 1 July, large public demonstrations against Arroyo appeared again, renewing calls for her resignation and impeachment. On 5 July, Garcillano resurfaced and claimed that the elections were clean; two days later, the president asked her entire cabinet to resign but reconfirmed her unwillingness to resign herself. The next day, 8 July, some of these former cabinet members and prominent political figures began calling for Arroyo's resignation.

In the next few days, the scandal continued to unravel, with the Catholic Bishops of the Philippines weighing in mostly against Arroyo, but on 12 July many prominent members of the political establishment, including the president's supporters and opponents and non-aligned persons, agreed on a major overhauling of the electoral commission and advocated a unified opposition to coups, rebellions, insurrections, and the like as solutions to political problems. Protests continued, and further procedures were set in motion to investigate the electoral fraud case, including possible impeachment proceedings. Toward the end of July, it was reported that Garcillano had disappeared to Singapore and then London. By early August, arrest warrants had been issued for him because he had ignored several summonses to testify to investigators of the case. Garcillano did not return to the Philippines until late November. In early December he turned himself in, and in mid December he admitted that his voice was on the tape. In early September, efforts by a minority of Congressional representatives to impeach Arroyo had been decisively defeated. That defeat was confirmed by the official approval of a congressional report ("Report 1012") that had first appeared in late August and had argued against her impeachment. By early 2006, Arroyo was taking increasingly extreme measures to quell protest. A State of Emergency in late February and early March[11] resulted in the suspension of state norms and in extrajudicial killings of leftist activists.[12]

Woven into the tale of the mid-2000s conjuncture in the Philippines is the unlikely appearance of the cell phone ringtone as a political agent. The story begins shortly after the first release of the audiotape recordings of the phone conversations between Arroyo and Garcillano. As the events of the scandal proceeded, popular acquisition of the sound files began to grow exponentially, all within the first few days of breaking. Before long, anonymous ringtone composers began posting free, downloadable ringtones online featuring various musical samples combined with a spliced conversation between the two wiretapped discussants from the original tapes, usually including Arroyo's anxious question "Hello? Hello Garci? So, will I still lead by more than one M?" and often with Garcillano responding "Hello Ma'am. We will try for that." Most of the ringtones featured some or all of these statements, which were effectively treated as "mash-ups," a specific type of quick-and-dirty remix form typically combining the vocal line or track of one song with the instrumental music of another track.[13] In the "Hello Garci" ringtone that initiated mass downloading, sometimes titled "Buking" (Tagalog slang for "someone who was found out"), a sample of the instrumental ending material of the American rapper 50 Cent's 2003 number-one hit "In da Club" was combined with Arroyo's voice to create a remarkably powerful political statement.[14] In part, the aesthetic and satirical impact of the ringtone derives from the listener's knowledge of the original song, which is an unapologetic celebration of dance-club hedonism and rapper/club entrepreneurship. The song ends with 50 Cent admonishing listeners with differing expectations by speaking the words "Don't try to act like you ain't know where we been either nigga / In the club all the time nigga, it's about to pop off nigga / G-Unit." Although the rapper's braggadocio and vaunting of his own financial success is a familiar theme in the history of hip-hop, he claims to be disinterested in musical-stylistic authenticity here ("I'm a tell you what [hip-hop producer Russell] Banks told me cause go 'head switch the style up / If the niggas hate then let 'em hate / Watch the money pile up") in favor of earning pay—which is, of course, another source of authenticity. Transposed to the context of the scandal, we can see that one might presume the cold calculation of financial self-interest now applies to the political self-interest of Arroyo, who indeed has effectively placed her own supporters in a political insiders' "club" and used these contacts to secure yet more political power.

The "Buking" ringtone (figure 6.1) was musically and perhaps politically effective for several reasons. First, Arroyo's voice stood out of the texture, with its insistent, imploring tone implicitly issuing a command—thus powerfully encapsulating the scandal in a single phrase ("Hello Garci"). Indeed, it is not a stretch to claim that the scandal is known as the Hello Garci scandal primarily because of the ringtones. Second, the music itself fits the atmosphere of protest in that it combines a party or club atmosphere with a certain degree of urgency, the latter created by the minor mode music and its ominous, minimal texture punctuated by the

Figure 6.1
The "Hello Garci" ringtone.

insistently repeated C♯ tonic notes in the guitar/plucked string instrument and the dynamic interjections on the "orchestra hit" synthesizer patch.[15] Third, the ringtone self-referentially cited the world of telephony itself. In this capacity, the old-style telephone ringer that begins the song refers both to the fad of using nostalgic ringer sounds as ringtones *and* songs that include the sound of a telephone ringing. Also, Arroyo's phrase would sound as if she were speaking to the cell phone's owner, thus interpellating the hearer and the phone's owner as Garcillano. And when we acknowledge that "In da Club" itself was a huge ringtone hit for a long time after the song's initial release (at least as late as the end of 2004), we might appreciate the fact that *every* sound in the ringtone has a direct association with the cell phone itself.[16] Fourth, the ringtone was effective *formally*. In taking what is an instrumental passage with a phrase ending or cadence, the ringtone constituted a well-balanced, three-part AAB form that begins and ends symmetrically with the same telephone ringer—essentially making the ringtone less an excerpt from a longer passage of music and more an independent, self-standing composition. And finally, the percussive textures and distinct sonorities within the ringtone made it capable of working well even on low-fidelity cell phone speakers (ca. 2005)—a crucial, if underestimated characteristic of any aesthetically effective ringtone.

The "Buking" ringtone was, however, only one of dozens of ringtones that were produced shortly after the scandal began, with many of them being posted on various websites for easy access. For example, at the website of the Philippines Center for Investigative Journalism, more than 30 versions of the Garci ringtones were listed—many of which are still available as of this writing.[17] Many of the recordings featured Anglo-American popular songs—not surprising in view of the ubiquity of Anglo-American popular music in the Philippines, as part of a broader American neocolonial market domination specific to that country, and the prevalence of "*videoke* [a]s a national pastime," according to Jan Maghinay Padios.[18] Several tracks stand out, including a bluntly ironic version of the ringtone using Billy Joel's "Honesty," featuring the chorus of that song. A few of the ringtones feature Beatles songs, including one ringtone that begins with "Do You Want to Know a Secret," then reveals someone asking "and how much?" [*at magkano*?], and then ending with the opening of "Don't Let Me Down," creating a narrative proceeding from a proposed secret-revelation, to a view into the illicit discussions, and then ending with an ambiguous exclamation asking not to be let down—conceivably either by a protestor who would need to fight the corruption of the Arroyo regime, or, ironically, one conspirator asking to rely on another. The song also ends up encapsulating a move from innocence to jadedness through the contrast of the early Beatles sentimental pop song with the later, rawer mid-tempo ballad.[19] We find several versions using the *Godfather* theme music, with quotations from Marlon Brando and others in the films; other ringtones use music and quotations from *Scarface*, *X-Men*, *Predator*, *The Last Samurai*, and several other English-language films, suggesting the prevalence of Hollywood in Manila. In many of these ringtones, Garcillano is spliced so as to blurt out "Ma'am" in response to Arroyo's "Hello Garci"; these exaggerated Filipino conversational formalities may also underscore the power relationship between the two speakers. Some of the most amusing ringtones sampled Arroyo's subsequent, noncommittal public apology, perhaps most impressively in juxtaposition with Tracy Chapman's "Baby Can I Hold You," which begins "Sorry is all that you can['t] say / years gone by and still / words don't come easily / like sorry, like sorry" and which finds Arroyo's "sorry" almost superimposed on top of the singer's.[20] And some of the remix/mash-up treatments of both sources (the Garci tapes and the apology) were remixed into full-length tracks that, due to their length, would not function as ringtones.

The appearance of the Garci ringtones played several roles within the scandal and political uproar. For example, they were directly inculcated in a pitched battle between the broadcast and print media outlets, who were attempting to disseminate recordings and transcripts of the conversations to the public, and the Arroyo administration, which, through the National Telecommunications Commission (NTC), initially banned distribution and rebroadcast of the tapes and transcripts. The International Federation of Journalists, an international federation of journalists unions and journalists' rights organizations based in Brussels, sharply criticized the absurdity of an administration banning distribution of materials that it, itself, had publicly aired in a press conference.[21] Apparently, one of the crucial moments in rescinding the ban was at another, subsequent press conference at which Arroyo defended the ban on the transmissions of the tapes and threatened the journalists present in the room with the NTC ruling. A journalist stood up in the crowd and played the Hello Garci ringtone on his or her cell phone, demonstrating that the issue was moot, since the tapes were already widely available thanks to the many websites distributing the ringtones. Soon after this was revealed in the press conference, sites serving as mirrors for redistributing the ringtones began crashing in response to an astounding increase in downloading, signaling the mass demand for the ringtones. Ultimately, as one writer put it, the effort to control distribution of the wiretapped conversations through an NTC ruling was a foolish maneuver, since it only increased the public's interest in the matter and therefore "was equivalent to throwing fuel onto a fire."[22]

In addition, these ringtones ended up serving as a form of political expression, sounding in the public sphere in the Philippines, especially in Manila. A mobile phone consumers' group called TXTPower played a particularly important role in fostering political ringtone use, since it was their websites that served as the main source of the more than a million Hello Garci ringtone downloads, including more than 500,000 from 17 June to

17 July 2005.[23] In addition to serving as a means for spreading rumors and information in a rather "viral," chaotic manner, ringtones became a central component of protest during the scandal, with coordinated playbacks of ringtones becoming part of the language of political protest.[24] In what must be numbered as one of the most remarkable and creative uses of the ringtone, many motorists wired their car horns so that, when honked, they would play one of the Hello Garci ringtones—a fact that was taken by the administration as being a blight on the culture.[25] Moreover, the ringtones were part of a broader set of protest-cultural forms that included screen savers, mobile phone wallpapers, and placards for street protest, many of which were also freely available on the TXTPower website. And yet, one would not want to overemphasize the role of these relatively new, technologically mediated forms in the creation of the protests themselves, since they functioned here—as do all political and politicized technologies—as a means of facilitating mobilization and participatory invigoration and not as a replacement for political organizing.[26]

Evaluating the political meanings and impact of the Hello Garci scandal is no easy matter, but we might draw some preliminary conclusions about it. It would seem that the class dynamics described by Rafael in the case of People Power II were still at play to some degree in the Hello Garci scandal, leading to continued divisions between middle-class supporters of Arroyo and the coalition of critics including the left, members of the elite (including Corazon Aquino), some former Arroyo supporters, and working-class factions still backing the circle of political figures including the ousted Estrada (who, early on, had volunteered to serve as an interim president if Arroyo were to resign or be impeached).[27] And yet, we should take note that the political demographic of cell phone users in the Philippines has changed somewhat since Rafael's original essay appeared in 2001. In particular, we might notice a growing trend in mobile telephone use that by 2006 covered more than 33 percent of the Filipino population, up from 10–11 percent in 2000.[28] This fact might suggest that the spread of cell phone technology had, in the Philippines, begun the process of moving from a semi-elite commodity to a mass-cultural one, and we might expect that more recent political uses of the mobile phone would reflect this demographic shift. But perhaps an even more significant factor is the precedent and historical burden of the earlier People Power revolutions, which on the one hand appeared to be highly successful and yet on the other hand have failed to realize the reformist aims sought by participants in those movements (such as a reduction in political graft and corruption), let alone providing more social provisions or employment for working-class

Filipinos.[29] Indeed, commentators on the Arroyo impeachment efforts have remarked upon the relatively low participation and apathy of the middle classes so invested in Estrada's impeachment—a marker both of passive middle-class support for Arroyo *and* of a disillusionment with the past results of mass popular demonstrations referred to as "people power fatigue."[30] Others have argued that the entirety of the present political class is "tainted" by scandals and nakedly partisan positioning and that therefore no figure or party has emerged that can claim unambiguous popular support.[31] But a more acute analysis is found in the writings of Amando Doronila, who argues that the politically active component of the middle class has essentially lost its hegemony (or moral leadership) over the impoverished majority of the populace, as a result of the latter's increasing awareness of being marginalized within the People Power movements.[32] Writing from the perspective of mainstream US political science, in the National Endowment for Democracy-funded *Journal of Democracy* no less, Paul Hutchcroft argues that the result of the Arroyo regime's heavy-handed efforts to maintain power has been a severe weakening of democratic institutions and structures in the Philippines. But it remains unclear how a country whose economy is overwhelmingly dominated by the informal sector (an estimated 60 percent) and by an entrenched political class discredited as incompetent and corrupt can expect to reproduce a developed-world model of democratic governance—let alone produce a stable form of governance—through Hutchcroft's gradualist proposals, which fail to address economic democracy in any substantive way.[33]

Toronto–Washington–New Orleans

If the political context of the Philippines bespeaks a popular disenchantment with a certain brand of mass collective action, in the United States the common goal of left-leaning political activists has been to counter widespread disempowerment and civic helplessness among working-class and, increasingly, middle-class Americans. Many of these activists are targeting the "millennial generation" of young voters who, as part of a ricochet or "echo" boom of the Baby Boom generation, constitute a large cohort—about 75 million people between the ages of 6 and 26—that is gradually coming of age politically.[34] It is with similar aims in mind that a group of mostly North American "mobile activists" has attempted to learn from and imitate the successes of political cell phone use found elsewhere in the world, notably the Philippines. A foundational moment in the development of a coordinated mobile phone politics might be found in the MobileActive conference held in Toronto in late September 2005. The conference was put together by a loosely affiliated political group called MobileActive, which describes itself as a "global network of activists and campaigners using mobile phones for civic action and engagement."[35] Funding for the conference was obtained by Katrin Verclas, the coordinator of the MobileActive website and e-mail list, primarily through a $100,000 grant from the Open Society Institute founded by the left-liberal philanthropist George Soros.[36] Having a longstanding background in the political applications of new technologies, Verclas was at the time a co-director of Aspiration: Better Tools for a Better World, a non-profit organization, based in San Francisco, that provides "software-related resources and services to other nonprofit organizations." For the conference, Verclas sought to bring together a transnational set of activists to have a fruitful conversation about the possibilities and difficulties involved in using mobile technology as part of their activism.[37] Located a couple of miles outside the city at an idyllically situated hotel, the MobileActive conference found itself populated with about 40 activists from Africa, the United States, Canada, and elsewhere, and included two members of TXTPower, Trixie Concepcion and Anthony Cruz. The latter were crucial in inspiring the other mobile activists about the power of the new technologies in political campaigns and mobilization. The conference itself was reportedly hectic but well organized, according to one participant, and involved speed-geeking sessions, presentation sessions, and breakout discussion sessions, as well as many moments for informal conversation.[38]

The impact of the conference was significant, in that it inspired North American activists to begin thinking more seriously about incorporating mobile technologies into their campaigns. One result of the conference was a growing awareness of MobileActive, whose network of participants had by 2007 grown to about 300 people located all over the world (but concentrated, for the most part, in the United States) and, according to Verclas, was part of a quickly expanding community of mobile activists perhaps numbering as many as 500–1000 participants.[39] More important, the conference gave the Filipino participants a platform for explaining the power of cell phone (and ringtone) activism by elaborately describing the activities of TXTPower during the Hello Garci scandal. As the activists from the United States listened, they began imagining similar sorts of campaigns in their own country but were quickly frustrated by the cumbersome, multiple-system mobile coverage there (which includes the use of GSM, CDMA, and other telephony systems, as well as monopoly control of ringtone distribution by each wireless carrier). The root of the problem, it appeared, was that individual carriers actually prevented the mass distribution of ringtones to all cell phones. One participant in the conference, Evan Henshaw-Plath, took an interest in this problem. Henshaw-Plath has a background in providing Web and tech support to Indymedia activists in Latin America, including Bolivia and Argentina.[40] He is also an experienced software and Web-application designer based in San Francisco and the former lead developer of Odeo, a software package that helps users organize their Podcasts.[41] A self-confessed "anarcho geek," Henshaw-Plath attempted to write a program, called Riot.Tones, that would facilitate easy distribution of free ringtones by accommodating the different communication systems and wireless carriers.[42] He began by learning about the international standards followed by wireless carriers and from other programs that had learned to distribute ringtones to all phones, across carrier networks, and from these observations, set out to make the first open-source, freely available ringtone distribution program. The online program would also include a sound editor, enabling the easy production of ringtones on the site itself. Unfortunately, Henshaw-Plath ran into problems during the coding process, and the project came to a halt.

Another American activist inspired by the activities and encounters at the conference was Eric Gundersen—Web programmer, co-founder, and later president of Development Seed, an organization, based in Washington, whose main goal was to provide "technical support and services for progressive organizations." Gundersen attended the Toronto conference; upon reaching the hotel, he ended up rooming with Anthony Cruz of TXTPower, who discussed his experience of distributing ringtones via the TXT website during the Gloriagate scandal and, according to Gundersen, demonstrated a great deal of knowledge about server structures and their use in this kind of application. Encouraged to make his own ringtone, during the conference Gundersen offhandedly remarked that, in view of the impact of Hurricane Katrina and the neglect of the federal government on New Orleans, it might be powerful to make a ringtone using President George W. Bush's much-ridiculed statement about the incompetent chair of the Federal Emergency Management Agency, Michael Brown: "Brownie, you're doing a heck of a job." Katrin Verclas encouraged Gundersen to follow through on his idea. Upon returning to work in Washington, he produced such a ringtone, which combined Bush's comment with Arlo Guthrie's famous cover of Steve Goodman's song "City of New Orleans."[43]

The ringtone itself is unusual, if not exactly aesthetically effective, in part because it juxtaposes the music with Bush's voice rather than superimposing the two. The song's instrumental introduction begins the ringtone but is interrupted by the word "Brownie," which happens a couple of times, before the opening line "riding on the City of New Orleans" is heard. At first, Bush's words merely interrupt the original flow of the song, but soon they begin to scramble or even recompose that flow. For example, immediately after the interrupted instrumental introduction, we hear the full statement, "And Brownie, you're doing a heck of a job," followed by a different section of the song's first verse that musically also functions as a pre-chorus: "Rolls along past houses, farms, and fields." Bush's statement appears again, stopping the flow of the music, and then the ringtone continues with a later point in the same verse/pre-chorus, "Freight yards full of old black men," and Bush's statement is heard for a final time. The effect is that of experiencing a detached ride offering a panoramic view of the city landscape after the disaster (with the "old black men" figuring the stranded, mostly black population in the city at the time), which is then cut to Bush's callous, ignorant statement about the state of the recovery efforts. In essence, it mimics the impact of watching the mass media at the time—a rare moment when, as Gundersen put it, "the press had picked up their heads for once"—and seeing the city's ruination starkly juxtaposed with the federal government's inadequate response. And by cutting away from the song mid-stream to Bush's comment, the ringtone highlights the simultaneity of the phenomena as they unfolded historically, as if occurring in two entirely separate realities, even while implying that the president himself impeded the city of New Orleans' own recovery.[44] The ringtone's reception was mixed: on the one hand, it gained a great deal of media attention in news outlets and the blogosphere (yielding more than 30,000 downloads); on the other hand, except in a few instances, it apparently failed to catch on as a ringtone. Gundersen attributes this failure to a poor choice and usage of the spoken text and to the choice of a song whose folk-pop style did not translate well into the ringtone context aesthetically or even sonically (sounding poorly on cell phone speakers). His awareness of the difficulty of creating an effective ringtone only gave him a greater appreciation for the communicative power of the "Buking" ringtone and the choice of speech fragments in the Hello Garci ringtones generally.[45]

Gundersen's experience with the composition of ringtones prompts the question "What makes an effective ringtone, and a political ringtone at that?" The veteran indie-rock band They Might Be Giants attempted to find out by trying their hand at writing a few, and, at least in one case, took inspiration from the political context in which they found themselves. The band members John Linnell and John Flansburgh took issue with the Bush administration's secret telephone-call monitoring program run by the National Security Agency, which was discussed in a long-withheld exposé in the *New York Times* late in December 2005.[46] In early February 2006, at the end of a recording session with producer Pat Dillett, the band decided to record three mini-songs, "Phone Phone Phone," "Ring Ring," and "Call Connected Thru the NSA," that could be used as ringtones or in other contexts.[47] The band then partnered with Xingtone to facilitate the sales of the ringtones themselves. Each of the ringtones is rather well composed, and suggests that one effective approach to the ringtone form is to actually write a miniature song that falls within the ringtone's usual 30-second time limit—a task to which the band was well suited, with its long history of writing short songs and distributing songs by telephone. In each case, the form of the mini-song includes some kind of short thematic statement, which is then repeated and can be prefaced by an introductory section ("Ring Ring") or can lead to an outro ("Phone Phone Phone," discussed in greater detail in chapter 7) or a continuation of the phrase that ends with a cadence ("NSA"). In addition, another element that makes these ringtones effective is a self-referential treatment of the telephonic context in which the ringtone would be heard—and in this way, the group's efforts are not unlike the "Buking" ringtone. The political song of the bunch includes a blunt and pointed set of lyrics—"Call connected through the NSA, complete transmission through the NSA, suspending your rights for the duration of the permanent war"—and evokes the old-fashioned telephone operator who would connect calls for dialers. The music, however, is what makes the song a little gem, with its hazily 80s-nostalgic, almost rock-anthemic character, which then morphs (on the cadence at "war") into a repetitive, minimalist texture not dissimilar to the sound of some ringtones.[48] Its only limitation perhaps lies in its lack of a sound sample directly stemming from the targets of critique—unlike "Buking" or Gundersen's ringtone—and hence not being a part of the mash-up "genre" of ringtones that has gained the greatest popularity and political use.

North American left-liberal mobile activists have not particularly concerned themselves with the aesthetics of ringtone composition. Instead, they have continued to expand the use of cell phones in political campaigns—for example, Amnesty International's "Close Guantanamo" campaign incorporated text messaging by participants[49]—with ringtones thus far playing only a very small part of a broader effort that probably will focus more on SMS text messages and cell phone photographs and videos. And yet, the trajectory that extends radially from the experience in the

Philippines to the rest of the world is by no means the only line of development for ringtone activism. An earlier, if less spectacular set of uses involved the presence of ringtones in electoral campaigns, and as many right-leaning movements as left-leaning ones have received media attention in this regard. Concerning electoral campaign ringtones, we might trace this use at least as far back as the earliest years of the new millennium. In June 2001, ringtones became popular in electoral campaigns in the United Kingdom. The Labor Party Member of Parliament and Deputy Prime Minister John Prescott hired the phone company Zed to design ringtones and text-message logos for his campaign. The pugnacious MP, who had brawled with a heckler in May of that year, soon became associated with a ringtone of the Rolling Stones' classic tune "Street Fighting Man."[50] Earlier in the year in Israel, the mobile communications firm Orange was approached by political parties to create ringtones for their campaigns. Soon enough, the musical version of an electoral slogan "Only Sharon will bring peace" became a very popular ringtone, advertising the conservative Likud Party Prime Minister Ariel Sharon's unilateralism that appealed to Israeli security fears and helped defeat the Labor Party candidate Ehud Barak in the February election.[51] In the intervening years, the trend has continued to grow, although its impact is not clear: in 2004, the Hindu-rightist Bharatiya Janata Party (BJP) unveiled an ambitious high-tech campaign involving ringtones of the BJP anthem, automated phone calls leaving messages spoken by Prime Minister Atal Behari Vajpayee, and text messages of Vajpayee's own poetry.[52] And yet, the election ultimately represented a referendum on the politics of the right represented by the BJP, with a tremendous upsurge in electoral participation by "Dalits, Adivasis, women, and the poor" that marked gains for leftist parties and the formerly dominant Congress Party and ousted the BJP from its former political dominance.[53] Ringtones and mobile media were certainly featured in the 2004 elections in the United States, as in the wireless firm Cingular's "Rock the Vote" campaign, which was also sponsored by Motorola and MTV. Although the campaign featured downloadable ringtones by artists such as Kinky Friedman and the Dixie Chicks, it would seem that the impact of the mobile media here probably was negligible—although the US youth voter demographic (18–29-year-olds), the only one to lean at all leftward, did increase its turnout substantially in that election (as well as in the subsequent 2006 and 2008 elections).[54] Perhaps the presence of Obama ringtones in 2008—some of which worked on the mashup/remix model combining samples of speeches and dance beats, although in a celebratory rather than critical vein—was one small component of that shift.

But explicitly political ringtones tied to mass movements have yet to make their mark within the bowels of the hegemonic state.[55]

Santiago–Madrid–Caracas

The third primary example under consideration confirms the development of the political voice-remix ringtone as a *global form,* exemplifying a partial incorporation of the unruly dynamics of mobile cultures into state and capital (their internal contradictions notwithstanding). Here, the scene is Caracas, Venezuela, the primary location of President Hugo Chávez's "Bolivarian Revolution," which in the mid 2000s was at the leading edge of left-statist politics.[56] The specific context is the December 2007 referendum[57] on a series of 69 constitutional amendments that would, among other things, reduce the maximum working week, lower the voting age to 16, extend social security benefits to the informal sector, and expand presidential power in various ways, all in the name of a state-socialist/social-democratic and arguably "post-neoliberal" project, one nonetheless structured by "magical" petrostate politics and by the presence of "compensatory" neoliberalism elsewhere on the continent.[58]

For our purposes, we take note of a minor but prominent component of the opposition's campaign, drawn from a media phenomenon that rippled through the entire Spanish-speaking world during November 2007, a month before the December referendum. Its origins lay in a particular event: the Seventeenth Ibero-American Summit, held in Santiago, Chile. A yearly gathering starting in 1991 that brings together the heads of state of Latin American and Iberian nations, the Ibero-American Summit is a public discussion forum whose goal is to increase collaboration and multilateral diplomacy between these nations, disarticulating them from their earlier Cold War allegiances, and to promote human rights, democracy, cultural diversity, security, and capital investment within this putative community of nations. As such, however, the gathering of former imperial powers and their colonies creates the potential for post-colonial and neocolonial flare-ups, the Seventeenth Summit providing a dramatic example. On November 10, the final day of the summit, a heated discussion concerning neoliberalism, privatization, and Spanish ownership of Latin American enterprises took place between the Bolivian president, Evo Morales, and the current prime minister of Spain, José Luis Rodríguez Zapatero.[59] Chávez followed this discussion by escalating tensions, in response to Zapatero's efforts to downplay these economic realities. Chávez repeatedly accused the former prime minister of Spain, the conservative José Maria

Aznar, of being a "fascist" and accused him of having supported the failed coup attempt against Chávez in 2002. Reacting viscerally to these remarks, Zapatero then interrupted Chávez, demanding that Aznar, as an elected official, should be treated with respect, even if they were "at antipodes in their [respective] ideological positions," and noted that he himself was very distant from Aznar's positions. Attempting to continue with his provocative remarks, Chávez attempted to reinsert himself into the conversation while Zapatero was still speaking, and protesting this counter-interruption, King Juan Carlos I of Spain uttered five words that became ubiquitous throughout the Spanish-speaking world: "¿Por qué no te callas?" ("Why don't you shut up?").[60]

As an awkward moment during an official discussion of present-day economic and political realities—one that dramatized the disparities within the "Ibero-American community"—it would have been quickly forgotten had it not been recorded and then seized upon by the Spanish-language media. First reported in the Spanish-language press and on Spanish television, the video clip of the exchange made its way to YouTube and became immensely popular, being viewed millions of times (in various versions) and thereby quickly becoming an Internet hit.[61] Commentary multiplied on the Internet through official media sources, blogs, and chat room conversations—one measure of this would be the 655,000 Web hits generated by Google or the 610 YouTube videos as early as 14 November 2007.[62] Soon, individuals across the Spanish-speaking world began producing cultural forms and saleable merchandise based on the phrase, including video commentaries, ringtones, T-shirts, and other assorted products. For example, Jesse Bravo of Big Springs, Texas printed the phrase on T-shirts with an image of King Juan Carlos wearing a jester-like crown and pointing at (indeed, interpellating) the viewer in the manner of Uncle Sam in James Flagg's famous World War I army recruitment poster, "I Want You." Gaining free advertising in an article by Simon Romero in the *New York Times*, Bravo sold the shirts globally and soon became a millionaire.[63]

As the phenomenon grew, the global Spanish-language press continued to report on and refer to the incident, with aftershocks lasting months after the initial tremor. But the media form most closely associated with the catchphrase, one might argue, is the ringtone. In a way, the immediate ringtonization of the event was unsurprising and even overdetermined: curiously, on video recordings of the event one can hear the "Nokia Tune" ringtone go off in the background 13–14 seconds before the king makes his infamous remark. Ringtones were also the most lucrative commodities emerging out of the Chávez-Carlos conflict; although the figures are unclear as to whether they are describing actual estimated sales or potential sales based on numbers of downloads, the Spanish newspaper *El País* claimed that "¿Por qué no te callas?" (PQNTC) ringtones were worth at least $2.2 million (1.5 million euros) by mid November.[64] Whatever the economic reality may be, there appears to have been a divide between ringtones sold for profit by companies like Jamba Spain, on the one hand, and ringtones that originally circulated freely on the Web, on the other. In the case of the first group, there was some concern that for-sale ringtones using the king's voice would actually violate his own copyright, and in this case, ringtone companies reportedly "avoided legal problems concerning breach of the king's image rights by using an actor to voice the line," according to the BBC.[65] It is difficult, though, to determine whether this is in fact true or not: the Jamba Spain website (jamba.es) as of 7 February 2009 still sold a number of different PQNTC ringtones,[66] including a number of them that appear to take actual recordings of the argument at the summit and speed them up slightly, giving them a "chipmunk" camouflage.[67] The site also boasted a ringtone by the group La Vaca Loca (The Crazy Cow)—perhaps inspired by Crazy Frog—set to a dance beat, in which the words are clearly enunciated by a performer.

As to the free PQNTC ringtones found on the Internet, Spaniards and other Spanish-speaking Web surfers gained access to them in numerous ways—through the plethora of blogs commenting on the phenomenon and providing sonic examples and links to the ringtones; through chat sites like Vagos.es, at which dozens of individuals posted MP3 files that could easily be converted into ringtones; and even through online news media sources providing access to various video clips that themselves could be turned into ringtones.[68] And whereas on the blogs and news sources, the files were embedded into the sites themselves, the chat rooms used file-sharing programs like rapidshare.com and megaupload.com for distributing ringtones. Although many different versions of the ringtone appear to have been popular, one in particular appears to have received a great deal of attention. (It is transcribed in figure 6.2.)

Somewhat less elegantly composed than the Buking ringtone, this PQNTC ringtone remix nonetheless shares some of the earlier ringtone's features. In particular, both combine recordings of politically charged speech with dance-music remixes, although the Buking ringtone uses a hip-hop track ("In da Club"), whereas the backing track here is a rather minimal techno arrangement (at 135 BPM). Both ringtones begin and end with a brief statement of a single textural element, framing the presence of the groove proper—in this ringtone, the element is King Juan Carlos' voice,

Figure 6.2
The "¿Por qué no te callas?" ringtone.

processed at the beginning to sound like a conventional rapid-fire snare drum fill on the /p/ of "por" (though perhaps also evoking a machine-gun or even a ringing telephone) and making a complete statement of "¿Por qué no te callas?" at the ending. Otherwise, however, the affective and communicative worlds of the ringtones are strikingly different. Whereas the Buking ringtone (and Hello Garci ringtones generally) emphasized the telephonic-conversational nature of the Arroyo quotation itself, leading to a high degree of self-referentiality within the ringtone, this PQNTC remix ringtone instead treats King Juan Carlos' voice in a direct, even assaultive manner, emphasizing the word "callas" (shut up) in particular (mm. 2–3) and then creating a call-and-response effect emphasizing the question "por qué" (why), which then receives the harsh answer "callas" (mm. 6–7). Hence, treated in an aggressive manner, the PQNTC ringtone overdetermines its gendering as male (male voice and "masculine" treatment), in contrast to the conversational and arguably feminine-gendered Buking ringtone (both through Arroyo's voice and through the telephone's evoking an office or even domestic space that could also be seen as feminine), a sense perhaps reinforced by the relatively relaxed groove (at least, for the context of this comparison). The affective intensity of the PQNTC remix ringtone is even augmented halfway through the track, as a result of the rhythmic break in the groove (m. 5) and added low-synth layer (mm. 6–8). One gets the sense, in the main, that while Arroyo was the butt of a (political) joke in the Hello Garci ringtones, Carlos was instead the mouthpiece for aggression directed at Hugo Chávez.

This supposition is borne out by an examination of the other ringtones and related digital miscellany associated with the "¿Por qué no te callas?" phenomenon. Perhaps chief among the offenders is Oliver Urdaneta, a Venezuelan singer-songwriter in his mid thirties whose work, according to a Wikipedia biography that he may well have written, has had a great deal of international success in performing the music for a *telenovela* on the decommissioned television channel Radio Caracas Televisión titled "A calzón quitao" (The Naked Truth).[69] Whatever the status of that putative success, his much greater notoriety owes to a parody-song composition "¿Por qué no te callas?" based on the song "Que Viva España" (Long Live Spain) in the *pasodoble* style (a two-step march associated with bullfighting), originally made famous in 1973 by the late-Franco-era singer, film star, and cultural icon Manolo Escobar.[70] Starting out with the inflammatory statement "Among parrots, petty thieves and filthy squalor / In Barinas a talking monkey was born" (Entre loros, ladroncito y porquería / Nació en Barinas un mico hablador), the song goes on to insult Chávez by using the racially charged epithet "mico" (monkey), cursorily describes King Juan Carlos' outburst as a scolding ("regaño"), and celebrates, as the statement of the people, Carlos' phrase—which serves as the song's main refrain (in place of "Que viva España!"). Urdaneta posted a video accompanying his tune on YouTube that reinforced the song's monkey-from-the-jungle stereotype and that as of mid 2011 had been seen more than 3 million times. In particular, the video features numerous images of monkeys and gorillas, some with red berets and all of which are clearly intended to portray Chávez.[71] Although the racial logics of this video and the political context are transparent, it is worth noting that another YouTube posting of the same video under the title "Racism on YouTube, Por que no te callas (cancion)" provoked heated responses claiming that the English-speaking poster, a "jimmart32," completely misunderstood the nature of Spanish "irony and sarcasm."[72] Fodder for such defensive claims might be supported by the fact that the video itself acknowledges (and lampoons) criticisms of its own racism, for instance, in a murder/betrayal scene parodying the "This is Sparta!" meme from the film *300* (2006), with the original film clip here featuring the collaged heads of Carlos (as Leonidas, King of Sparta) and Chávez (as the ostensibly Moorish Persian messenger, played by the Ghanaian-English actor Peter Mensah).[73] Chávez's protests that his impending murder is racist—a possible reference to past and possible coup attempts against him—are met with the king saying "¿Por qué no te callas?" and then giving Chávez the boot while uttering, roughly, "shut your deformed mouth [or] I'll smack you!" (¡Cierra el boquino, hostia ya!). Combining the crude humor of the video and lyrics with the pro-Spanish nationalism (and fascist overtones) of "Que viva España!" Urdaneta's effort seems tailored perfectly to the demographic of the ethnically and ideologically "European" anti-Chávez opposition in Venezuela (see below).

Other PQNTC ringtones and videos include the more generic, if distinctly video-gamish techno/trance-style remix with King Juan Carlos' voice by DJ Pichu—who elsewhere boasts being one of the few "raperos" from La Gomera (of the Canary Islands)[74]—whose more amusing aspects include a helium-voiced advertisement for himself and beginning and ending with the Windows XP startup and closing sounds, respectively.[75] A similar, but more intense remix was produced by DJ Devil, whose YouTube video emphasizes Chávez's devil-like characteristics, at one point displaying an image of the Venezuelan president's face morphed into a devil-face with the numbers "666" appearing on the forehead.[76] Yet another remix, called the "Disco Trance Mix" featured Zapatero's voice saying "Presidente

Hugo Chávez," followed by the king's "¿Por qué no te callas?"—with the two effectively doing a kind of "good cop/bad cop" routine; the minor-mode seriousness of the remix gives the voice samples a certain quasi-cinematic urgency lacking in some of the other remixes. The equally minor-mode-inflected "Reggaeton" Mix, based on the chart-topping Latin American reggaeton hit "Dale Don Dale" by Don Omar, is treated in audio and video form like a stylized *telenovela* dialog between Chávez, Zapatero, and Carlos." Making the remix especially amusing is the fact that the chanted chorus, "Dale Don Dale" (which roughly translates to "Go, Don, Go!") is now reinterpreted. Originally referring to Don Omar himself, through a kind of claim to musical (especially reggaeton) royalty ("Don" roughly meaning "lord"), in the new context the chant arguably refers to the king as well, seemingly lending him the support of a male reggaeton crew and thus framing the dialog as one in favor of the king. To reiterate, were we to take away one point from the meanings of the ringtones and their popularity, it would be that there is widespread, but by no means universal, resentment against and annoyance with Chávez in the Spanish-language world—a fact quickly confirmed by the numerous blog comments accompanying the online links to the videos and ringtones.[78]

Wherefore all the uproar surrounding *Presidente* Chávez? The answer to this question is complex and cannot be fully addressed here. Instead, we might make four points on the events in question.

First, noting that a large portion of the mainstream Spanish-language global media was very critical of Chávez's behavior, we might first emphasize that attacks on this behavior are largely symptomatic of their non-acceptance of his political project of Bolivarian revolution and twenty-first century socialism.[79] In this regard, the comments of the blogger Okrim Al Qasal are quite apt:

> When a leader, in this case Chávez, strays even minimally from the script of the puppets that appear on the screens of CNN, with their language of pseudo-tolerance of the "I'll respect you if you stimulate free markets, and if not you are a tyrant dictator"-type, then controversy erupts. If someone thinks that the King told Chávez to shut up only because of his correct and fair description of [the People's Party's] Aznar, then he is wrong. The Bolivarian Revolution and the reform proposals are touching upon very, very strategic interests. You can smell it in the air.[80]

For his part, Chávez has played up his own combativeness, at points matter-of-factly suggesting that diplomatic and economic relations with Spain could be harmed, rhetorically asking the king if he signed off on the 2002 coup attempt, and generally drawing on a political practice of the insult characteristic of Venezuelan politics but not necessarily the rest of the Spanish-speaking world.[81] This disjunction in political-social mores is an advantage that he exploits in the theater of international politics.

Second, although some responses to the diplomatic row explicitly criticized Chávez's politics and favored King Juan Carlos' comments as a rejection of that politics, a large swath of the international media response to the event offered its criticisms in a more muted way. In particular, drawing on a peculiarly liberal logic of legal form and geopolitical convention, many of the criticisms of Chávez emphasized the degree to which he violated all manner of diplomatic, political, and social-conventional norms.[82] Of course, these formalistic critiques are not proffered with any sense of symmetry—a fact that unmasks their putative objectivity. One can see this, for example, in the way that Zapatero's interruption of Chávez garnered much less attention than Chávez's subsequent interruption of Zapatero's prior interruption.[83]

Third, the event pointed to contemporaneous developments within the Venezuelan mediascape that have proven both alarming to Spanish-language media-capital in Latin America and supplied yet another pretext for criticizing Chávez. In the aftermath of the 2002 coup attempt, the alleged complicity of specific media organs like Radio Caracas Televisión in the coup itself and the 2002–2003 oil strike led the Chávez government to undertake various new media regulations and interventions, culminating in the non-renewal of the station's broadcasting license (then relegating it to cable and satellite distribution only) in May 2007, and to further promote politically neutral or state-run media outlets.[84] The last of these have been exemplified both by state-imposed broadcasts or *cadenas* ("chains") that often feature presidential speeches and by Chávez's own television show, *Aló Presidente* (1999–present), a strange and unprecedented project by any measure. The show features the president himself speaking on various subjects, making impromptu policy, singing songs, interviewing citizens, receiving screened telephone calls, introducing media clips, etc., in an unscripted and unpredictable format broadcast on Sundays and lasting between four and eight hours per show. Although reception of *Aló Presidente* has been predictably mixed, the show is successful to the extent that it has become essential viewing for journalists and pro- and anti-Chávez interests, given the spontaneous decision making it documents; at the same time, its host's sheer garrulity could easily be seen as both tiresome and trumpeting victory in the president's ongoing battle with corporate national media. Both the *cadenas* and the show provide specific targets for desires to silence Chávez, and perhaps not only from his detractors.[85]

As can be seen above, the critiques of Chávez, as opposed to the critiques of and meditations on the behavior of the king largely originating in Spain, took on an especially polarized cast within Latin America and Venezuela. In part, this can be understood through my fourth point: the particular materialist and political-economic histories of Spain's relationship to Latin America, both within the last 20–30 years and as part of the longer history of Spanish imperialism and colonialism, were seen as relevant frames for the media event in Santiago. Regarding the shorter time frame, the significance of the previously mentioned recent history of privatization in Latin America and the role of Spanish multinational firms in buying up privatized public goods and services must be reiterated here. In the course of the 1980s and the 1990s, Spanish foreign direct investment, most of it in Latin America, increased from $1.8 billion in 1980 to a peak of $92 billion in 2000 before tapering off somewhat starting in 2001, perhaps on account of the crash of the Argentine economy and the global economic slowdown following the deflation of the US tech-speculation bubble.[86] Spanish firms invested more than $30 billion in Argentina alone during this period, and by 2000, the two largest banks in Spain had gained control over 20 percent of banking in the country.[87] By the same time, Spanish interests had secured majority ownership of several energy and telecommunications firms, including the three largest firms in the Chilean stock market.[88] The degree of Spanish control over the Latin American economy, unprecedented since colonial independence, was a direct result of the IMF "structural adjustment" policies that leaders of indebted developing nations agreed to in order to secure loans from the Fund, and the results of these privatizations in general were rather mixed, in many cases resulting in significantly higher prices and decreased access to services that were delivered more poorly than before, as well as triggering both severe impoverishment and mass anti-privatization organizing—with the Bolivian "water wars" and "gas war" serving as examples of the latter.[89] For this reason, a group of Latin American leaders at the 2007 Ibero-American Summit, including Evo Morales of Bolivia, Carlos Lage of Cuba, Daniel Ortega of Nicaragua, and Hugo Chávez of Venezuela all "called for the establishment of a new economic model as an alternative to the brutal neoliberalism that has ravaged the continent since the 1980s." Evo Morales himself stated at the summit that "basic services like potable water, sewer systems, telephone or energy cannot continue being a private business."[90] Also, Ortega, in particular, complained about a "Spanish utility with a virtual monopoly on power distribution in Nicaragua."[91] The seriousness of these conditions continues to be underestimated by liberal-centrist writers like Simon Romero, Caracas correspondent for the *New York Times*, who assessed the complaints as follows: "Spain's investment in the region has slowed drastically, a result of a surge in trade within the region, new investment from China and India, and complaints against some Spanish companies. So it has become easy for a populist to make points with poor voters by bashing Spain."[92] A journalist whose opinions, particularly on Chávez, conveniently coincide with the mainline positions of the US Department of State, Romero cynically misreads the earnest hopes of progressive political leaders to create possibilities for change—possibilities, they hoped, would be received sympathetically by the socialist Zapatero. As it turned out, however, the Spanish prime minister responded with feel-good abstractions and didactic history lessons about the human condition, ideas and civilization, poverty, racism, absolutism, and so on while dismissing nationalization as a useful political-economic strategy ("nationalization is not an end")—a move one might read as a post-leftist, third-way-esque compromise with capital or, more cynically, as an implicit nationalist defense of Spanish multinationals.[93]

And yet, the ringtones and popular-cultural/new-media forms of expression discussed above did not address the problem at this level of economic reality, but rather from the perspective of a broader historical framework, one that pits Europeans and Latin American whites who view themselves as European against indigenous, African-diasporic, and *mestizo* populations, a racial dynamic that by and large is reproduced in the class structure of Latin American countries. It is at this level of analysis that notions of "imperial nostalgia" come into play, with the confrontations of colonizer and colonized bubbling just below the surface during the Summit and erupting with full force into the (digital) public sphere as a result of the king's outburst. One could see the potential conflict at the summit itself, certainly, with a telling moment being Zapatero's smug comment that "Even Karl Marx was a European," the punctuation mark to a broader excursus on the role of European history and the French Revolution in the development of ideas beneficial to humanity.[94] These racial dynamics certainly were present in both Spain and in Latin America. In the former context, we might note the importance of Latin American immigration to Spain, which increasingly beginning in the 1990s became an alternate destination to the United States for Latin Americans escaping hardship and seeking economic opportunities. By 2005, more than a million Latin

American immigrants—hundreds of thousands of Ecuadorians, Colombians, and, because of the 2001 economic crash, Argentines—moved to Spain, with immigration clampdown by the Bush administration after 11 September 2001 exacerbating the tendency.[95] And though there have been relatively few pronounced incidences of violence against immigrants in Spain, "'low-level' racism" remains a reality on the Iberian Peninsula.[96]

In Venezuela itself, as elsewhere in Latin America, the racialized class divide is stark, particularly in the wake of the mass impoverishment of the population during the neoliberal reforms of the 1980s and the 1990s. With the largely white elite sequestered in particular enclaves, protected by gated communities, security fences, and private bodyguards, they would appear to lead a rather different life from the Venezuelan poor—who, in Caracas, populate the hillside shanty towns or *ranchos* that surround the city. And yet, one might immediately assume that cell phone access and ringtone use was, *ipso facto*, limited to the elite, largely white opposition to Chávez. In fact, this is not the case, although the PQNTC phenomenon did translate into Venezuela in specific ways, along class and race lines. Most important, the altercation at the summit in November was parlayed into mobilizing the two sides in December 2007. On the opposition side, the slogan "¿Por qué no te callas?" became one of the major rallying cries during the referendum campaign, with the ringtones described above turning out to be ubiquitous on opposition phones, at rallies, etc. Sometimes, groups of students from particular universities—such as one group from the Caracas branch of Andrés Bello Catholic University, a private university—would download and use the PQNTC ringtones in order to signify their political allegiances. According to a philosophy major named Laura Solórzano, "It's a form of protest. It's something that a lot of people would like to tell the president. Now, whenever we call each other, that's what we hear."[97]

On the other side of the coin, a young pro-Bolivarian Venezuelan, Elena Pereira, alerted me to a series of ringtones available on the Venezuelan government's own website, which included folk songs by the Venezuelan Communist activist and protest singer Alí Primera, a Venezuelan representative of the "Nueva canción" movement; other ringtones at the website include several by the Venezuelan artists' collective Lloviznando Cantos, who are attempting to revive the "Nueva canción," and an Afro-Cuban son/salsa-style number by Grupo Madera accompanying the words "¡Uh! ¡Ah! ¡Chávez no se va!" (roughly, "Chávez is not going away!" originating in the 2002 anti-coup protests), which may have been the most popular of the bunch.[98] Ease of file sharing and exchange facilitated the dissemination of the ringtones, with Bluetooth technology playing a major role in that process. Pereira also noted that the ringtones were available when the ringtone fad came to Venezuela (around 2005–06) and that they were widely used during the presidential recall referendum of 2006, which was won handily by Chávez (62.9 percent).[99]

These ringtone postings seem generally in line with the Chávez government's proactive but open-ended approach to cultural policy—in this sense, making a break with efforts to control cultural production characteristic of self-proclaimed socialist countries (though perhaps less so in Latin America—with Cuba being an intermediate example). On the one hand, as Elena Pereira noted, "Many artists and singers are openly criticizing Chávez and the Government does not pay attention to them," but on the other hand, the government actively promotes independent media through the Law on Social Responsibility in Television and Radio and pro-government cultural producers through the Ministry of Popular Power for Culture.[100] Additionally, successful cultural programs that began prior to the Chávez administration, including "El Sistema," originated by the musician and economist José Antonio Abreu and source of the acclaimed Simón Bolivar Symphony Orchestra—conducted by Gustavo Dudamel since 1999—are being further promoted and funded by Chávez.[101]

That the same pro-Chávez ringtones were used again during the reform referendum in 2007 might, however, be one small indicator of the lack of organization of the Bolivarians' campaign. And it can certainly be said that the PQNTC phenomenon and its ringtones gave the opposition a certain boost of morale, even if its absolute numbers did not increase by any substantial margin. Whatever the case may be, the results were disappointing for the government: on 2 December 2007, the reforms were defeated narrowly, 50.7 percent to 49.3 percent, with a large percentage (44 percent) of the electorate abstaining from the elections—the vast majority of which were Chávez supporters in past campaigns.[102] Significantly, and in the face of the US-dominated media that claims Chávez is an anti-democratic dictator, the government accepted the results and went through a process of self-critique, examining the causes of their failure. With the subsequent investment of resources in the United Socialist Party of Venezuela, though, the Bolivarians have taken small steps toward weaning the revolution from its excessive reliance on a single individual, which is particularly urgent given Chávez's recent battle with cancer.[103] All this notwithstanding, Chávez appears to have gotten the last laugh—on 15 February 2009 he won a new referendum vote limited to abolishing term limits with 54 percent of the votes (and about 70 percent of the electorate voting).[104]

A Wireless Politics?

The politics of ringtone use—itself a subset of the wider problematic of a *wireless* or *mobile phone politics*—is a complex phenomenon, one whose outlines are only partly discernible in the foregoing case studies. We might formulate an assessment by framing the problem in two parts: the successes and failures of ringtones within particular social movements, and the role of ringtones in the successes and failures of particular social movements. Considering the first part in light of our case studies, one element of a successful political ringtone becomes remarkably clear: its ability to articulate critical aspects of the sociopolitical situation or crisis in question with the sociocultural (indeed, aesthetic) forms of expression particular to the constituency responding to that situation or crisis. In other words, the ringtone has to encapsulate the political issues of the moment in a way that makes sense and appeals to those who are interested in doing something about those issues. Of the "Hello Garci" ringtones, Gundersen's and other US-based political ringtones, and the pro- and anti-Chávez ringtones, those that gained traction within their political contexts (especially the "Hello Garci," "¿Por qué no te callas?" and Bolivarian ringtones) engaged directly with the political constituency's *aesthetic* forms. In the "Garci" case, American and Western popular musics and, to a lesser extent, Filipino popular musics, made their way successfully into ringtones, the 50 Cent hit "In da Club" being particularly effective and circulating widely. With the PQNTC ringtones, various popular music genres within the Spanish-speaking world, including reggaeton and especially electronic dance musics (techno, trance, etc.), as well as the kitschy use of *pasodoble* in the Urdaneta parody-song, pointed to the imaginative affiliations with a largely white European popular culture among the relatively affluent Venezuelan youth using the ringtones. With the forms of politicized popular culture used in the Bolivarian ringtones, especially with success in the 2006 recall referendum, we find that the aesthetics of Cuban musics and "Nuevo canción" central to the Latin American left since the 1950s were reproduced and articulated with the new social movement (particularly clearly with the "¡Uh! ¡Ah! ¡Chávez no se va!" song and ringtone). Gundersen's "City of New Orleans" ringtone, on the other hand, failed to connect to the popular musical forms that might have resonated with a large population, in part simply by choosing a Baby Boomer-era "folk" song that, despite its many merits, both did not appeal *en masse* to cell-phone-bearing constituencies and did not translate well in the new, limited-fidelity medium (at the time). And here, we might legitimately ask who these constituencies are. Although one might be able to extrapolate it from the above examples, the work of Manuel Castells, Mireia Fernández-Ardèvol, Jack Linchuan Qiu, and Araba Sey emphasizes clearly that global mobile communications cultures are predominantly *youth cultures*, whose interests in mobile "entertainment, games, and media" are significantly greater than corresponding interests among older mobile phone users.[105]

Evaluating the role of ringtones within these social movements, however, is much more difficult. But, we might begin by noting that the new kinds of "networked" social movement activity that function through mobile communications do appear to favor younger constituencies and hence overlap significantly with the "mobile youth culture" described by Castells et al. Indeed, it is difficult to separate the new technological media, which are assimilated by younger users more thoroughly than by older users, from musical styles (such as hip-hop, reggaeton, and electronic dance music) whose very production practices are similarly dependent on widespread microelectronics technologies. (Indeed, the presence of hip-hop in the Buking ringtone and electronic dance music and reggaeton in the PQNTC ringtones would seem to mark those organizing constituencies as particularly youth-driven—"youth" being broadly understood, extending even to young professionals in their thirties.) That said, in none of the cases above did ringtones significantly serve to mobilize new constituencies or produce spontaneous mass gatherings in analogous ways to the celebrated uses of text messaging in the Philippines, Spain, and elsewhere—although these earlier uses arguably paved the way for later politicized ringtone use.[106] Although the ringtones themselves were often part of wider efforts to disseminate political messages and information through alternative, independent, and relatively autonomous means—this is particularly the case in the "Hello Garci" ringtones—for the most part the ringtones seemed to serve as forms of individual, group, and mass-popular expression. In a sense, the problem would merely seem to return us to instrumentalist arguments over the role of music within social movements regarding its effectiveness in increasing and maintaining participation, although, following Eyerman and Jamison, the widespread use of political ringtones clearly demonstrates that the social movements of the new millennium are both meaningful "spaces of cultural growth and experimentation" and that they, like earlier social movements, rely on the "mobilization of tradition" as a part of their overall dynamics.[107]

That said, the PQNTC case may be viewed as the (temporary) endpoint of a trajectory in the brief history of the political ringtone, which is now marked by two developments. First, the political ringtone, which was once free of charge, soon became, at least in part, a lucrative commodity—a move reinforced by the appearance of remarkably successful political ringtones within the developed world (Spain). Arguably, one might claim that this development signals the further incorporation of cell phone politics back into the global economy and, hence, as part of what Guy Debord ruefully described in his last years as the "integrated spectacle."[108] Second, the Spanish-language case demonstrates the extreme fungibility of political sound file ringtones, to the point that they may be exchanged throughout an entire global circuit—in this case, the circuit bound both by a globalized, mostly conglomerate-owned news media and by an internationally accessible online media system (the Internet). Certainly, the situation was a relatively rare one, originating as it did in a geopolitical context relevant to a wide array of Spanish-speaking nationalities and thus being best understood within an international conjuncture instead of national ones alone. Nonetheless, that such transnationalization became articulated with an essentially liberal-bourgeois campaign in one particular state perhaps both confirms Rafael's assessment of cell phone politics generally and attests to the extreme spectacularization of global politics today. In light of such speculations, particularly when compounded with the growing obsolescence of the ringtone itself, we may well wonder how much of a future remains for the political ringtone.

What, then, will become of the political voice-remix ringtone? On the one hand, as part of the extraordinary growth of global cell phone access, political organizing and mobilization efforts will undoubtedly continue to involve cell phones and perhaps ringtones of this sort in the short to medium term, gradually invalidating Rafael's claims of its class and political specificity.[109] On the other hand, countervailing pressures against political ringtone use can be found in the multi-billion-dollar global ringtone industry's decline, as discussed in chapter 1. The ringtones discussed above at once point to both tendencies through their transnational distribution and considerable but inconsistent capitalization. The political voice-remix ringtone is, of course, not only a global form in itself, as mentioned above, but also one variant of the broader global practice of sample-based digital music, and on this score, we might take counsel from Ian Baucom, whose comparative analysis of the global forms of world literature and the corporation draws its cyclical logic from the work of the world-systems theorist Giovanni Arrighi. Noting the ways in which the joint-stock chartered company of the seventeenth and eighteenth centuries prefigures the transnational corporation of the twentieth and twenty-first, or the ways in which the conservative and Romantic historical particularism of the late eighteenth and early nineteenth centuries is recollected in anti-poststructuralist efforts at literary-theoretical historicism, Baucom succinctly describes the logic at work in capitalism's cycles of accumulation: "expansion contracts, contraction enriches, and enrichment haunts."[110] Whether or not we identify similar parallels between the public-sphere-generating print explosion of the eighteenth century, including the often, but by no means exclusively, explicitly political broadsheets or broadsides of the period, on the one hand, and the present abundance of digital information commodities, including ringtones both explicitly political and not, on the other, it is surprising the degree to which Baucom's formula elucidates the fiscal and cultural fortunes of the ringtone, which, having expanded, is now steadily moving out of the enriching contraction phase. How will the ringtone, once having disappeared, haunt *us* in global capitalist modernity? Although the much-publicized fantasies of technological "convergence" might provide us with some clues, as might Baucom's suggestion that dead genres haunt living ones, perhaps the most politically valuable act would be to abstain from prognostication, leaving it to the dubious interests of the self-appointed gurus of the mobile entertainment and digital industries.

7 A Spectrum of Forms: The Aesthetic Logic of Original Sound-File Ringtone Composition

I basically use my ears, and my creativity and do it over and over again.
—Brice Salek, Ringtone Records[1]

What does it mean to compose a ringtone? The question is not trivial, for there are few universals associated with mobile phone ringtones themselves now that they are digital sound files (usually MP3s) often of any length. (They are, nonetheless, conventionally around 15–30 seconds long and often subject to length restrictions. Apple's iPhone formal maximum for ringtones is 40 seconds, but this restriction is easily bypassed.[2]) Certainly, original monophonic and polyphonic ringtones were composed (semi-) anonymously as part of the labor of ringtone production (as discussed in chapter 2), and labels such as the German company GoFresh produced original polyphonic ringtones for sale outside of established ringtone-industry contexts (i.e., wireless providers and mobile entertainment firms selling popular song-based ringtones).[3] Moreover, experimental algorithmic-compositional projects for creating new ringtones existed in the polyphonic period: perhaps the best-known is Wolfram*Tones*, the MIDI file generator inspired by the work of Stephen Wolfram and produced by algorithms realized using the software package Mathematica.[4]

But the creation of new ringtones took a qualitative and quantitative leap forward as a result of the possibility of placing potentially *any* sound file on a cell phone, to be played back on the phone's ambient speaker, which is usually used for ringtones and other alert-signal sounds, as well as its speakerphone function.[5] As I argued in chapter 4, precisely on account of their simplicity or primitiveness, depending on your viewpoint, the earlier forms of the ringtone—especially the monophonic ringtone—provided the most distinctive sonic material that artists and composers could employ in the context of new concert music compositions, interactive multimedia performances, or art installations. The dialectical closure of this creative possibility was signaled by the emergence of the sound-file ringtone—which could either merely mimic the older phone models through prerecorded sounds or transform mobile phones into mobile audio playback and multimedia devices, thus rendering their ringtone-specific features less central.[6] However, the new technological moment of the sound-file ringtone also resulted in a dialectical *opening* of new possibilities for original ringtone composition, with original ringtone albums increasing in number and availability from the early 2000s, just as sound-file ringtones were starting to appear on higher-end phones: in particular, as the ringtone became less sonically distinctive, it transformed into a more capacious vehicle for a wide variety of sounds. Of course, new constraints imposed themselves, the primary ones being the influence of preexisting music and sound recording genres (including various music genres—songs, long-form instrumental compositions, electroacoustic music, etc.—and comedy recordings), which were absorbed into the new medium in a partly parasitic manner; the technological constraints of duration during the early sound-file ringtone moment (roughly between 2003 and 2007) and the resulting temporal limits residually remaining as a convention of the "ringtone genre"; and the constraints imposed as a result of scenarios for the use of mobile devices—typically, that ringtones can expect to be interrupted, and that they are usually reproduced on relatively low-fidelity ambient cell phone speakers (improvements in the quality of speakers in the 2000s notwithstanding).[7] In the following discussion, I first review some of the aesthetic qualities of the sound-file ringtones treated in earlier chapters of this book, developing some categories to discuss these. Then, I examine a body of originally composed ringtones that add to these categories, focusing mainly on albums by Ringtone Records. My hope is to provide some preliminary terminology for discussing the musical-formal aspects of ringtones; for, despite ringtones being of such short duration, it should become clear that their formal properties vary widely.

Three Tendencies: Excerpted Songs, Phonomontage, and Original For-Sale Ringtones

A few different aesthetic possibilities emerged within the new dispensation of the sound-file ringtone, some of which I have already discussed. At least three dominant tendencies in sound-file ringtone creation or composition can be noted. In the first, the vast majority of ringtones on the market are adaptations of preexisting popular songs and are intended to provide a (sometimes unwelcome) representation of music to phone owners and passers-by, as well as, in Christian Licoppe's terms, a familiar "treat" that compensates for the mobile phone's imposition of perpetual availability characteristic of the present-day ideological-labor formation that Luc Boltanski and Ève Chiapello term "the projective city."[8] The primary aesthetic logic at work in the creation of such ringtones is one of *excerpting*, as discussed in chapter 2. The labor involved in creating such ringtones is predominantly a matter of *hook-finding*, which often amounts to excerpting a song's chorus, but not always (for example, the ringtone for Deep Purple's "Smoke on the Water" would instead focus on the distinctive guitar riff that starts the song).[9] As a representation of other music, this type of ringtone does not constitute a complete, self-standing statement,[10] and so in some sense is open-ended, leaving the listener hanging, as it were; only its indexical function as a marker of taste or fleeting musical interest causes the ringtone not to sound incomplete, in the sense that, once recognized, the labor of identification and indexing is completed.

A second tendency, found in originally composed "political" ringtones—i.e., articulated with social movements—frequently seem reminiscent of the traditions of committed visual art (especially photomontage) by superimposing recordings of political figures' speech in combination with sound effects and preexisting songs. Exemplifying a crude mash-up/sample/quotation-based practice, one may characterize such work as *phonomontage*, borrowing the term from the sound poet Larry Wendt and the film scholar John MacKay.[11] The examples discussed in chapter 6, the "Hello Garci" ringtone and the "¿Por qué no te callas?" ringtone, are paradigmatic in form and function: these ringtones offer amusing commentary on the political imbroglios from which they emerged and feature musically closed, coherent statements with clear beginnings, middles, and ends.[12] The beginnings of both present some kind of introductory sound alone (a ringing phone or an altered voice), the middles in both cases essentially coincide with most of the grooves of the instrumental musical material, and the endings exhibit a closing gesture characteristic of certain kinds of pop cadence formulas (in these cases, instrumental breaks after final groove articulations, sometimes reinforced by pitch arrivals, which are followed by unaccompanied sounds filling in the empty sonic space). Heard in both unmarked public situations as part of an expression of political identity/position or amid collective playback at a political rally, such ringtones are intended to be heard from beginning to end and constitute more or less discrete, closed statements.

The third tendency concerns the creation of original audio file ringtones composed for sale (as opposed to the free distribution characteristic of most political ringtones or to the for-sale ringtones derived from popular songs). With the expansion of the ringtone market in the mid 2000s, a number of artists sought to exploit the possibility of building a niche market for originally composed sound-file ringtones. Although composing music based on 30-second forms may seem a quixotic task, in one sense it participates in the long history of compositional miniatures within Western classical and popular music. The former has a long history, although one celebrated point of origin might be found in various nineteenth-century art musical tendencies (bagatelles, character pieces, art songs, preludes, and so on), the aesthetics of the Romantic fragment and celebration of "open" forms being crucial here, even if closure was attained to widely varying degrees in such pieces.[13] The formal dynamics of musical flow and structure were common to longer and shorter compositions, and thus even in the shortest works, which Douglas Green includes in his category of *one-part forms*, a single musical phrase or pair of phrases—musical units that are found in symphonies and single-page songs alike—would necessarily constitute the entirety of such a work.[14] But with the arrival of sound recording, the locus of miniaturization shifted from that of the space on a printed page to absolute temporal durations, the 78-rpm record being about three minutes (or three and a half, depending on the record size, groove spacing, and other factors) in the first decades of the twentieth century and that duration determining the length of most popular songs and even some classical compositions composed for records.[15] As the storage capacity of sound recordings increased—first, in the years shortly after World War II, with the long-playing record (roughly 45 minutes), then in the early 1980s with the compact disc (80 minutes), and most recently with hard drives that can store days, weeks, months, and even years of recorded music—contrarian tendencies emerged reacting to musical practices exploiting such length. For example, in the late 1970s and the early 1980s punk rock and then hardcore punk's disdain for art and progressive rock's distended songs and albums resulted in an abundance of two-minute and one-minute songs; the derivate "grindcore" genre not long afterwards produced so-called microsongs, the most extreme being Napalm Death's 1.6-second-long "You Suffer" (1987), which apparently holds the world record for shortest recorded song.[16] Similarly, minute-long contemporary classical compositions, particularly in the world of electroacoustic music, have offered opportunities for composers to gain attention in a music market saturated by new offerings; one well-known vehicle for short compositions is the 60×60 project, which began in 2003 and features concerts (and sometimes albums) by sixty composers of works each one minute in length.[17] In contrast to these clearly experimental endeavors, short-form composition appears in more functional contexts as well: advertising jingles have historically been compressed into standardized spans ranging between two and sixty seconds, the thirty-second ad being quite common on television. Jingles, which would fill in the spaces of these ads, could range from short musicalized mentions of a brand to short songs. More recently, sonic brands that draw on the aesthetics and history of ambient music have arguably also compressed musicalized ad time toward the few-seconds end of the spectrum (Brian Eno's work for Microsoft, including his "Windows 95 Sound," is a classic example of sonic branding, as is the Intel sound).[18] A similar and often closely related functionalism can be found in film and television music, with the seconds-long use of short cues, brief theme songs, and bumper music (bumpers being short ads or announcements for a show itself played during transitions between show time and advertisement time).[19]

In light of this cursory history, the ringtone appears as merely one particularly functional variant of a broader trend toward the production of music of short duration. How, then, should we understand the aesthetic dynamics of original ringtones? We might be tempted to take our cues from the ringtone production practices discussed in chapter 2, in which excerpted popular song ringtones are to be understood as formally *open*, whereas the political phonomontage ringtones are formally *closed*, to invoke that standard binary opposition in Western music theory.[20] But upon closer examination the opposition is not entirely adequate for addressing the formal dynamics of original sound-file ringtone composition. If we sublate the dynamic to a higher level of generality, though, we might see that both the phonomontage and hook-excerpt ringtones rely on a certain degree of syntactical definition, with both (usually) working with phrases and/or quadratic syntax or at least two-measure hypermeter, as well as relatively well-defined grooves and musical textures. To reiterate the above in the most obvious terms, both types of examples considered here draw largely on familiar popular musical styles within the West.

The lineage of original ringtone composition, on the other hand, lies on a continuum between those demonstrating the structural and stylistic consistencies described above and those more amorphous in form. In presenting a rough classification of types of musical sound in originally composed ringtones, I will largely be drawing from the catalog of recordings released by Ringtone Records, a small company started by Brice Salek in 2007 in Santa Monica, California. Salek, an artist trained as a designer, hit upon the idea while conversing with friends, trained himself in sound design and electroacoustic composition, pursued the production of ringtones through 2012 with an admirable single-mindedness, and continues to produce recordings to the present day; the current catalog lists more than 30 separate albums. I also consider the Touch Ringtones recording of 2002, a compilation with contributions of a large number of fairly well-known composers, performers, and sound artists (among them Evan Parker, ikue mori, Ryoji Ikeda, fennesz, Phil Niblock, Monika Nyström, Atau Tanaka, and Brandon LaBelle).[21] Six categories that can be divined from an examination of this corpus (and a few other ringtones) are miniature songs, through-composed miniature compositions, aestheticized signals, repetition-variation forms, sound and music fragments (often conceptualist), and text-based or speech-based ringtones.

Miniature Songs

The miniature song is a popular song form, with vocals set against a musical accompaniment or backdrop, spanning roughly 30 seconds or less.

As was mentioned in chapter 6, They Might Be Giants released a series of three ringtones in May 2006, each with a phone-themed concept ("Ring Ring" and "Phone Phone Phone") and one of them additionally referring to the wiretapping of civilians undertaken by the Bush administration ("Call Connected through the NSA"). Although They Might Be Giants were said at the time to have "embraced the ringtone as a stand-alone medium" (whereas other musicians typically turn their existing songs *into* ringtones), they are known for writing short songs—and for their "Dial-a-Song" answering machine of 1983–2006, on which callers could hear a short song (often changed daily) as the answering machine's message.[22] Figure 7.1 shows a transcription of "Phone Phone Phone." It cleverly incorporates something like a phone signal (the three-note riff on the words "phone phone phone") into a very simple melody. Formally it is quite similar to the political ringtones mentioned above—particularly the Gloriagate one (figure 6.1). It begins in G major, without a separate introduction, and produces an eight-bar phrase ending on V, which is repeated; the second time, an extra measure is appended. The second phrase then cadences on I and leads to a small coda featuring hemiola bass and drum patterns and a dissonant, even Bartókian piano riff.[23] Although the proportions are different from those of the Gloriagate ringtone, especially in that the ending of "Phone Phone Phone" is rather distended, the formal structure is quite similar: both are based on a repeated unit (a refrain-like phrase in "Phone Phone Phone," a subphrase in the Gloriagate tone), then

Figure 7.1

They Might Be Giants, "Phone Phone Phone" ringtone (2006). Reproduced by permission of They Might Be Giants.

a cadential or ending figure, after which the musical "work" of the ringtone is done. Thus, upon comparing the miniature song with the longer, standard compound AABA or verse-refrain song forms, one appreciates that in the miniature song each component of the larger form is *not* shrunken down to fit smaller proportions. Rather, the basic unit of form changes to accommodate the scale change. Hence, instead of the form featuring two contrasting sections each including one or two phrases, here the refrain phrase (or a pair of them) *is* the form itself.[24]

Other examples of miniature ringtone songs can be found on three albums on Ringtone Records by Brice Salek titled *Ringtone Songs* (September 2009), *Ringtone Songs, Love and Chords* (December 2010) and *Ringtone Research* (February 2010). On the former two recordings, Salek opted for simple folk-style guitar strumming and an out-of-tune, quasi-Dylanesque drawl, conveying romantic and existential sentiments while at times implicitly referring to the absent presences implied by the telephonic scene itself, as in the amusing, somewhat Lou Reed-ish "Toes Are Getting Cold

Ringtone," the lyrics of which read "My big sweet love / Where you at? / Like I lost my hat / I go soft into sleep, lookin' down below the sheets / I don't see you there, just my toes, and they're gettin' cold." In the third record's songs, Salek uses heavy processing on his voice and instrument, perhaps to hide recording imperfections or to extend the range of possible vocal characters and styles. In the later recordings, the effect is much more neo-psychedelic and twee, with the lyrics generally far more minimal and functionally oriented, as in the "Wake Up Ringtone Song" and the "[illegible] Are Here Ringtone Song." Each of Salek's ringtone songs presents a single, distended musical phrase (or, in the case of the more minimal songs, a phrase fragment) during its 30-second duration.

Miniature Compositions (Through-Composed vs. Textures/Gestures)

Like the miniature ringtone song, the miniature composition is similarly tied—if not as extremely so—to an ethos of closure and completion, even over rather short durations. On the shorter end are those examples on the Touch Ringtones CD that seem to present relatively complete utterances. Consider Oren Ambarchi's "Skadooor," a five-second work that packs in a great deal of textural variety and complexity into its brief duration, ending with a punchy, almost door-slam effect in the right channel close to the end. An Australian of Iraqi-Jewish descent and an experimental electric guitarist and drummer who has released numerous recordings on the Touch and Tzadik labels, Ambarchi adopts a straightforwardly electroacoustic compositional approach, with extreme disjunctions between the left and right channels. The complexity of the parts might lead one to describe the work as "through-composed," in that it doesn't exhibit any clear formal repetition. Admittedly, achieving a sense of closure in five seconds is difficult. Most of the through-composed miniatures on *Touch Ringtones* are at least a little longer, ranging from around ten seconds (Atau Tanaka's "Therering") to about twenty seconds (Andreas Karperyd's "Accelerate"). Other miniature compositions on that recording may either present a single, mostly consistent and even seamlessly repeating texture (e.g., the white-noisy channel oscillations of Cecilia Heisser's 14-second "Molar") or demonstrate a relatively coherent gesture (e.g., Tanaka's "Therering" features a wobbly, almost Xenakis-ish synth or Theremin-like glissandi rising to a shriek, recalling a boiling kettle). There are also longer form tracks on the Touch recording, but because they extend far beyond the average ringtone range—up to three minutes—I'm excluding them from consideration.

In contrast, the Ringtone Records examples of through-composed miniatures reach beyond the aesthetics of electroacoustic music and experimental electronica to include older-fashioned forms. One subtype of these is the *moto perpetuo* ringtone, which usually includes a monophonic (or sometimes two- or three-voice polyphonic), homorhythmic stream of notes that sounds somewhat like a subclass of "effect"-oriented preset ringtones, and are often rather minimalist or sonorist in effect. "Frantic Ringtone" on *Crazy Phone Ringtones* produces a quasi-toccata replete with Bartókian and Ligeti-esque inversional symmetry and chromaticism in ways reminiscent of the vast, if recent, sonic legacy of video game "boss music," featuring the appearance of (and a fight with) a significant enemy at a particular level of play.[25] Another compositional type includes the pastiche composition, such as the "Classical Ringtone" on the album of the same title; in "Classical Ringtone," the angular string line sounds more neoclassical than classical. Also, through-composed electroacoustic efforts are quite common, as in "Tesla Ringtone" (also on *Crazy Phone Ringtones*), which seems to imitate "classic" music from that genre like Varèse's *Poème électronique* and compositions of a similar vintage. (Indeed, in terms of the processing and effects palette in Ringtone Records albums, Salek—being largely self-trained in musical performance, recording, and production, and working on a more limited budget without expensive computing and studio facilities—seems generally inclined toward a kind of "classic" electronic music sound, or the now-straightforward forms of sound synthesis generating the blips and bleeps of much postwar tape music.) The examples here follow nonstandardized syntactical patterns, and so the dynamics of closed and open endings are debatable—as in the case of the Ambarchi composition—and perhaps, in the end, less important. Apparently far more important are the music-generic and stylistic features on which these ringtones draw, which provide sufficient familiarity to situate and cognize these examples as particular kinds of musical utterances.

Aestheticized Signals

In stark contrast to the musically generally closed forms described above—or, if not exactly closed, then tapping into familiar musical genres—perhaps the most distinguishing feature of the ringtone is the fact that at one level it is precisely to be understood not as music, but functional sound. As Christian Licoppe eloquently put it, "The choice of musical ringtones reflects an acute sensitivity to this tension between a ringing sound and music, between its function as an invitation to answer and as an opportunity for sensorial experience."[26] Licoppe situates the old-fashioned phone ringer squarely within the functional, rather than the aesthetic, but it is clear that the old ringers have peculiar sonorous traits—in North America, the standard ringing cadence is called "2-4," or two seconds on, four seconds off—and that these traits could serve as the basis for what we might understand as an *aestheticized signal*; in other words, that the sound that beckons an answer would maintain its general rhythm or periodicity (on the order of several seconds), be framed by some amount of silence, and repeat exactly, but that this rhythm or periodicity may also be shaped and designed for certain aesthetic ends (beauty, complexity, etc.)

Some of the most compelling recordings on the Ringtone Records label are aestheticized signals, such as "Rising Bell Ringtone" on *Ringtone Research*. A delicate, simple monophonic functional-musical utterance, it repeats roughly every seven seconds, so more or less on the scale of the generic phone ringer in North America. Using an E-Phrygian collection, it interleaves an ascending cycle of perfect fourths (E4-A4-D5) with a descending cycle of diatonic thirds (D6-B5-G5), before ending at a convergence point using the two remaining pitches between the final notes, E5 (returning to "tonic") and F5 (one half-step above).[27] (See figure 7.2.) In another example, "Siren Blast Ringtone" on *Phone Ringtones* (February 2010), elegant delay effects are applied to a note sliding upward from E♭5 to F5. In both cases the dynamics of openness and closure do not work in the same way as in the examples in which musical phrases are normative, or even those that hew to more recent music-generic practices (as in electroacoustic music). For, in a sense, the repetition of a figure *guarantees* its role—as a signal to alert a caller—and so whatever is placed within that framework, as long as it is audible, will perform its role. This means that all of the shortest musical figures on the *Touch Ringtones* CD—and there are a number of them that are from one to two seconds long—as well as even some of the longer ones discussed above as miniature

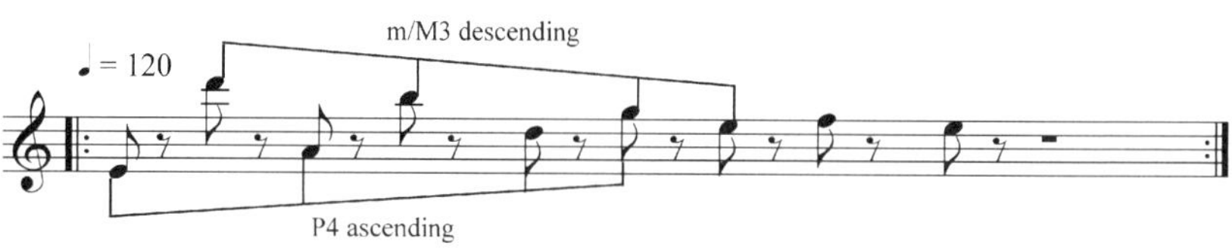

Figure 7.2

Interleaved interval cycles in Brice Salek's "Rising Bell Ringtone," *Ringtone Research* (2010). Reproduced by permission of composer.

compositions could likewise potentially be understood as repeating sounds within a signaling system, particularly if one's phone is properly set up to repeat that signal a sufficient number of times.

Salek draws on a wide range of signal genres to create ringtones of this sort. The two examples above draw on the beeps of electronic music and the sound of sirens, respectively, but several other genres are invoked in the Ringtone Records catalog. For example, *Sonar Ringtones* (2008) is devoted to the imitation and variation of echo/delay-inflected sonar signals, submarine pings, and aquatic sounds. Functional sounds such as Geiger-counter clicks and nuclear-themed quasi-cinematic audio effects such as bomb whistles appear in *Nuclear Ringtones* (2010); sci-fi sound effects are also heard in *Ringtone Spaceship* (2008) and *UFO Ringtones* (2010). Imitation video game sounds are found in *Game Players* (2008) and *Balls of Jesus* (2007), the latter of which also features doorbell sounds and various other signals. And, of course phone ringers themselves are imitated or sampled throughout the catalog, including on *The Phone Ringtones* (2008). In surveying the plethora of aestheticized signals on Ringtone Records, one is struck by the sheer variety of sonic signals that are found in media representations and in the course of everyday life in contemporary modernity. That they are not understood *as* aesthetic attests to the fact that they are not really listened to within their functional contexts, and yet those very contexts create codes for the signals that can, in turn, serve an aesthetic purpose.[28] But insofar as many of the sounds on Ringtones Records albums (particularly the synthesized sounds) are actually produced by similar sound effects, one may come to suspect that, at least in some cases, the effect is driven less by acuteness of listening and imitation and more by aesthetic predilections or limitations. Of course, the notion of the aestheticized signal itself mediates between these two positions.

Repetition-Variation Form

The stark contrast between the miniature compositions and signal-based ringtones hinges on whether literal repetition does not or does feature in them. But many compositions, even if they seem through-composed, feature elements of repetition and recurrence as a matter of course. The issue that appears to be at stake for present purposes is the degree to which the forms of repetition under consideration work in such a way as to signify a "signal" rather than a "piece of music." But perhaps the greatest achievement of the Ringtone Records recordings is the presence of a formal type that seems to blur these two categories together in what I would describe as a kind of "repetition-variation form." Although other, preexisting musical practices like minimalism or terminologies such as "developing variation"[29] may account for the interaction of musical transformation and repetition—with respect to the variability of goal-directedness, the postmodern problematic of "recombinant teleology" may be relevant[30]—the key difference here is that we find a form of sonic practice in which the boundary between functional signal and music is simply not so clear and is made even more unclear precisely by the context of its use as a phone ringtone. The contours of this practice are hard to describe and easily blur into the aestheticized signal, on the one hand, and the through-composed electroacoustic piece, on the other, but its most distinctive feature is the repetition of a signal-like sound, which is then contrasted and complicated by other sonorities, textures or effects that disrupt the possibility of precise, exact repetition. The other important aspect of this formal type is the way in which elements seem to vary and change on roughly the same scale as the six seconds of the 2-4 cadence in the standard North American ringer signal—that is to say, both the repetition of events and the appearance of new events operate at the resolution of a few seconds, rather than at a substantially faster or slower pace. In some cases, entire genres of semi-functional music and sound may be invoked, providing a reference point suitable for this ringtone form.

On this last point, "The Cute Elf Trail Ringtone" (figure 7.3), from *Ringtones Brighter Side* (December 2008), features a number of distinct musical units that appear with some regularity, following a general order. The units themselves are four repeating mid/high-frequency figures and a series of yet higher-frequency chime washes that cover over them. On the basis of the labeling in figure 7.4, the order of events is roughly AE/AE/AE/ABE/'/C/'/ED/EA/AED/EA/EA/BE/'/C/'/E/DE/AE/AA/EA/EA/EAB*/'/C/'. (The asterisk indicates that the last B is shorn of its final C8. Apostrophes signify short pauses.) These elements do not repeat exactly and in order, but the appearance of C three times at roughly equal spacings suggests a degree of signal-like regularity. Moreover, the aesthetic material itself must be considered. We have very high-frequency music that sounds like a video-game accompaniment—precisely, a combination of "trigger music" (the music/sound caused by a player's action) of different types (A seeming like a character's "step" motion, E the result of some general scoring action, B a more significant scoring action, C a signal that some relatively new and important status has been reached, as suggested by its implied dominant-tonic harmonic closure, and D possibly being a negative scoring action).[31] The point of this exercise is that the music sounds like a series of repeated

Figure 7.3

Musical Units in Brice Salek, "The Cute Elf Trail Ringtone," *Ringtones Brighter Side* (2008). Reproduced by permission of composer.

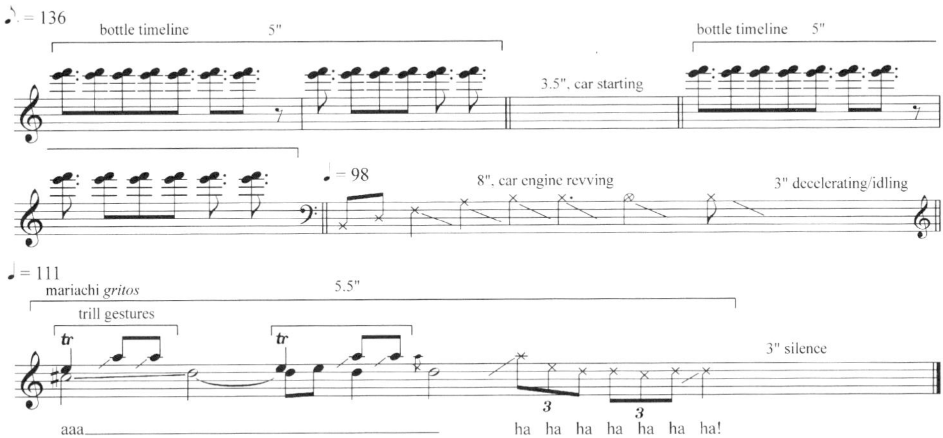

Figure 7.4

Brice Salek, "40 Oz. Tequila Driver Ringtone," *East L.A. Ringtone Ringtones* (2008). Reproduced by permission of composer.

signals, and takes place at a pace roughly equivalent to phone signaling—C occurs about every nine seconds—while also clearly drawing on the sonic aesthetics of video games. This meeting point of a semi-functional musical aesthetic with the periodicity of the phone is what makes these ringtones particularly striking, generating a creative form that nonetheless immanently emerges from its telephonic context. In a very different example, we can hear similar kinds of inconsistent repetition in "40 Oz. Tequila Driver Ringtone," as part of the *East L.A. Ringtone Ringtones* album (from November 2008). Here the aesthetic is completely different—all prerecorded sounds, no synthesizers, comprising a synecdochic sequence portraying working-class Latino masculinity. Yet the tapped tequila bottle that opens the recording, providing a neo-African timeline pattern, returns after the first started car, at about 8 seconds; the second car then extends from 13 seconds to 21 seconds, after which we hear a series of echo/delay-processed mariachi *gritos* or yells and trills (presumably from Latinos in East Los Angeles).[32] Now, again, for this example, which is at the outer limit of this type of form, the pacing of events still roughly follows the scale of the 2-4 signal, which gives it, even here, something of a phone-signal-like quality. Moreover, even the source materials used in the ringtone—the bottle timeline, the once-repeated phone-like *grito* trill gestures, even the rhythms of the revving engine (see figure 7.4)—are to different degrees reminiscent of telephonic and functional signals in timing, in gesture, and even in sonority.

Sound Fragments and Concept-Art Ringtones

Ringtone Records tracks are consistently about 30 seconds in length, whereas the contrasting *Touch Ringtones* recordings vary more widely and include a number of very fragmentary sounds, such as found sounds from Bulgarian radio, English television/radio programs, and other unknown sources. Such effects go a long way to making the Touch recording sound "sophisticated," like an album of contemporary experimental electronic/electroacoustic music, of the sort its niche-market demographic would appreciate. Overt conceptualism is found in a ringtone by the Stockholm-based art duo Bigert & Bergstrom called "Four Seasons." Bigert & Bergstrom take short fragments of Vivaldi's famous violin concerti *The Four Seasons* and subject them to extreme abbreviation. Offering a predictable if entertaining critique of the classical music industry and its kitsch forms, as well as the processing of time under late capitalism, the "Four Seasons" ringtone aspires to a level of conceptuality and engagement with the kinds of traditions of artistic institutional critique that are central to the established, global art market at present. This kind of conceptuality, which privileges incompleteness and openness as an aesthetic and even ideological virtue, is notably absent from the Ringtone Records recordings.

Ringtones Based on Text and Speech

The presence of recorded speech in ringtones is attributable to the rise of the sound-file ringtone. Since various music-industry stars and media figures have used their voices to produce promotional ringtones (e.g., the Green Day ringtone mentioned in chapter 2), it is not surprising that others would take advantage of the sound file's capacity to accurately reproduce human speech by placing it at the center of the ringtone, rather than only using it to reproduce song lyrics—which were by and large absent from polyphonic ringtones. Indeed, it is this capacity that appears to have generated a welter of comedy/novelty speech-based ringtones, in which jokes, effects, short phrases (often directed toward phone owners) serve as the content of the ringtones themselves; perhaps most original ringtones available for purchase, if not consumed, are actually of this type. In the catalog of ringtones examined here, such approaches are also present: a good example would be the that of "porntones," which range from pornographic film music styles to (simulated) audio clips from porn films and which are represented on the Ringtone Records album *Sex Ringtones, Volume 1* (2011), although these are often defamiliarized in the ways discussed below.[33] More aesthetic interest emerges when recorded speech interacts with the other genres of ringtones described above. Obviously, lyrics feature in miniature songs, but in addition to being sung, lyrics can also be spoken (as in the "Too Sexy for Your Call Ringtone" on *Ringtone Songs*, which references Right Said Fred's 1992 novelty hit "I'm Too Sexy"). Conceptualist speech-based ringtones surface, not surprisingly, on the Touch Ringtones recording, particularly on the tracks made by the politically conservative conceptual/performance artists Gilbert & George (three numbered tracks titled "The World of Gilbert & George"). These tracks include stilted, highly formal monologues or conversations between the two, sometimes punctuated by sound effects, effectively translating the extreme formality of their attire in their visual work and performances, as well as presenting another instance of their life-as-artwork philosophy. Aestheticized speech, similar to aestheticized signals, and speech-inflected repetition-variation forms feature prominently in the Ringtones Records catalogs. Among the former, one might consider the "Look Mommy Ringtone" on *UFO Ringtones*, in which a voice that sounds like that of a little girl (but probably is Salek's voiced, altered) presents mostly discrete pitches while saying "Look, Mommy, a U.F.O.," making the ringtone almost melodic in character. (See figure 7.5.) Among the latter are the strange "I Wanna Beer Ringtone" on *Must Have Ringtones* (2009) and the perverse

Figure 7.5

Brice Salek, "Look Mommy Ringtone," *UFO Ringtones* (2010). Reproduced by permission of composer.

"Aye Papi Ringtone" on *East L.A. Ringtone Ringtones*; in both, the title phrase is repeated and varied, sometimes accompanied by sound effects and/or drumbeats. The presence of recorded speech as an interchangeable element within the aesthetic frame of composed ringtones breaks down the division between language, sound, and music in a way that has long been characteristic of electroacoustic music.

The Ringtone Record Market and the 30-Second Form

The many forms described above, with their varying degrees of closure and proximity to the sonic practices of traditional telephony, exist in a marginal sector of the ringtone market, which is still dominated by excerpts from popular songs. The stakes in composing original ringtones are, therefore, very low, and most ringtone albums seem to be one-off projects, like the Touch recording, or are part of entire catalogs made by small production firms. Indeed, in examining the recordings of ringtone albums on offer at Amazon.com's MP3 store, one finds 1,236 albums of ringtones for sale (by Ringtones Friends NYC, Beat Boutique, Readymade Ringtones, On-The-Go Music, and Hahaas Comedy, among others).[34] As was mentioned above, most are outright novelty recordings, cheaply made and generic in form and content, often repeating tracks verbatim on different albums; interspersed among these ringtones are the recordings by Ringtone Records, which are at the upper end ($8.99) of the pricing spectrum (which can go as low as $5.49 and even less for particularly short albums of ringtones). Salek, who runs Ringtone Records, is in a difficult situation. Not having access to the major experimental art, sound-art, and music markets that made their presence felt on the Touch recording, he finds himself locked into a field that he, from what I can gather, simultaneously seeks to conquer (as an entrepreneur) and to supersede (as an artist). There is something admirable, even mildly heroic, in attempting to produce something with a semblance of aesthetic sophistication in a market nearly devoid of it.

The major difference between Salek's project and the one-off, star-based, even quasi-curatorial nature of the Touch recording, as far as I can tell, is that Ringtone Records demonstrates a commitment to the 30-second sonic form in a remarkable and unparalleled way. Salek says he was "thinking about the time of 30 seconds as an important interval, which is about the time that a standard song often has a chance to keep the listener interested." He continues: "I would hypothesize that a song changing up at around the 30 second interval might have something to do with media and the length of commercials subliminally dictating the attention span."[35] The 30-second form is, of course, not unique to the ringtone and advertisements; it is found commonly in audio samples available at iTunes and at other sites selling digital sound files. On YouTube, a number of songwriters have posted 30-second original songs, which seem more likely to grab a casual viewer's attention. Professional songwriters have been increasingly encouraged to create music in 30-second snippets, rather than in larger durations.[36] To the extent that Salek is correct, in the distracted society of mobile and digital media, the 3-minute, 78-RPM-record-dictated duration of the minimal sonic form during much of the twentieth century may have contracted rather significantly, providing yet another instance of capitalism's production of social acceleration (as discussed in the introduction to part II). If 30 seconds is all we are left with, we had better make good use of it.

Epilogue to Part III

The mass use of cell phones and ringtones at present is, no doubt, a striking characteristic of the new millennium, one that is transforming social mores, consumption habits, aesthetic ideologies, political mobilization, and more. Near-universal use of cell phones will surely have more unforeseen effects of great moment—in dialectical fashion, quantity will surely again and again transmute into quality. Yet, in view of the immanent necessity for the intertwined mobile communications and media sectors to promote the cell phone, one must be skeptical toward claims, implicit or explicit, about the scope of historical change attributable to the new mobile device regime. In the case of the political ringtone and more general political use of cell phones, one wonders to what extent these uses appear with such frequency in the global media because they serve a need and an interest: more precisely, demonstrating the promise of globalization and capitalist modernity's technological advances, especially in the context of anti-developmentalist and market-based solutions to global economic injustice (for example, the recently touted notion of microcredit).[37] To the extent that civil society organizations or non-governmental organizations promote the cell phone in situations favorable to the global liberal common sense (such as elections) while shying away from their promotion in more radical social movements—such as anti-imperialist resistance struggles or battles over the privatization of the global commons—these organizations may ultimately play into the surveillance-control-consumption agenda of the neoliberal state and multinational corporation, the declining public image of the latter two notwithstanding. But, once ultimately shorn of its status as a leading-edge commodity of the moment, the mass potential of the ringtone and the cell phone might be realized in ways both powerful and unremarkable. The making *mundane* of the cell phone—a process well underway and already realized within the majority of the world—would seem to be a prerequisite for its unfettered use in the service of human emancipation. Indeed, one might further argue that such mundanity is a requisite for transforming the ringtone into a site for aesthetic experimentation, as demonstrated in the original ringtones discussed above. Yet the limitations of the system of cellular telephony, characterized aptly by Jonathan Zittrain and Daniel Miller as being representative of a "proprietary model" of digital media, in contrast to the flexible and "generative" character of the original Internet, would appear to place fundamental boundaries on its liberatory utility.[38] Perhaps, then, the degraded pre-adolescent utopias of Crazy Frog are nothing more than harbingers of the universal subscription system sought by the global media conglomerates, with resistance to that system reduced to consumers evading digital police by way of imaginary conveyances.

IV Revivals and the (Universal) Particularization of the Ringtone

The dialectical processes of decline (part II) and reversal (part III) in the sociocultural transformation of the ringtone cannot account for every cultural or national register in which the ringtone appears. Indeed, in some contexts the shifting formats of the ringtone provided a recharging or reviving effect, in which the new modalities of the ringtone fostered new enthusiasm and opportunities for its use. Hence, the contexts to consider below are roughly coterminous with—and sometimes exceed—the ringtone periodization discussed variously throughout this book (roughly, 1998–2010), the decline of the industry either having a delayed effect or becoming downgraded in price and normalized—and thus conforming to the notion of industrial "death" outlined at the end of chapter 1.

The domains of ringtone use surveyed below—the worlds of film and television (chapter 8) and popular music (chapter 9)—are, perhaps even more than the rest of the book, focused on cultural products and forms indelibly tied to specific nation-states and geographic regions: the UK, the US, and China. This is particularly the case in the chapter on (mostly) black popular music, whose emphasis on English-language and especially American music is perhaps most evident, but in the television/film chapter the shadow of Hollywood also looms large, whether as a direct financial/creative foundation, a structuring influence, or a source of opposition—the last as might be seen in the extraordinary work of the Chinese filmmaker Jia Zhangke, discussed at the end of chapter 8. In this sense, we might view the relative geographical and even national specificity of the material under consideration as an example of *particularization*: as the ringtone has spread pervasively throughout the world, it has been taken up within national and regional markets not discussed in this book and imbued with specific regional and national significance, surely to varying degrees. For example, one might investigate the role of the ringtone and mobile phone more generally in Bollywood cinema or Spanish-language *telenovelas*, to take two of the most important and popular forms of moving image-based media in the world today, and arrive at different conclusions from those drawn below. The particularization process of cultural forms is reminiscent of what Fredric Jameson considers to be the particularization of the dialectic itself. This is to call attention to the decreasing favor that the notion of the universal has had in the postmodern period, in which difference is reflexively privileged and allegedly championed over similarity, even if the implicit universals of capital and liberalism (as in the discourse of human rights) still somehow manage to undergird—and even produce—the wealth of diversity on offer in the world.[1] From such a perspective, "the dialectic, which formerly reigned supreme, is reduced to a local law of this or that corner of the universe, a set of regularities observable here or there, within a cosmos which may well not be dialectical at all, but rather one of sheerest chance, of virtualities, say, or chaos theory, when not some altogether untotalizable and non-theorizable noumenon."[2] The ringtone, from this perspective, would encompass the entire world of music, the global regularities of which may be a chimera that the totalizing researcher hunts in vain.

But the implicit universals mentioned above, which undergird the present status of the capitalist world system, also have a geographical and national-cultural centricity, which is to say that the priorities of the system still reflect an extant, if waning, hegemony of the United States and its relationships to the previous global hegemon, the United Kingdom, and to its apparent successor, the People's Republic of China. The extensive debates over whether the globalization of cultural production is, in fact, the Americanization of global cultural production is a complicated, much debated matter, one that cannot be settled here; but it suffices to say that to the extent that hybridizations feature a Western, and often American, set of source materials—"pure" South-South hybridization being as poorly defined and elusive a category as the "unspoiled," "authentic" pre-modern cultural practice—the matter is often clearest at the economic level. That is to say it concerns a relationship between different nations' rates of cultural export and import and a question of where the lion's share of surplus value goes within the international network of cultural commodity chains, on account of the differential and geopolitically conditioned degrees of protectionism and market access.[3] The power of Hollywood and of the global music conglomerates as cultural exporters, whatever their actual degree of dominance in local markets and subjectivities, cannot be denied; hence, the subjects of the following chapters cannot be simply reckoned as examples of dialectical particulars, tempting though such a prospect might be theoretically and politically. To the extent that it represents a privileged or even hegemonic cultural force, whether "American" or "global," the ringtone as filtered through Hollywood or the global music industry is also a vehicle for—and perhaps even a type of—"world music," which is to say a marketplace that in principle includes every (musical) sound recording but is nonetheless dominated by certain artists, genres, aesthetics, structural materials, and production techniques.[4] Still, the predominance of the United States within global cultural markets (film and, to a lesser extent and in a more complex way, music) can only remain a hypothesis to be tested and critiqued; the ringtone-dialectic's totalization is necessarily arrested halfway, as it has not yet consumed all of the material necessary for the completion of its process—which could only achieve sublation in a more expansive, and ideally collective, research project.

8 Personalization and Spectatorship: The Ringtone's Narrative Functions in Cinematic and Televisual Media

If you have attended a Hollywood blockbuster horror film or thriller in the last decade, you probably have witnessed a scene involving a cell phone failing to function at a critical moment during the narrative—often with a character ominously saying "There's no signal." The video blogger Rich Juzwiak (of the blog FourFour) noticed this cinematic trend and produced a video sequence titled "No Signal (And Other Cellular Drama),"[1] which captures what has evidently become a marked cliché in the world of cinema today—one that plays on popular perceptions of mobile phones as guarantors of personal safety—and builds upon a small body of research examining the recent history of the cell phone in cinema and television.[2] For scholars of sound and music, however, the "no signal" trope is somewhat vexing in that it typically presents the narrative function of mobile-telephonic sound as a kind of charged *absence*, rather than making meaningful use of the sonic and musical features of ringtones as aesthetic objects. In steering clear of possible arguments about the image of the cell phone's efficacy or its failure, this chapter instead attempts to chart out another stream of televisual and cinematic representation of mobile modernity, one that depends on the logics of *personalization* and *actor-object identification*, themselves rooted in numerous points of determination. After examining the ideological sources of these problematics and extending them with respect to mobile phones and ringtones, I will in turn examine the particular uses of ringtones in two television serials: *Coupling* (UK, 2000–2004) and *The Sopranos* (US, 1999–2007). I continue with a consideration of ringtones as sonic brands, with attention paid especially to the use of the iPhone "Marimba" ringtone in three productions. I end the chapter with a series of reflections on the use of ringtones and other mobile musical practices in the films of Jia Zhangke, one of the most remarkable Chinese directors of his generation. At times, Jia's work takes us beyond the dynamics of personalization, actor-object identification, and sonic branding to show the ways in which ringtones participate in acts of communication between characters, through degraded sonic tokens of an idealized music that those characters are rarely, if ever, allowed to hear.

This chapter undertakes two tasks. The first is to offer a modest cultural contribution to the broader conjunctural analysis of political and economic conditions prevailing in during the 2000s in the United States and United Kingdom, and to contrast these to the contemporaneous situation in China—with the capitalist dynamics of personalization and sonic branding linking the two within a singular, internally differentiated modernity.[3] The second is to begin to answer the following question: What does the ringtone (and, in some cases, comparable sonic forms) *do* in films and television shows? In the course of this exploration, it should become evident that the answer to this question is complex—even contradictory—and involves two interconnected threads. The ringtone participates in a form of cinematic and televisual realism, in which characters and commodities (i.e., ringtones) are portrayed as having relationships to one another in ways not dissimilar to real individuals' relationships to their commodities or as advertised quasi-propaganda about consumer-commodity relations, with the latter not necessarily being dependent on explicit product placement. Moreover, the fact that movies and television shows are made *for* viewers leads the relationship between character and ringtone-commodity to convey information to viewers in ways uncharacteristic of experience in unmarked, everyday life; sometimes this information is conventional, at other times it is metatextual. Hence, ringtones offer a means of reading or interpreting moving-image/audiovisual media. The logic of personalization discussed in what follows provides a tentative hermeneutic, which proves relevant to varying degrees.

Actors and Objects: Personalization, Film and TV Props, and the Mobile Telephone

In popular media and advertising discourse, the ringtone is a facet of cell phone "personalization," or customization of the phone's variable digital features (ringtones, ringback tones, wallpaper) and display arrangements (especially through carrying cases and the increasingly popular "skins"). As was discussed in chapter 1, personalization is in actuality a marketing system and ideology of tailoring advertising and products to individuals, one aim of which is to encourage personal expression through consumption. When transferred into televisual and cinematic media, the ringtone often appears to maintain elements of this personalization-ideology. The functional role of personalization vis-à-vis the mobile telephone would appear to concretize a relationship between a single object (the phone) and a human individual (the owner) that is powerfully resonant with longstanding conventions involving the relation between actor and movie prop. This dynamic is perhaps best articulated by Walter Benjamin, in a perceptive footnote to his "Artwork" essay noting cinema's transformation of relations between actors and stage property in theater:

> Searching for . . . [film] actors [is] like hunting for rare props. . . . If the actor thus becomes a prop, the prop, in its turn, not infrequently functions as actor. At any rate, it is not unusual for films to allocate a role to a prop. Rather than selecting examples at random from the infinite number available, let us take just one especially revealing case. A clock that is running will always be a disturbance on the stage, where it cannot be permitted its role of measuring time. Even in a naturalistic play, real-life time would conflict with theatrical time. In view of this, it is very revealing that film—where appropriate—can readily make use of time as measured by a clock. This feature, more than many others, makes it clear that—circumstances permitting—each and every prop in a film may perform decisive functions. From here it is but a step to Pudovkin's principle, which states that "to connect the performance of an actor with an object, and to build that performance around the object . . . is always one of the most powerful methods of cinematic construction." Film is thus the first artistic medium which is able to show how matter plays havoc with human beings. It follows that films can be an excellent means of materialist exposition.[4]

Reading the work of Arnheim and Pudovkin, Benjamin extrapolates from what appears to be a general principle of cinema: namely, that any self-contained entity in film can almost function as a kind of equivalence for another, on account of cinematic conventions (e.g., the close-up shot) and editing techniques.[5] In many classic writings on cinema, the prop—a physical object touched by characters—is often understood as a concept extending beyond its literal meaning to encompass actors (Arnheim), music (Balázs), and even a more abstract thematic idea (Bazin).[6] Regardless of whether or not we might want to read this equivalence process as a general cinematic allegory for capitalist exchange, we may further appreciate that this exchangeability leads to the possibility of pairings, as in the actor-object dyad described by Benjamin and Pudovkin. A part of the story of such pairings concerns affective relationships: Deleuze refers to them as *emotion/object pairs* characteristic of a certain cinematic realism.[7]

The formal similarity of the actor-object and consumer-personalized commodity dyads is made more meaningful when the latter is considered in light of the mobile telephone. For the commodity in question is a leading-edge technological item whose form and content are constantly in flux and therefore articulated with contemporaneous marketing fads and new trends (such as the upsurge in app sales on account of the iPhone and other smartphones).[8] Thus, a case can be made that the presence of the mobile phone in cinema, when paired with a character, becomes a form of advertising for mobile telephony itself, particularly in its contemporaneous guise, and that the character-phone dyad functions as an implicit *advertising testimonial* or *transfer*—in the propagandistic senses of the terms, the former involving a human being vouching for some object or cause, the second being a simpler mapping of associations from one entity onto another.[9] Moreover, if the character is played by a famous actor, the phenomenon arguably transmutes into a generalized form of celebrity branding.[10]

The potential intertwining of the actor-object dyad with that of the consumer-personalized cell phone exemplifies the ability of the moving image to depict the commodity lifeworld of the present—offering the possibility of a "materialist exposition," even if it isn't one that simply displays the power of the commodity over humans. Indeed, the prop's ability to convey a materialist history and function as represented commodity isn't unique to cinema and television, but originates with the early modern theatrical stage—the role of the prop (*Requisit*) on which Benjamin was preoccupied in his early masterpiece, the *Trauerspiel* book.[11] Indeed, a part of the prop's materiality, according to Andrew Sofer, is its featuring of motion and manipulability—and often, but not always, a kind of portability or mobility—and the prop is furthermore marked by a degree of displacement, or what Frances Teague calls its "dislocated function."[12] Beyond whatever materiality the prop object or commodity possesses outside the performance context, the materiality of the prop also involves its narrative functions of "visual shorthand" and "characterization," demonstrating the importance of the actor-object dyad on the theatrical stage as much as in television or cinema; in some cases the theatrical prop, too, presents the possibility of substituting for the actor, in a process that Sofer terms "surrogation."[13] But, the prop also plays a concrete, material role in motivating action on the stage or screen.[14] Moreover, when appearing within the moving image, the prop can also provide a degree of continuity that may be lacking in the collaged aesthetic of film and video, as well as generate new layers of meaning as a result of the cinematographic process.[15]

The foregoing excursus provides some insight into the potential meanings and roles of the cell phone in film and television, but with the introduction of the ringtone as a significant narrative entity, our understanding of these issues must be revised somewhat. One such aspect must deal with the fact that the prop in question—a mobile phone with a (usually musical) ringtone—takes on a more complex identificatory function, on account of the ringtone itself. Although it is obvious that the ringtone—as the best-known example of mobile phone personalization—should serve as a means of identifying a character, through the actor-object relationship described above, we also should take note that this capacity intersects with the longstanding musical tradition of character identification using the leitmotif. Emerging, in part, from Richard Wagner's operas and the interpretive discourse surrounding them, Hollywood film composers imported the technique into their work in a somewhat simplified form as early as 1910 and would employ it, as well as other techniques, to produce the "classic" sound of American cinema in the 1930s and the 1940s.[16] However, the simplistically illustrative nature of the technique—derided even in their original operatic form by Claude Debussy as a "calling card" for a character—led Adorno, Eisler, and other critics to suggest that its simplified version in cinema was a means of aiding film consumption.[17] Since ringtones rarely exhibit the capacity for transformation, especially of a sort that maintains their thematic material (as would be the case with Wagnerian leitmotifs), it is more typical that they serve just as this sort of "calling card"—though that card can present information to viewers that characters themselves do not "hear" and it can be transformed in the course of a film through its appearance in different arrangements or formats (i.e., not just as a ringtone). The effect is a concrete example in which music itself transforms into something like "a sort of acoustical stage property."[18]

Moreover, the fact that the mobile phone in film or television has this surplus capacity to "speak" (or "sing"), as it were, makes it a special sort of prop, one that is relatively *animate*. The leitmotivic and musical function of the ringtone, or its "prop music," is therefore only one type of a number of active, agential behaviors of a mobile phone—which can ring like a traditional phone or make unusual sounds (using nonmusical ringtones, for example),[19] sound or move about while on vibrate mode, and present the speech of a voice on the line—Michel Chion's "acousmêtre" or acoustic being.[20] The mobile phone is, unlike the fixed-line telephone, also self-evidently *mobile*, giving it a certain flexibility in depiction, given that it can show up in multiple places—in hands, pockets, or purses, on tables, desks, or chairs—and ring in various locations, including spaces of transit (like moving vehicles). We might view such objects as puppets operated by a puppeteer or dummies by ventriloquists, particularly in light of Andrew Sofer's warning that "props remain objects, not subjects" and that "stage props are 'motivated'—literally put into play—by actors but are not themselves animate."[21] But, the combined effect of moving image representation, the ideology of personalization, and the cell phone's own noisy motility arguably combine to give it the semblance of what actor-network theorists term "symmetry" between human and nonhuman actors.[22]

As is evident from the above, there are clearly two elements involved—the mobile phone as an object or a prop and all that it facilitates, and the ringtone as a sonic (and often musical) entity. Though these elements often combine to create a particularly animate prop, the actor-object dyad (and consumer-personalized commodity) can also be split into a triadic relationship of actor, cell phone, and ringtone, creating a more complex set of identifications that can be exploited in televisual and cinematic contexts. Again, since Balázs and Adorno and Eisler argue that music can serve as a kind of prop, we might posit that the diegetic sound or music of the ringtone can function as a prop that, due to the differences in conveyance of sound and image, can be split from the phone as prop.[23] Though such a possibility exists for any noise-making prop (a horn, a gun), which would then to some degree necessarily induce such a triadic relationship, the uncanny degree to which cell phone ringers can be difficult to localize—exacerbated by the ubiquity of its sounds, its relatively low sonic fidelity, and, with sound-file ringtones, its tendency to blend in with other ambient music—perhaps renders the ringer and phone more autonomous than might be the case for other sound generators on stage, particularly from the point of hearing of the audience member. On the other hand, taking a cue from Michel Chion's perceptive claim that "sound in film is voco- and verbocentric," on account of film's human spectatorship, we must see the ringtone as providing a kind of ancillary, and potentially ironic, function, interacting with other sounds on the screen but never occupying a dominant role.[24] Keeping these issues in mind, we should nonetheless consider the problematic of identification (and its ideological confirmation in marketing personalization) in mind as a point of departure when considering the appearance of ringtones on the screen.

Ringtones in *Coupling* and *The Sopranos*

The BBC-sponsored sitcom *Coupling* and HBO's epic Mafia drama *The Sopranos* make for an odd pairing. They are treated in depth on account of the relatively early adoption of ringtones in television shows—most of the ringtones under consideration here are monophonic—and hence may have served as inspirations for the generally later ringtone uses described in subsequent sections of this chapter; they are also considered for what they are able to tell us about the conjuncture of the early 2000s in the United Kingdom and the United States. Among the numerous ways, however, in which the two shows differ—and they do so in almost every conceivable way (genre, degree of cinematic production, duration, regional and class origins of characters, uses of music and sound, thematic foci, depictions of sex and sexuality, existing scholarly research, etc.)—are their respective uses of ringtones, both in degree and kind. *Coupling*, written by Steven Moffat and airing on BBC2 from 2000 to 2004, features a decidedly white, middle-class, heterosexual, and urban characterscape, set in London and featuring the developing relationship between a writer named Steve Taylor (played by Jack Davenport) and Susan Walker (played by Sarah Alexander), apparently a mid-to-high-level employee at an accounting firm. Indeed, the working labor of the main characters is hardly ever mentioned or depicted, and when workplaces do appear they merely provide pretexts for various shenanigans. Instead, the show focuses on the leisure activities of the couple and their circle of friends—all white, attractive thirty-somethings and describable as what were once called "yuppies," lending comparisons to the demographically similar American sitcom *Friends* (though in discursive tone it is closer to *Seinfeld*). These leisure activities are, for the most part, those oriented around sex and dating (or the lack thereof), usually generating more talk than action. Often the characters spend gender-segregated time in a pub (the men) or a restaurant (the women) recounting and analyzing the sexually charged and awkward moments of quotidian interaction—though the men appear to have more sexual hang-ups than the women, who are generally more sexually confident and aggressive. A much-acclaimed production that was remade into an unsuccessful American series, *Coupling* is difficult to describe in unique terms. What makes it compelling is not its conceptual or aesthetic originality, but rather Moffat's whip-smart writing and the crisp delivery of the cast.[25] The show is, like many sitcoms, theatrical in structure and tone, and it features long monologues on the relations between the sexes by Steve or bizarre theories about sex and sexual interaction by his closest friend, the amusingly awkward Welshman Jeff Murdock (who is also one of Susan's co-workers).[26]

Ringtones—particularly melodic or musical ones that function "in excess" of the signaling function of more generic phone ringers—appear only twice in *Coupling*. But they are used in a consistent manner, with variants of the same ringtone, and they are integral to the episodes' plots. Both instances occur in the show's second season. In the first instance, in season 2, episode 2 ("My Dinner in Hell"), Steve and Susan are planning to have dinner with Susan's parents, who make Steve uncomfortable with their frank discussions of sex. Socializing before dinner with Jeff and the show's other major male character, a bumbling, conservative playboy named Patrick Maitland, Steve recounts an awkward conversation with Susan on the topic of male masturbation, during which his sexual fantasy involving the British TV journalist and personality Mariella Frostrup is revealed. Meanwhile, Susan and her friends—a ditzy radio traffic reporter, Jane Christie, and an aging-obsessed beautician, Sally Harper—discuss the same conversation between Steve and Susan. Jane also mentions her techniques for stalking celebrities, which include slipping a mobile phone into the celebrity's pocket, then calling the phone to retrieve it and scheduling a meeting with the celebrity. The plot thickens when, at the bar, Steve runs into Mariella Frostrup, who spills a drink on him and causes him to drop his phone. Jane inserts Steve's phone into Mariella's pocket. Great discomfort ensues during the dinner with Susan's parents, who Steve mistakenly thinks are ridiculing him for masturbating too much. The culmination of the episode arrives when Susan attempts to find Steve, who has stepped out in anger, by calling him from her land-line phone. Just as Steve returns, the television features Mariella, whose pocket now contains his ringing cell phone—offering a remarkable moment in which the mobile telephonic triad is fully co-present and yet spatially separated. In the second case, in season 2, episode 9 ("The End of the Line," the season finale), the plot is built around a comic series of exchanges on telephones. Susan pretends to be Giselle, a French client of her employer, and Steve adopts the persona of the Australian bartender Dick Darlington, who works at a bar called Susan's (allegedly in honor of Susan's wild sexual romp through the country during her college years). At one point, Steve calls Susan from a land-line phone in an attempt to clear up confusion, but ends up calling Jeff, whose phone looks similar to Susan's and which he accidentally picked up at work.

Aside from their amusing phone-switch gags, which lead to all manner of confusion and narrative conundrums, these two scenes are somewhat strange in that the sounding ringtone is the "Nokia Tune," probably the best-known ringtone in the world.[27] Though it is not at all unusual for Steve and Susan to have the Nokia tune as their ringtone, it is strange that the characters in both cases react to the ringtone by so clearly identifying their phones as *theirs*, in view of the commonness of the ringtone. (Whereas Steve's is the standard version of the ringtone, Susan's is transposed to F major from A major, and thereby perhaps further feminized.) The gap between personalization (via actor-object identification—the object in the first case initially being the ringtone alone, separated as Steve is from his phone in physical space) and the relatively impersonal nature of the ringtone might be interpreted in a few ways. The first and most likely explanation is that the ringtone's sound itself is an arbitrary sonic prop that merely functions to impel the narrative forward. In this reading, the audience takes it as a matter of faith that the ringtone is identified in a simple sense with a character. Another reading might view this use of the Nokia ringtone as a form of auditory product placement, though effected through the sounding ringtone rather than verbal mention of a product (as is typical in the auditory case).[28] In this view, Steve and Susan are something of a branded couple, with matching his-and-her versions of the Nokia ringtone. A third interpretation focuses on the "newly dominant" distribution of the mobile phone in British life, with penetration rates only at 25 percent in 1998 and 46 percent in 1999 (that year being the "tipping point" according to one reporter), and by 2001 hovering around 75 percent.[29] Arguably, the mobile phone would in 2000 and perhaps into 2001 have retained a trace of relative class affluence (if surely mitigated by high rates of youth culture penetration and decreasingly conspicuous consumption), such that a group of thirty-year-old mobile subscribers at that time were still somewhat less likely to be classified as working class or poor. Operating at the crossroads between cell phone owners of middle-income status and those lower on the income scale, *Coupling*'s central couple pose, through their geographic, racial, class, and age specificity, as representatives of the whole British populace, as a hegemonic particular representing the universal of the national-popular.

The dynamics of ringtone use in the highly acclaimed series *The Sopranos* are rather different, and far more complex.[30] A sprawling, epic portrayal of the suburbanization and decline of the Italian-American Mafia featuring James Gandolfini as Tony Soprano, the show is the brainchild of producer and writer David Chase, who drew on his overbearing Italian-American family and personal problems with depression and mental illness in order to produce the Tony Soprano character, with his chaotic and schizoid life revolving around his therapy sessions with a psychiatrist, Dr. Jennifer Melfi (played by Lorraine Bracco), in the course of the series. *The Sopranos* has drawn a large number of scholarly treatments, extending from the straightforward and informative compendia of Maurice Yacowar to studies of the serial's ethnic, psychiatric, geographic (dealing with the urban and suburban context of North New Jersey), and racial aspects; it has also elicited a range of theoretical and philosophical assessments interpreting the show's nihilism, existentialism, postmodernism, post-Cold War paranoia, and much more.[31] A number of writers have pointed to Chase's innovative uses of music, especially his avoiding non-diegetic film scoring in favor of diegetic source music—a feature that, aside from the opening and closing credits music, is dispensed with only rarely.[32] Another facet of Chase's use of music is a kind of punning or semiotic wordplay, in which song titles and lyrics (as well as episode titles) take on new meanings in the context of the show, such as in the uses of the song "Return to Me" to comment on loss, and the absence of non-returning characters, but in other contexts the song's inclusion of the word "return" can signify returns as earnings.[33]

For all of the extant discourse on *The Sopranos* and music, little, if anything, has been written with regard to the ringtones that frequently appear on the show. Table 8.1 summarizes these ringtone appearances in cursory format, and allows us to make a number of observations based on examination of specific scenes in which these diegetic sounds appear. First, perhaps the most salient feature of the ringtones in the show is their consistent use for ironic commentary, which is not unlike how songs themselves can be used in auteurist films and television shows, including this one.[34] The technique also draws on a practice continued from theatrical presentation; as Sofer notes, "Often props convey information of which the character himself may be unaware and become vehicles of dramatic irony."[35] To the extent that the "awareness" part of this statement is true of ringtones in TV and film in general, a corollary is that characters in these cases do not really hear their own ringtones, or, taking Claudia Gorbman's cue that "diegetic music functions first and foremost as sound," at best they hear them as functional signals—cues for action or behavior, rather than music for listening.[36] The best example might be the somewhat shocking moment in season 4, episode 3, when Janice Soprano, Tony's ne'er-do-well sister, is seen anally penetrating Ralph Ciforetto, a widely despised mob captain who is cheating on his wife and prefers to engage in heavily masochistic sex (as we learn here for the first time). Ralph's ringing cell phone interrupts the proceedings, playing a monophonic version of the theme song from the film *Rocky*, "Gonna Fly Now." The manifestly masculine associations with the song (especially Sylvester Stallone/Rocky Balboa) contrast strikingly with the feminine-gendered or "bottom" sexual position that Ralph takes here—reinforced both by Janice calling him a "slut" and "whore" and by the diminutive effect of the monophonic ringtone itself. The transformation of Ralph's character as feminized and "deviant" may elicit sympathy for this character but also serves as a pretext to discredit him within the world of the show's diegesis—a world captured, for that instant, by the (hetero)sexually normative Italian-American working-class world of the referenced film.

Table 8.1
Important ringtones in *The Sopranos*.

Year	Season	Episode	Phone-owning character	Ringtone (usually monophonic)	Context
1999	1	13 (13)	Mikey Palmice	Ringer signal (digital)	Shot to death by Chris and Paulie in exchange for Brendan's death
2002	4	3 (42)	Ralph Cifaretto	*Rocky* theme song ("Gonna Fly Now," Bill Conti), 1976	Being anally penetrated by Janice Soprano
2002	4	3 (42)	Bobby Baccalieri	Verdi, *Rigoletto*, Act III, "La donna é mobile," 1851	Complaining about wife, unaware that she is dead
2002, 2004	4, 5, 6	4/13 (52), 5/2–3 (54–55), 6/21 (86), others	Tony Soprano	"Melody 1," Ericsson T39 preset [applied to Motorola phone], late 1990s	Numerous, usually functions as an annoying interruption
2004	5	2 (54), 6 (58)	Tony Blundetto	Queen, "We Are the Champions," 1977	Awakened by Tony S. while sleeping in mother's basement; call from Gwen, received while Tony B. is painting his new massage studio
2006	6/I	11 (76)	Dominic "Fat Dom" Gamiello	Beethoven, "Für Elise," 1810 (polyphonic)	Killed by Silvio and Carlo after making jokes about Vito and Carlo's sexuality
2007	6/II	21 (86)	Paul "Paulie Walnuts" Gualtieri	Simon and Garfunkel, "Cecilia," 1970 (polyphonic)	Called back by Tony, concerned that Carlo isn't there and might be dead

Other examples of this sort of dramatic irony are relatively common. In another example, when Tony Blundetto, Tony Soprano's cousin, appears in the fifth season of the series, having been released from prison on a mob-related conviction, and has now decided to avoid the Mafia life and become a massage therapist, one distinctly gets the sense that the character's sexuality comes into question by Tony's group—though such a claim is never verbalized. However, when Blundetto is awakened late at night in season five, episode 2 by a phone call from Tony Soprano, we hear his ringtone as Queen's "We Are the Champions," a song that, in view of the well-known queerness of Freddie Mercury, might be said to allude to the ambiguous sexuality of the character. (However, the manifest content of the song, with its claims of the narrative subject having "paid my dues," and "done my sentence," resonates rather explicitly with Blundetto's release from prison after almost seventeen years.) Later in the season (in episode 6), we hear the song as a ringtone again when his girlfriend, Gwen, calls him as he is painting his new massage studio—a venture that will soon fall apart—both representing the career struggle the character faces at that moment and providing confirmation of his now-demonstrated heterosexuality (and refutation of potential assumptions from the first appearance of the Queen ringtone). In yet another example, the death of Bobby Baccalieri's wife Karen in season four, episode 3, is foreshadowed through the call that Bobby receives from his son, conveying his wife's request to run some errands—at which Bobby vents frustrations about his wife. The ringtone, the classic *canzone* in praise of women's unpredictability from Verdi's *Rigoletto*, "La donna é mobile" (Woman is fickle), comments on the situation in a number of ways: it speaks immediately to Karen's changing her mind about the errand (leading to Bobby's frustration), it foreshadows the news of her death in suggesting that her condition on earth (and perhaps life in general) is a fickle matter; and, through the use of the false cognate "mobile," reflexively points to the use of the cell phone in conveying this news.

Another characteristic of *The Sopranos* is a ringtone's repeated, consistent appearance that sonically links different episodes and provides a certain auditory continuity. Although this potential is suggested by the cross-episode transformation of associations with "We Are the Champions" as Tony Blundetto's ringtone, it is most strikingly realized in the symmetry between Mikey Palmice's death (shortly after which his cell phone rings with a simple digital ringer signal) in the season 1 finale and Fat Dom Gamiello's death in the penultimate episode of season 6, part I (shortly after which his phone rings with a polyphonic version of Beethoven's

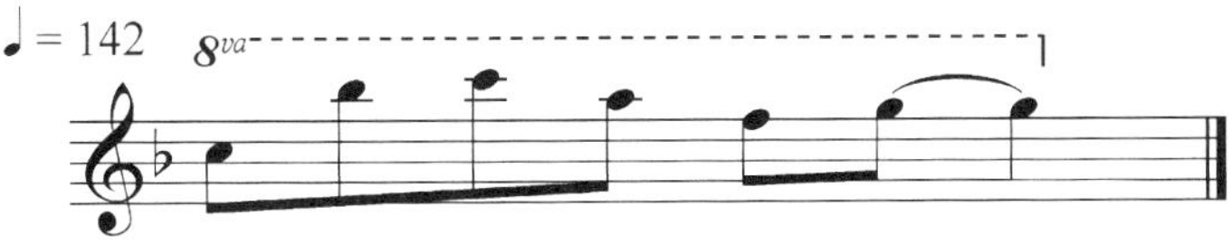

Figure 8.1
Tony Soprano's ringtone (Sony Ericsson, "Melody 1").

"Für Elise"). Here, the symmetry not only spans most of the entirety of the series, but also demonstrates the technological contrast between the functional, non-melodic ringer signal of the earliest digital and mobile phones and the MIDI-synthesizer technology of the polyphonic ringtone seven years later (although by this point sound-file ringtones were becoming dominant in the United States).

A more involved example of a ringtone that spans multiple episodes in *The Sopranos* is the particular tone used by Tony Soprano, which is the Sony Ericsson preset monophonic ringtone called "Melody 1," appearing on phones such as the T39 (in 2001) (see figure 8.1)—a striking inconsistency, since Soprano appears to use a clamshell phone, probably one from Motorola's V series.[37]

Attaining an almost leitmotivic or *idée fixe* condition through its repeated appearances, the ringtone's musical features are remarkably annoying—a trait not lost on Tony Soprano, who seems exasperated every time he has to answer its call and attend to some "work" or underworld-related responsibility. In addition to its cheery, singsong homorhythms and trebly frequency range characteristic of many ringtones, this one additionally features a harmonic-functional twist: a dominant prolongation that doesn't resolve, and an implicit melodic line that is perpetually frustrated from descending to the tonic. Although this ringtone appears most strongly as a punctuating irritant in the season 4 finale, it returns early on in season 5—despite the implied two-year time lapse—as if to haunt and nag Tony, embodying in simple sonic form the burdensome responsibilities of his illicit work and multiple kinship networks (Mafia, family, mistresses).

Taken together, these techniques serve an auteurist agenda in *The Sopranos*, bringing a cinematic aesthetic to the serial televisual form as a result of a large budget enabling high production values and the relative freedom from broadcast restrictions granted by HBO as a premium cable channel. The effect of these techniques and this auteurism, when heard at ringtone-level, would appear to present the alienated consciousness of the powerful creator himself, the producer David Chase, whose omniscience lends a certain distance to the oscillating poles of an often numbingly horrific violence and a stifling domestic life found in Tony Soprano's daily rounds and routines.[38] But Soprano's own, inescapable ringtone may also serve as a cipher for the nagging, deep-rooted problem of his own murderous sociopathology, to which he cannot fully admit and awareness of which he has therefore partially repressed—perhaps because both the pathology and its cognitive repression provide the sources of his own material sustenance, the psychological means of reproducing himself as an underworld capitalist but otherwise living an ordinary suburban life.

These two television series, when perceived with their ringing cell phones firmly in mind, resonate with the broader Anglo-American political conjunctures in which they were temporally located. The conjunctural situation, understood as a long moment with internal contradictions and dynamics, can be characterized in terms of the conjunction of a number of factors and events: the defeat of the Clintonian agenda in the wake of the impeachment proceedings in late 1998 through early 1999; the election and judicial emplacement of George W. Bush to the presidency in late 2000; the formative moment for neoconservatism in the 9/11 attacks in New York in 2001; the quagmires of the Middle East wars extending from 2002 to the present; the extension of Tony Blair's New Labour agenda through its articulation with the Bush administration's neo-imperial Project for a New Century; and, perhaps most significantly, the economic crash of 2000–2002. In response to the latter, the result was the reflation of new speculation bubbles in land and commodities, the Bush administration's abandonment of the "strong dollar" and the strengthening of the European economy relative to the United States, whose trade currency had slowly been increasing relative to the euro since the "Reverse Plaza Accord" of 1995. Since this accord made US exports less competitive (and European ones more so) and left consumers with lower spending power and less valuable holdings, the situation in Europe and North America was now effectively reversed. If we consider the ringtone itself—largely a phenomenon of Asia and Europe, rather than North America, when interpreted from a tripolar perspective—as partly a product of what Robert Brenner has called the "hydraulic dynamic" of the leading capitalist economies in the wake of the long downturn since 1970, then the moment of the late 1990s and the early 2000s appears as one in which the cylinder of the world system felt the pressure of the piston's movement with particular force.[39] In light of such a context, the buoyant, almost giddy quality of *Coupling* reflects a British consumer economy on the make—even though it was never exactly that when understood in context[40]—whereas *The Sopranos*, with its grim dual allegories of the struggling working-class man and the dictatorial leader (embodied in the character of Soprano himself), encouraged a certain detachment with which to view the rapidly deteriorating conditions of accumulation and governance in the US marking this moment.

The Nokia Tune, the "Marimba" Ringtone, and Apple's Product Placement

As can be ascertained from the ringtones in *Coupling*, the most problematic relationship between personalization and ringtones emerges when the ringtone is profoundly generic. In other words, the use of such a ringtone to identify a character is compromised when the purview of the generic ringtone seems to extend beyond the individual character in question, on the one hand, or when the ringtone's sonic content is sufficiently indistinctive to identify *any* character, on the other. Prime examples in this regard are those involving some of the best-known preset ringtones in the world—especially the previously mentioned Nokia Tune—and the United States (the iPhone "Marimba" ringtone). Since an estimated 40 percent of all mobile phone users are either uninterested in changing or don't know how to change the default ringtones on their phones, specific preset ringtones (e.g., Motorola's "Hello Moto," or the two mentioned above) have become profoundly important if underappreciated sonic presences throughout many global soundscapes—particularly since handset manufacturers don't seem to adjust ringtones much, if at all, for specific national, regional or linguistic markets. Of course, that sonic presence is significantly a brand presence, even if the ringtone's function as a corporate sonic logo, a branded "vessel for associations," may not be as explicitly articulated as in traditional advertisements; in contrast, one might consider the famous Intel logo (see figure 8.2) in television ads or the MGM lion's roar at the beginning of a film produced by that studio, for example.[41] (Of course,

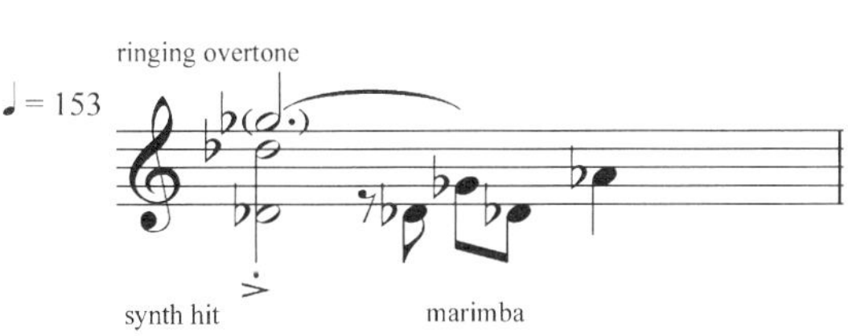

Figure 8.2
Sonic logo of Intel Corporation.

the advertising brilliance, or insidiousness of the ringtone, depending on your viewpoint, is that it isn't generally understood as an advertisement, though it often functions as one—especially in the case of default/preset tones.[42]) Though minimal in content, sonic logos are subject to copyright protections when registered as legal "sound marks," as is the case with Intel's logo—which is copyrighted "in the United States, Australia and New Zealand, but interestingly, not in Europe."[43]

Nokia Tune examples are extremely common, and appear as early as the *Coupling* episodes in 2001, but have appeared sporadically since then. Two sets of straightforward instances include those in the American film *Cellular* (2004) and in the UK TV mini-series *Mobile* (2007). Whereas in the latter the Nokia Tune is used as a pretext for cell phone rage (featuring a yammering tourist in the Lake District and an angry engineer with his family), in the earlier case the Nokia Tune appears twice: the first time, we hear it as the kidnapped Jessica Martin (Kim Basinger) rewires a smashed telephone and reaches a random stranger named Ryan (Chris Evans), a frustrating character who seems to be an irresponsible and self-serving young man in his twenties—a trait perhaps confirmed by the annoyingly familiar Nokia Tune itself. However, toward the middle of the film, after Ryan has become fully involved and invested in rescuing Jessica and her family, he attempts to bargain with her kidnapper, a corrupt policeman, in order to facilitate an exchange: a piece of incriminating evidence to be traded for those held hostage. In a tense, even melodramatic scene, Ryan waits for the kidnapper's return call after bargaining with him for the Martins' lives, calling the kidnapper's bluff; when the Nokia Tune returns, its narrative urgency has now overcome whatever reservations we may have had about it (and Ryan's character) previously, and we are relieved to hear it. A third example depicts the Nokia Tune as persisting two centuries into the future, in a scene in the recent *Star Trek* (2009) film prequel to the original 1960s television series. A young James Kirk is portrayed as absconding with his father's vintage Corvette, taking it on a joyride while appropriately listening to the Beastie Boys' rap-rock classic "Sabotage"—the video for which featured a mock credit sequence parodying those in 1970s serial crime dramas and involved car chase scenes and action stunts. The version of the Nokia Tune that appears, signaling Kirk's father's phone call chastising and threatening his son with punishment, is slightly sped up and rendered with a more clearly synthesized timbre, and it features a pitch-bend effect on the final A (though it remains in the original key of A major and clearly recognizable as the Nokia Tune—a fact reinforced by the "Nokia" brand label appearing on the futuristic car-phone/stereo system).

With the iPhone's revival of the ringtone, a similar use of its most familiar preset ringtone, "Marimba," took place in 2009–10, when the brand had an outsized presence in American film and television. Though in the early 2000s the Nokia Tune may have been associated with relative class elites, in the course of the decade its status seems to have become generally less specific (if still reinforcing the class prejudices of the culture industries producing filmic and televisual representations). In contrast, and on account of its relative novelty and high cost, the iPhone may still signify a certain class status for characters, who are usually within the overrepresented middle-to-upper part of the class spectrum, thereby overtaking the role of the Nokia Tune, especially in US contexts; however, the iPhone's trendiness also appears to make it seem appropriate for those in the years of emerging adulthood (their twenties) and slightly older. Accenting the youth associations with the iPhone, appearances of the "Marimba" ringtone are found in 2009's failed second-person-based series *Secret Girlfriend* ("your" phone is an iPhone), and in the cop comedy-drama *Psych* (especially in 2010); in the latter, the not-quite-grown-up series protagonist and pretend-psychic, Shawn Spencer, owns an iPhone, which usually features the "Marimba" ringtone and often goes off at inopportune times. Far more interesting, however, are two Hollywood feature films that implicitly draw upon the ringtone's class-based meanings. Consider, for example, *I Love You, Man* (2009), a "bromance" comedy ostensibly about male homosocial bonding set in Los Angeles. A multi-talented but strait-laced and somewhat dorky real estate agent, Peter Klaven (played by Paul Rudd), has growing material success and owns a house in the wealthy Los Feliz neighborhood but lacks male friends and hence lacks a best man for his wedding. Peter also possesses an iPhone with the "Marimba" ringtone, which is heard periodically throughout the film.[44] After a series of misadventures resulting from "man dates" to procure new male friends, Peter meets Sydney Fife (played by Jason Segel) by chance at an open house for the former's client, the actor and bodybuilder Lou Ferrigno. Peter and Sydney hit it off, bonding in part over the Canadian prog-rock band Rush, whose libertarian politics understatedly confirm the fact that both characters seem to be aspiring capitalists: Sydney claims to be an investor (although we are led to be somewhat skeptical of this claim, in part due to his shabby wardrobe), and Peter wants to buy some land to build a commercial development, though he needs to sell Ferrigno's house to secure the necessary capital. (Peter's fiancée, Zooey Rice, played by Rashida Jones, co-owns and runs a boutique called ZHD with her friends Hailey and Denise.) After spending a great deal of time together, some of it at Sydney's house in the bohemian Venice district, Sydney asks Peter to lend him $8,000 for an investment, since his money is "tied up in equity" and he is therefore "cash poor" at present. In a moment of sentimental appreciation for Sydney, Peter agrees and also asks him to be his best man at the wedding. It turns out that Sydney has spent the money on a series of somewhat outrageous (and eventually successful) billboards intended to help with the Ferrigno house sale, though this precipitates a falling out between Peter and Sydney. Shortly before the wedding, however, Zooey calls Sydney and invites him to the wedding to be Peter's best man, shortly before the couple's vows, Sydney enters and presents Peter with an $8000 check, stating, "you know, I know you don't believe me, but I'm actually a pretty successful investor, so . . . the billboards were my wedding gift to you guys." Although the moment provides the fictitious class resolution to the class-based friendship conflict, we might have surmised it a little earlier, since Sydney answers Zooey's last-minute call on his own iPhone, which features a ringtone of Rush's song "Tom Sawyer."[45] Personalized and less foursquare than Peter's phone, Sydney's prop demonstrates that he is nonetheless in the same tax bracket as (or even a higher one than that of) his seemingly more responsible "bromantic" partner.

The "Marimba" ringtone and iPhone also feature in another class allegory semi-disguised as a clash of personal qualities: the 2010 film *She's Out of My League*, directed by Jim Field Smith. The beautiful and successful event planner Molly McCleish (played by Alice Eve) loses her iPhone, which results in an unlikely romance with the awkward, working-class TSA guard Kirk Kettner (Jay Baruchel), all taking place in Pittsburgh. First heard when she calls her lost phone (which is found by Kirk), Molly's "Marimba" ringtone figures in another crucial moment in the narrative, when her friend Patty (Krysten Ritter) makes a "bad date bailout call" and Molly decides not to answer, signaling that the date is going well—thereby rejecting the upper-middle-class privilege that the ringtone seems to represent. In contrast to Molly, Kirk is revealed early in the film as owning an older-fashioned flip-phone with a hardcore rap ringtone (or parody thereof); the ringtone song's lyrics "smack her butt, and call the bitch 'slut' / likes to fuck when she wants, uh-huh" wafts conspicuously over the sounds of tumbling pins and rolling balls in the bowling alley scene, during which Molly asks Kirk out on a date. Given that Kirk is known to be both a particularly nice and unaggressive (read: "wimpy") person, the ringtone produces a moment of dramatic irony—its absurd, overcompensatory presence and "bad" (or inappropriate adoption of black working-class) taste implicitly ridiculing him in the way that other characters (including his own family members) explicitly do in the film. As is the case with *I Love You, Man,* various narrative elements in the film disguise or even mystify the stark class divide between Molly and Kirk—perhaps most powerfully revealed when Molly comes to a Kettner family lunch and is subjected to a barrage of stereotypically white working-class tastes and perspectives (such as the NASCAR wedding planned by Kirk's brother and future sister-in-law and their upcoming family vacation to Branson, Missouri). Chief among these mystifications is the one embedded in the film's title: the notion of an individual's "league," computed in part through markers of class status (e.g., car ownership and education) and physical attractiveness (which, as normatively imagined, is not entirely separable from class status). And although, as in most romantic comedies, everything works out in the end, Kirk getting Molly *and* achieving his dream of becoming a pilot, it is by the self-nominated natural aristocracy's uplift by the bootstraps (in Kirk's words, "I *am* good enough") that he, with his friend Stainer's encouragement, overcomes the realities of his class background that do not always explicitly signify class identity in the film and are therefore mystified. As a result, particular props (mobile phones, ringtones, cars, clothes) are never understood as narrative elements in a totality elucidating that one's "league" has quite a lot to do with one's economic quintile and class upbringing.

If, on one level, the appearance of the "Marimba" ringtone supplants the Nokia Tune's original function as a generic ringtone with a tinge of class elitism, the conspicuous presence of Apple products in both of these Hollywood comedies suggests a broader dynamic at work: product placement. For example, at multiple levels, *I Love You, Man* seemed to function as an opportunity for an Apple product placement deal—Peter's "Marimba" ringtone is heard numerous times in the film; shots of iPhones ringing and used as cameras are prominent throughout; the iTunes software is even mentioned by name at one point; and the film also features Matt Costa's "Mr. Pitiful" (*Unfamiliar Face,* 2008), which was used in iPhone 3G commercials in the summer of 2009, a few months after the film's release. Although Apple product placement is less extreme in *She's Out of My League,* the few shots of the iPhone and the central role it plays in the narrative of the film—as the primary prop first initiating a connection between Molly and Kirk—likewise grant Apple pride of place within its consumerist horizon.

However, these deals, like most of Apple's transactions and surely like some of the other examples cited above, probably were results of free exchange, the placement being facilitated by free Apple goods for the film and with the company making few stipulations as to how the products should be depicted.[46] Some startling statistics indicate the extreme dominance of Apple products in Hollywood today. Abe Sauer describes the situation:

> In the 44 films in 2009 that topped the box office for at least one weekend, an Apple logo or device could be seen in at least 18 of them. (That's almost 41%.) In some, Apple products even eclipsed their human scene partners. This high appearance rate does not include the heap of mass-market films from 2009 that did not own a weekend but also featured Apple product placement. [This latter group would include *I Love You, Man*.] . . . For one month in 2009, product placement tracking service Nielsen IAG noted 62 total "occurrences" of Apple products, more than any other brand, including AT&T (59), Coca-Cola (55) or Ford (41). Playboy, even with a whole show crafted around its brand name, logged a comparatively paltry 21 occurrences. This is a downturn from 2008, when Apple appeared in exactly half of the 40 #1 films. This achievement comes at the end of a decade-long climb for Apple. . . . Last year, Apple's product placement plan finally topped the heap as well-with Apple appearing in more #1 films than any other brand, including Ford and Pepsi, which had held the most appearances for every year going back to 2001 (except 2003, when the #1 was Cadillac).[47]

Sauer goes on to note the "deeply inauthentic nature to many of Apple's appearances, particularly films set in administrative hells where the brand is rarely found, whether hospitals (*Scrubs*) or government offices (*24*, *Law Abiding Citizen*)."

Why have Apple products been highly successful in these spaces? Nominally it has to do with the ease with which Apple grants clearances to use its products on sets, but some other reasons can be identified. For example, Apple's visibility has increased tremendously in the last decade as a result of the iPod and now iPhone, leading to a spike in sales across the brand's product line. Moreover, the fashion-item status of Apple products, owing to their clever product designs, and the quasi-religious zeal of Apple and Mac fans seem to make the company immune to various forms of consumer criticism, political and otherwise, and to lend the company a politically progressive association favorable to Hollywood.

In addition, Apple's avoidance of payment for product placement makes it unclear when an appearance of an Apple product is or isn't an actual product placement, lending all brand appearances in film and TV a naturalized veneer. (At most, the company gives its products away for use.) Finally, the overrepresentation within the media industries of Apple computers, which are often seen as the best for editing music and video, perhaps leads those working on film and television shows to be either confirmed Apple and Mac fans or habitual users of the company's products.[48]

Insofar as Apple has become a nearly ubiquitous brand in a context of declining success of product placements and has been able to make its products central to cinematic narratives (i.e., the "story-point product placement" coveted by brand marketers, as is evident in *I Love You, Man* and *She's Out of My League*), the company has had success with an audience increasingly hostile to branding and advertising during a major economic recession.[49] However, in view of the national specificity of the mass media under consideration here, one might propose a different explanation: that Apple, particularly for the zealous (and predominantly US-based) 2 percent of the global market that consistently uses its most expensive products, represents a strategically competitive American corporation (the other one playing a similar role currently is Google) that intervenes in markets (mobile telephony, hand-held music devices) long dominated by Asian companies (Sony, Samsung, LG), European companies (Nokia, Ericsson), and cross-regional joint ventures (especially Sony-Ericsson after 2001).[50] Representing the "good" corporation and therefore "good" American intervention in the global marketplace, Apple's corporate "halo effect" may be doing all sorts of cultural work in assuaging fears of national decline and stoking a mild form of consumer nationalism, at least among the Americans purchasing the company's products.[51] The rationale for such digital chauvinism is, like all chauvinisms, illusory: the company seems merely to have facilitated a degree of technological parity with the rest of the developed world, particularly where smartphones and mobile Internet access are concerned.[52]

The unstable dynamics of what Apple signifies outside the United States are perhaps best expressed in the UK television serial *The Trip* (2010), starring the comedians Steve Coogan and Rob Bryden. Working as restaurant critics writing a series of articles for *The Observer*, Coogan and Bryden play semi-fictionalized versions of themselves and venture on an eating tour across the north of the country, all the while engaging in somewhat antagonistic conversations and competitive impressions of famous actors (Michael Caine, Anthony Hopkins, Sean Connery). The plot of the series concerns the contrasting love lives of the show's two protagonists: whereas Bryden is depicted as a contented and loving family man with a wife and child who otherwise has little luck with the opposite sex, Coogan is seen as a rather lonely ladies' man whose estranged American girlfriend, Misha, is currently in the United States. Coogan's iPhone ringtone figures rather cleverly in this story. In the first episode of the series, we first hear his text-message ringtone, the familiar "Marimba," which seems to represent the two actors' yuppie class status (which itself is reflected in their easy comfort in the upscale restaurants they visit).[53] In the course of the series, however, the "Marimba" ringtone takes on greater geographical specificity, in part through the use of contrasting ringtones. In the second episode, Coogan receives a call from within the UK with news about his son, the ringtone here being a generic digital ringer; similarly, another call from within the UK (from his assistant/secretary) is signaled by an old-fashioned mechanical ringer. But elsewhere in the same episode, Coogan receives a "Marimba"-producing text message from Misha. In the fourth episode, the "Marimba" ringtone is heard once more, this time (rather confusingly) heralding a call from Matt, Coogan's American agent. Thus, in the course of the short, six-episode series, "Marimba" has shifted from a marker of class and status to being associated with the United States and thus reinforcing notions of Apple Inc. as a signifier of its national location. But in *The Trip* this association is also imbued with a sense of loss: Steve's relationship with Misha seems fairly well doomed throughout, and at the end of the series, Steve turns down an offer to be in a US-based TV show that would have offered him a chance to be in the same country as Misha. Reflecting Steve's attachment to his own country (and his children who live there), the cheerily tropical "Marimba" ringtone here subtly reenergizes the old-fashioned signification of the United States as a land of opportunity and, with it, a certain erotic wholeness, even as its actual status within the world economy continues, in many ways, to dwindle. What is more remarkable in the series, then, is that Steve Coogan rejects the United States, though he continues to long for his erstwhile lover located there, poignantly watching videos of himself with her on the very same iPhone just before leaving his definitive message on his agent's answering machine.

Jia Zhangke and Mobile Music in Postsocialist China

In the examples outlined above, the ringtone generally serves as a cue for the audience to pay attention so that they might follow the narrative shifts arising from mobile communication and/or, as I have tried to show, come to some greater understanding about the characters who possess certain ringtones. If this is clear, then it seems to be the case that, as noted above, ringtones typically go unheard by characters themselves, at least in the sense that they do not receive sustained hearings of a sort that is somehow profoundly meaningful for those characters. Our meditations here might take as their point of departure Claudia Gorbman's description of "narrative cinema's frequent use of diegetic music for ironic effects, [in which] such music is not only 'unaware' of the dramatic situation, but 'indifferent'—or, as [Michel Chion] puts it, 'anempathetic.'"[54] Gorbman also emphasizes that apparently coincidental diegetic music can be "made to match the scene's mood and pace with an uncanny consistency."[55] From a comparison of Gorbman's descriptions and the foregoing discussions of ringtones, it would seem that the ringtone's cinematic and televisual treatment must be inserted into a long tradition of diegetic music use in these media. However, Chion's notion of "empathy" pertains to the affective quality of the music (and its relationship to onscreen activity) from the point of view of the spectator, and the gap between a character's affective state and the music has, for him, "the effect not of freezing emotion but rather of intensifying it, by inscribing it on a cosmic background."[56] But what if the analytic orientation is shifted from marking empathy (or its lack) for characters themselves to a perspective that the viewer is not fully capable of adopting sympathetically? Hence, instead of considering examples of music's empathy in which we hear the affective responses of characters through a ringtone in a "metadiegetic" sense, if the spectatorial frame eliminates or diminishes opportunities for such empathy, we might see the viewer as a voyeur who cannot truly "feel" or understand what the characters are going through *unless she has gone through such an experience herself*. Put another way, might we draw on Michael Fried's famous theatricality/absorption opposition to describe the subjective agency of character depictions in art (and which could easily apply to film and television—the breaking of the fourth wall signifying the presence, for Fried, of theatricality) and consider *the subjective agency of the music itself*, so as to imagine an "absorbed" diegetic music that is more important to (absorbed) cinematic characters than to spectators?[57]

With the above in mind, let us consider a rather subtle treatment of the ringtone in the "Asian-digital age," as found in the work of the remarkable young filmmaker Jia Zhangke (b. 1970).[58] A member of the so-called sixth generation of Chinese filmmakers who has drawn inspiration both from the new documentary movement in China and, in a far more limited sense, from aspects of fifth-generation directors (e.g., Zhang Yimou and Chen Kaige), Jia Zhangke has sought to document the economic, physical, and social transformations taking place in post-reform China and to produce realist narratives of ordinary Chinese citizens navigating the perils of the country's unstable condition of "postsocialism."[59] The conjunction of Jia's interest in documenting the profound social changes in China from the 1980s on and his longstanding and acute understanding of popular music and music's relationship to film would seem to have facilitated his observations of music in its mobile electronic and digital forms—on mobile phones, from portable stereo systems, with digital musical lighters, and through wireless microphones in karaoke parlors, for example.[60] But one might additionally suggest that—in contrast with urbanite directors from Japan or even South Korea—the striking contrast between Jia's relative (if decreasing) freedom from culture-industry commodities while growing up in his somewhat remote hometown of Fenyang in the 1970s and the 1980s and his being thrust into the increasingly high-tech lifeworld of Beijing (first in college in the early 1990s, at the prestigious Beijing Film Academy), as well as the rapid marketization of China in the wake of Maoism, have sensitized him to the effects and meanings of chirping digital devices. Indeed, their cinematic treatment has accorded well with the director's "critical realist" aesthetic.[61]

Arriving at an understanding of Jia Zhangke as a "mobile music" director requires some prior sense of his approaches to both music and to digital media used to (re)produce music. With respect to music, one can appreciate that Jia has developed a relatively consistent aesthetic of music and popular song in his cinematic *oeuvre* that includes a number of features. First, Jia's films, whose sound design is often very noisy, are typically saturated with music and musical references, at times exhibiting an almost bewildering degree of density.[62] Many of the musical references stem from Mandopop (a commercial popular music genre largely based in Taiwan and legally available in mainland China only as a result of state reforms and economic liberalization) and Cantopop (which is comparable to Mandopop but features the Cantonese language and is based predominantly in Hong Kong), but, as Tonglin Lu has suggested, the director tends to combine musical forms and practices, arguably lumping them together into a general popular culture that can include regional pop forms, Western (especially English-language) popular music, and even Western classical chestnuts.[63] Second, Jia tends to home in on particular songs, even on rare occasion reusing them in different films (for example, Sarah Yeh's theme song for the 1989 John Woo film *The Killer*[64]), and often treats them as having a kind of thematic function, threading their way through his films—such as in the use of his third feature film's (*Unknown Pleasures*, 2002) title song "Ren xiaoyao," which appears periodically in *Unknown Pleasures*. Third, Jia favors the presentation of songs through various modes of delivery and with varying degrees of sound fidelity, often favoring low-fi and ambient versions of songs or pieces presented in public or semi-public spaces. Such "performances" range from songs played on small, cheap speakers and recorded by an ambient microphone to the inclusion of the song directly on the film soundtrack, in a relatively loud mix; they also include accompanied or unaccompanied singing by the actors, a favorite technique of Jia's. Fourth, a thematic song can be heard in multiple formats (combining the last two techniques described above), allowing the song to undergo a transformation through the film. Again, *Unknown Pleasures* offers an instructive example. As Michael Berry has argued, the song, sung by Richie Jen and a hit in China in 2001, is first heard in a room playback of the recording, as an accompaniment to a dance performance introducing Qiaoqiao (Zhao Tao); in this context, the song features an extra percussion track (performed live or added onto the recording) and is sped up to 110 BPM (from 97 to 98 BPM), raising it a whole step in pitch. Later in the film, in a parlor room, Binbin and Yuanyuan argue, listen to the song, and then sing along with the recording; again, the music's fidelity is low, the result of an ambient recording of the television, but it is heard in its original guise (without extra percussion, and at the slower tempo and lower pitch) and, being relatively closer-miked and without any room reverb, it is of significantly higher fidelity than in its first appearance. At the very end of the film, after Binbin has been arrested for attempted robbery and ordered to sing while in custody as an act of humiliation, he offers up a movingly artless *a capella* version of the song. Finally, during the closing credits, immediately after the forced-singing scene, we hear the previous version of the song again, with Binbin and Yuanyuan's voices singing along with the chorus.[65]

Extrapolating from the above, one might say that Jia Zhangke's musical aesthetic in his films seems in part to be a matter of "recoding"—a metaphor meant with semiotic codes and software code in mind, but also with its biological use to describe transformations of RNA molecules that had been altered from an initial transcription from a DNA sequence. Here, "recoding" denotes a kind of quasi-leitmotivic recontextualization of a musical number or figure, which is transformed not only through harmony or instrumentation, but rather through different renderings that can range from re-instrumentation to karaoke-ization to different playback media and acoustics (as suggested above).[66] Crucial to Jia's aesthetic of recoding is that there are clearly less and more "adequate" or "competent" performances or instances of a particular musical excerpt and that the director finds a great degree of pathos in quotidian renditions that might otherwise be thought of as substandard or unsatisfactory.[67] As described in *Unknown Pleasures*, the effect of this technique can structure the narrative arc of an entire film—in this case, the transformations of "Ren xiaoyao" making the film a rather bathetic tragedy—and can work in both disconnected appearances separated over long durations and in immediate juxtaposition (the latter featured at the end, when Binbin's *a capella* performance is immediately followed by the sing-along version by Binbin and Yuanyuan heard earlier). Jia's predilection for the everyday and his realist aesthetic also appear to lead him to present musical material in various states of inadequacy, using "idealized" versions of that material infrequently (when appearing in full fidelity, it is usually in a non-diegetic soundtrack) and even then usually contrasting them with non-idealized versions. In part, this may have to do with Jia's interest in representing the unbridgeable distance between characters' aspirations or desires and their unsatisfactory environs; in Jason McGrath's terms, the director's films often thematize "the gap between the imaginary of popular culture and the intractable reality that the characters actually inhabit."[68] Perhaps it is noteworthy that Jia does not seem to favor scenes of headphone listening, which has played a certain politicized role in American scholarship on East Asia. Jia avoids scenes merging source scoring with non-diegetic soundtracks, which could facilitate a banal treatment of idealized listening, on account of their stereotypical appearance in Hollywood and world cinema (think of "rebellious teenagers" listening to music on headphones—examples in *The Sopranos* and in the 2007 film *Persepolis* come to mind). But his eschewal of such solutions may be primarily realist in nature and may also have to do with the relatively slow adoption of headphone listening itself in most parts of the world until the advent of the portable MP3 player and music phones in the last few years.[69] Indeed, instead of being represented as resolutions to the contradictions of popular culture in Jia Zhangke's films, digital and mobile musical devices instead seem to contribute artless musical performances themselves, in some cases substituting for human voices in the process. As such, mobile music for Jia contributes to the dialectic of mobility and immobility that he seeks to portray in Chinese contemporary life, with bodies that move through and between spaces, and even to new cities, but nonetheless are never capable of going anywhere or achieving fulfillment in a more tangible sense.[70]

Jia's preoccupations with popular music, digital music, and mobile music extend as far back as his first full-length feature film, *Xiao Wu* (Pickpocket), which was released in 1997—shortly before the era of the customizable ringtone. That film has been lauded for its innovative use of sound and music and has already come to be viewed as a classic of contemporary Chinese cinema. *Xiao Wu* portrays three failed-relationship scenarios involving the protagonist, the film's eponymous pickpocket played by Jia Zhangke's frequent collaborator and "alter ego," the actor Wang Hongwei. In the first scenario, Jin Xiaoyong, Xiao Wu's former pickpocket comrade, has become an entrepreneur and now is trafficking in stolen cigarettes and the sex trade. In the second, the relationship is with a karaoke girl named Hu Meimei. The third scenario involves the protagonist's relationship with his family. By the end of the film, the protagonist has been arrested. In the concluding sequence, he is inadvertently humiliated in public by an accumulating crowd of staring bystanders while temporarily handcuffed to a telephone pole. Concerning the music in the film, I have only three main points to add to the excellent discussions of the subject by Tonglin Lu, Michael Berry, and Jason McGrath.[71]

First, there is extensive use of source scoring and diegetic music, much of which emanates from human bodies (singing and whistling) and mobile and portable musical devices of various sorts (portable radios, hand-held electronic gadgets, and karaoke machines with wireless microphones). By a cursory reckoning, there are almost thirty distinct musical episodes in the film, ranging in length from a few seconds or less to a minute or two, and the number would increase if we did not allow for some degree of continuity (for example, not counting the individual musical passages that are all part of a collaged soundtrack from John Woo's *The Killer*, which appears during the middle of the film).[72] In a reasonably conventional example of the source-scoring effect, we might consider the pop song based on the story of the King of Chu, as discussed by Lu.[73] As Xiao Wu walks down the stone-walled alleys of Fenyang early in the film, we hear the song, which seems to function as a non-diegetic soundtrack; but once the protagonist has exited the alleys and stands in the open street, in a cut coinciding with a pause in the song (probably just before the bridge), we see a table with some small, single-unit home stereos and boom boxes for sale. As if to confirm the (newly) diegetic status of the song, the saleswoman presses stop on the tape on one of the boom boxes, causing the music to stop, just as the music crescendos on the dominant chord before transitioning to the next section (possibly a return to the verse). Shortly afterwards, the saleswoman hits play on the boom box, and the song starts up again (though possibly at a different point).[74] Important here for present purposes is the effect of the low-fidelity portable reproduction system, whose lack of sonic fidelity might have something to do with the poor sound quality of this song and whose wobbly and drifting pitch seemingly intentionally produces a disconcerting effect.

Second, one song in the film that does not seem to have drawn much attention is Michael Jackson's "Heal the World," from *Dangerous* (1991), which sold hundreds of thousands of copies in China and helped contribute to Jackson's popularity in that country.[75] Like the King-of-Chu-based song, this one also hovers unclearly between diegetic source scoring and non-diegetic soundtrack music, and is heard during the scene in which Xiao Wu watches Meimei get a haircut. Shortly after the hairdresser turns on the radio, which plays a fragment of one song and then some radio chatter, Jackson's song is heard—seemingly from the radio itself, but then transforming into the scene's soundtrack. Taking place during a formative moment in the couple's abortive relationship, the song's romantic-ballad R&B style (with sensitively sung vocals, slow tempo, and smooth seventh chords) would seem to be foregrounded by the scene, inflecting the sentimentally humanitarian ethos of its lyrics in a more personal, individualistic manner, perhaps. Moreover, the song provides one moment of auditory respite from musical references that seem to link Xiao Wu and Xiaoyong and the (failed) homosocial bonds and new economy-produced disjunctions in class status that seem to structure many aspects of the film (including Xiao Wu's relationship with Meimei).[76] Yet, in addition to the fact that its location in China is a result of global capitalist exchange, Jackson's song, which in its video featured representations and critiques of military violence linked to the political upheavals of the end of the Cold War, indelibly marks a moment of global transformation in what one might call "the long 1989"—a moment that includes postsocialist economic globalization in China, accelerating in the years following the Tiananmen protests and massacre of 1989.[77] As such, and despite the rare hopeful tone in an otherwise mutedly bleak film, the broader structuring circumstances producing the rift between Xiaoyong and Xiao Wu still appear to be at play even here.

Third, and perhaps most important for the study of ringtones, is the role of the musical lighter that plays a single-oscillator, monophonic version of the right-hand part of Beethoven's piano bagatelle "Für Elise"—a clear predecessor of later monophonic ringtones using the same melody. Lu skillfully demonstrates the ways in which the lighter serves as a surrogate voice for the reticent Xiao Wu, both in an awkward conversation with Xiaoyong (in which the latter offers unconvincing excuses for not inviting Xiao Wu to his wedding) and in a touching scene in which Xiao Wu "sings" to a bedridden Meimei by using his lighter as a surrogate for his voice.[78] Through the frequent appearance in the film of this piano-lesson chestnut (which is surely by now an item of global European classical-music kitsch), Jia's recoding technique is made particularly clear in two moments. At the very end of the first section of the film, just as a sorrowful, drunken Xiao Wu hears a song dedication on Fenyang TV to Xiaoyong and his bride, the protagonist semi-consciously drowns out the song with the musical lighter, which ultimately runs out of battery power, in the process playing a grotesquely slowing-down and wobbly version of the Beethoven melody (itself underscored by a cut to black). Thus, the already unimpressively tinny, even annoying version of the tune becomes rather lurid and ever more inadequate, providing a powerful moment of deflation for an already dispirited persona. Then, toward the very end of the film, after Xiao Wu has been arrested for theft, the arresting officer handcuffs him to a motorcycle, allows him to watch a television, and encourages him to reflect on his crimes. On the television there appears a commercial for a shoe company with an arrangement of "Für Elise" for several instruments, including a plucked or hammered string instrument (perhaps a highly distorted and/or synthesized piano, perhaps a *guzheng* or a modern Chinese zither), some inner-voice winds (perhaps from a keyboard synthesizer), and an electric bass.[79] This version of the tune, already overdetermined as kitsch, is particularly garish, the bass line filling in the blank spaces of the left-hand part with harmonic support and chromatic passing tones absent in the original and, moreover, doing so on the fourth half-beat of the triple meter, creating the effect of an unflattering pop-style syncopation or polyrhythm. (See figure 8.3.) If we read the lighter

Figure 8.3
Xiao Wu (1997), 1:39:11–1:39:30, TV commercial, "Für Elise" arrangement.

(and its version of "Für Elise") as standing in for Xiao Wu's voice (and the protagonist himself, more generally), we might, in turn, view this inflated arrangement of Beethoven's composition as a proxy for the post-reform capitalist state, with its increasingly uncensored embrace (or at least tolerance) of popular music.[80] Yet both the digital lighter and the advertisement are products of global capitalism, underscoring the naiveté of Xiao Wu's clinging to notions of an older brotherhood facilitated by a non-capitalist world—notions that are figured by a similarly backward looking discourse in *The Killer*.

In the years since the release of *Xiao Wu*, Jia Zhangke has continued to make use of similar film-musical techniques that he developed in his first full-length film while also adopting new elements into his practice. Perhaps the most important of the new elements is scored soundtracks, appearing primarily in more recent films such as *The World* (2004) and *Still Life* (2006), both of which feature soundtracks composed by the multi-talented Taiwanese musician Lim Giong. Beginning in the early 1990s, Giong, whose music was associated with the New Taiwanese Song genre, recorded a number of albums marked by a certain eclecticism and incorporating folk-pop, pop-rock, "alternative" rock, soul/R&B, synth-pop, and even a "world beat" ethos evocative of David Byrne or Peter Gabriel. During the 1990s and the 2000s, he became known as an electronic dance music producer and film composer, working first as an actor and performing musician and then as a soundtrack composer with the great Taiwanese director Hou Hsiao-Hsien. Partly through Jia's admiration of Hou's films and through a newfound interest in artifice and virtuality (complementing and rounding out his realist aesthetic when trained upon metropolitan life), Lim Giong's dance-electronica seemed to provide an effective counterpoint to the consumerist spectacles in *The World*.[81] *Still Life*, in contrast, is marked by the absence of thumping bass-drum beats, and instead features Lim Giong's synthesized ambient music (at times accompanied by vocals), producing a hazy atmosphere of ethnic fusion, particularly during transitional and connecting scenes in the film. Additionally, the film highlights the quasi-musical sounds of repetitive labor—particularly the massed sound of building demolition by hammering men—and the plaintive moans of ships' foghorns as they traverse the Yangtze River. However, Jia's familiar technique of recoding still acts to drive the narrative diegesis along as it does in his earlier films, and he continues to favor mundane forms of music making, such as unaccompanied singing, ambient recordings of musical performances, and even the sounds of digital devices.

The World is best known for its emphasis on text messaging, but ringtones do make an occasional appearance. Specifically, the female protagonist, Tao (played by Zhao Tao), has a mobile phone with an arpeggiated preset ringtone, the bubbly activity of which is consonant with the other treatments of digital life (including the Flash animation sequences). Notably, we hear this ringtone for the first time as Tao rides on the Beijing World Park monorail, which transports her between various parts of the park for her work; we hear it again as she rides the bus and yet another time on her second monorail trip. Linked to the mobility of the cell phone and moments of transit that ultimately lead nowhere, Tao's ringtone is arguably a little sonic lie about freedom in the face of the deadening routine that the park's performers must endure in order to produce its dream-factory spectacles; that lie, however, is inseparable from the utopian horizon of an outside (and better, less alienated) world which the mobile phone seems to offer.

In *Still Life*, by contrast, mobile phone ringtones are treated as central to the narrative in a couple of ways. The first involves a remarkably touching scene with the main male character, Han Sanming (played by Han Sanming, a cousin of Jia's who also appeared as a similar character in *Platform*). A miner having traveled from Shanxi province to find his estranged wife of 16 years, Han Sanming temporarily resettles in Fengjie, a city being flooded, destroyed, and rebuilt on higher ground as a result of the massive Three Gorges Dam project. To make a living, he finds work as a manual laborer in building demolitions, and befriends Brother Mark, a petty swindler and low-level underworld figure who lives in the same boarding house as Han Sanming. Unlike the soft-spoken Han, Brother Mark theatrically sympathizes with (and divines) his interlocutor's plight—he has paid for a wife from Fengjie, she has left him, and he wants to retrieve her. After remarking that Han is "a nostalgic," Brother Mark acts out a scene from *The Killer* in which that film's protagonist (played by Chow-Yun Fat) laments the decline of the old traditions of honor in the Hong Kong triad to a trusted underworld brother and wistfully notes that the two of them are too nostalgic. After Mark affirms Han's request for help (translated as "I hope I can count on you" or "If something comes up, please help me"),[82] the two characters share mobile telephone numbers, calling each other's phones and listening to their ringtones in the process. First we hear Han Sanming's ringtone, a monophonic rendition of a melody titled "Bless the Good People," which Ackbar Abbas suggests has old-fashioned, Communist-era overtones but which is actually the theme song of a 1989 television soap opera called *Yearnings* or *Aspirations* (Kewang), which deals with the history of the Cultural Revolution and its aftermath. Then, Brother Mark proudly holds up his phone, playing a sound-file ringtone of "Shanghai Bund," the theme song from the 1980 Hong Kong TV series "The Bund" (Shàng Hǎi Tān) about the triad (and which also featured Chow-Yun Fat).[83] The moment is telling in a number of ways. Mark's youth and up-to-date familiarity with Fengjie is affirmed by his relatively advanced ringtone technology, in contrast with the older, rural Han Sanming's old-fashioned monophonic phone. And, the thematic references of their ringtones, in addition to underscoring their status as underworld member and laborer, respectively, also highlight the two characters' different temporalities and historical trajectories (the triad underworld of Mark's ringtone being the condition to which mainland China might be said to aspire, as opposed to the world of the Cultural Revolution of Han's ringtone that China has sought to escape), though Mark's is actually the older song. Even more remarkably, the film cuts during Mark's ringtone to a shot of a nearby television showing an arrest, a tearful bystander, a tourist boat on the Yangtze, and a man on a hilltop waving his jacket—possibly scenes referring either directly or indirectly to the destruction caused by the Three Gorges Dam project. In the immediately following scene, as the film cuts to a boat-level point-of-view shot (showing a board on the mountain reading "THIRD PHASE WATER LEVEL: 156.3 METERS") and then a shot tracking in on a barge on the Yangtze River, the music transforms from the ambient recording of Mark's ringtone to a direct reproduction of the song, which is now part of the film's non-diegetic soundtrack. A subtle version of Jia's recoding technique, the sonic ideal of "Shanghai Bund" presented in the soundtrack is contrasted ironically with the hulking barge, a symbol of the priorities of industrialization transforming the area with devastating consequences.[84] In particular, the song is split between ringtone and soundtrack rather smoothly. In cutting from the television, it moves past the lo-fi reproduction of its opening two phrases, which are in turn declamatory, even martial, and then rather spare and more clearly orchestrated (with a flute and a wood block entering), and settles into the river scene just before the fuller soundtrack version of the song is heard. The easy-listening/light-rock drumbeat that is heard mid-song effects a particularly smooth transition to higher-fidelity sound, moving gently in tandem with the rolling movement of the landscape as captured by the on-boat camera—arguably transforming the previously unmarked diegetic music of a ringtone into empathetic music, reminiscent of editing techniques found in many Hollywood films. (See figure 8.4.)

Later in the film, as Han Sanming attempts to get in touch with Brother Mark (with whom he was to have dinner), the camera pans slowly from

Han on his cell phone to a table featuring three strange men dressed in operatic garb, engrossed in their cell phones (playing video games or text messaging). This apparently surreal moment is then followed by a shot of Han calling Mark again the next workday at a demolition site, and we hear a trebly version of "Shanghai Bund" in the distance. As Han searches for the sound source, finding it in a pile of rubble, he calls the other workers over to clear the debris, realizing that Mark is dead, either by accident or (more likely) as a result of the underworld conflicts and paid thuggery in which he was involved (and from which Han abstained).[85] In a fitting memorial to Mark, Han Sanming conducts a solitary wake ceremony for the body of his deceased friend: he lights three cigarettes, in place of the traditional incense, and places them at a makeshift altar that includes a photograph of Mark and, of course, his mobile phone. Soon afterwards, Mark's corpse is sent out into the water, hence providing another interpretation of the river's waves mentioned in the song's lyrics, which initially seem placed to refer to the flooding caused by the Three Gorges Dam.[86] Retrospectively, the scenes of tragedy on the television during the characters' earlier ringtone conversation may have foreshadowed Mark's death, but the contradictions of unbridled industrialization and the working-class conflicts and underworld activity it produces may actually offer the more sobering and coherent explanation for this character's fate.[87]

Were we to attempt an explanation as to why Jia pays such close attention to these mundane musical matters in his films, we might posit that it is precisely because they matter to their characters. As meaningful but dilapidated tokens of sonic ideals represented only outside cinematic diegesis, if ever, snatches of sung melodies or their tinny reproduction on low-fidelity speakers and synthesizer systems are typically the only musical materials that Jia's itinerant and mobile characters make use of in their daily lives. Yet, although the gap between aspiration and reality (a theme in his films generally) is revealed in such disparities of musical fidelity, the director's humanism allows for these moments to be genuinely meaningful in characters' lives.[88] Most film directors and television producers are typically content to have ringtones convey meanings about characters to audience members (whether in a general sense or as providing ironic commentary), and this is not entirely untrue of Jia's work. Moreover, Jia's occasional use of non-diegetic, high-fidelity versions of a character's "own" song provides information to audiences—even as it places those audiences on the opposite, idealized side of a divide between the space of utopian spectatorial omniscience and the "real" obscurity of the diegetic world.[89] But for Jia Zhangke ringtones also convey meanings to and from

Figure 8.4
Still Life (2006), 38:42–39:09, "Shanghai Bund" ringtone/soundtrack.

Figure 8.4
(continued)

characters themselves, depicting moments of communication immanent to the diegesis itself and not primarily designed to structure the spectator's affective responses in a traditionally "empathetic" way or even as a cue for (postmodern) irony. In emphasizing music's ability to generate empathy between characters, Jia's work offers a powerfully realist riposte to the reifications of personalized consumerism, even if by no means simply or simplistically so—after all, his characters are often associated with specific ringtones.[90] Nonetheless, a great many characters in film and television do not really hear the ringtones they purport to possess; in this director's films, however, they are accorded the rare decency of being genuinely able to listen to and even commune with them.

9 What's in a Name? Race and the Ringtone's Revival in (Un-)Popular Music

On January 12, 2010, Berin Szoka, a blogger for the Technology Liberation Front—a libertarian media commentary site advocating against Internet regulation—posted a satirical report titled "We Must End the (Reverse) Digital Divide!" The occasion for the post was the recent release of statistics by the Pew Research Center's Internet & American Life Project showing that African Americans and Latinos had surpassed white Americans in wireless Internet use. Szoka, a Senior Fellow at the market-oriented Progress & Freedom Foundation (PFF) and director of its Center for Internet Freedom, ended his brief recapitulation of the Pew report's observations as follows: "Congress must act to correct this clear racial travesty. . . . Only then will digital Racial Justice be achieved for *all* Americans."[1] While the post's subtext is that left-liberal petitions on behalf of goals such as "Racial Justice" are pointless, even dangerous exercises—requests for government intervention obviated by the guiding hand of the all-wise market—it points, perhaps unwittingly, to a transformation in understandings of black America in the age of Obama. For while racial justice certainly continues not to be served, a certain "normalization" in whites' perception of African Americans, in part a cumulative effect of the expansion of the black middle class and a product of the "post-Soul" era, can be attributed to the transformation of material conditions in black life that extend beyond the election of a black president.[2] Indeed, the far right's complaints of (reverse) racism, with the rhetoric of which Szoka's post flirts, seems largely to be a minority, if certainly disturbing reaction partly stemming from the statistical fact of the declining white majority in the American populace.[3]

One element of that broader set of material transformations in black life is technological, but Szoka's post only points to it obliquely. Particularly, mobile telephony appears to have assumed a significant role in black society, crossing the class division between the black middle class and the black working class and transforming African Americans into a lucrative market of mobile consumers. The cell phone ringtone plays a prominent role in this story, resulting in the ringtone's "racialization" in the United States—and with popular music as the primary terrain of the transformation.

Popular Music and Black (Telephonic) Modernity

The study of ringtones and their racialization can, in part, be understood as an addendum to an important argument made by Alexander Weheliye. He argues that the notion of "the 'digital divide,' while surely drawing much needed attention to certain politico-economic inequalities, cannot but reinforce the idea that Afro-diasporic populations are inherently Luddite and therefore situated outside the bounds of Western modernity."[4] Yet such assumptions patently ignore the ways in which these populations engage with the technological, for example through the domain of popular music. According to Weheliye, "popular music offers one of the most fertile grounds for the dissemination and enculturation of digital and analog technologies and has done so at least since the invention of the phonograph at the close of the nineteenth century. Pop music also represents the arena in which black subjects have culturally engaged with these techno-informational flows, so that any consideration of digital space might do well to include the sonic in order to comprehend different modalities of digitalness, but also to not endlessly circulate and therefore solidify the presumed 'digital divide' with all its attendant baggage."[5] Consider the work by Eszter Hargittai and other sociologists on a "second-level digital divide" produced by the variable acquisition of skills, in contrast to a divide based on technological access. It may reinforce racialized conceptions of technological access—surely class and education are the primary markers of such asymmetries, rather than race—and yet also speak, if obliquely, to aspects of present social reality—precisely, in part, on account of the racialized asymmetries of class and educational access.[6]

These two arguments operate at different, seemingly incompatible registers—one on the plane of cultural theory, the other of empirical sociology—but something of a sublation of the two can be effected. For a host of complex reasons involving differential access to mobile and digital technologies, working-class African Americans have engaged with them in ways both under-recognized and articulating cultural/musical habits of everyday life with the production of new modes of consumption. Specifically, as working-class African Americans gained access to mobile phones largely through the rise of prepaid payment plans, they soon transformed into a lucrative market for digital goods, a market that was in part fostered by relatively public approaches to music listening extant in other black diasporic contexts (boom boxes, sound systems, etc.).[7] Here, skill development can be understood as tailored to the particular devices of mobile consumption, though it is surely a more complex matter than simply mastering the requirements of using those devices.

The lynchpin to making this mobile entertainment market a reality was a preexisting phenomenon that one might cautiously term "black telephonic modernity": the favoring of telephonic communication among African Americans (and perhaps the black diaspora generally, insofar as the tendency is not unique to African Americans by any means and is characteristic of many status groups' mobilization and maintenance of kinship relations and social networks).[8] This tendency was further transformed by mobile phone use in the everyday lives of black consumers. Three articles of "evidence" might be examined in this context. First, the following comments from Robert L. Johnson, founder and former CEO of Black Entertainment Television (BET), counter some of the stereotypes about African Americans and the digital divide:

> You look at cable penetration in African American households. Look at the VCR, telephone: African Americans probably spend more for long distance telephone per capita than any other population. Beepers, cell phones, the technology has to serve their needs. The African American people love to communicate, and they do it orally. They don't do a whole lot of letter writing, but they will definitely go to the long distance phone and check in with their extended families, distant relatives, and friends. The same thing with the beeper or the cell phone.[9]

Although Johnson maintains a certain essentialized view of the "African American people," particularly when it comes to writing and literacy, his point about black telephonic communication merits attention. As an entrepreneur cashing in on an African American target market, Johnson appreciates its lucrative potential and describes companies' avoidance of it as "economic racism":

> African Americans spend on average 30 bucks a month on cable, 50 bucks a month for long distance service. . . . What you have to do is demystify the computer, and then let them know that there's content on the Internet that's specifically geared and tailored to their interests, and you've got to promote it. . . . So what I'm saying is the Internet people have yet to wake up to the fact that African Americans spend $500 billion on consumer goods, and a lot of those dollars go into entertainment and communications. The long distance people got it.[10]

The conjunction of entertainment and communications that Johnson describes is pertinent here: the mobile phone has transformed as a vehicle for both markets, not only juxtaposing them in the same device, but also combining them, such that entertainment goods (e.g., music) are used as a means of telephonic communication (e.g., ringtones, ringback tones) and communications goods (e.g., text messages) transform into or articulate with significant entertainment goods (e.g., text-message promotion campaigns and mechanisms, such as phoning and texting votes for competitors on *American Idol*).[11] Concerning African Americans in particular, that conjunction has been fruitful for the mobile telecommunications and entertainment industries. According to a 2010 Pew Internet & American Life Project report written by Aaron Smith, African Americans and Latinos outnumber white cell phone users in wireless Internet use, confirming a trend first noted in 2009, *and* they surpass white mobile phone ownership and functional use.[12] The numbers, though perhaps skewed by lower, class-determined rates of African American life expectancy, point at least to equivalent if not much greater use when corrected for age demographics, and suggest that American minorities are largely accessing the Internet through mobile phones—although African American laptop ownership is also growing.[13]

Perhaps the most important evidence comes from the domain of popular music, in which mobile phones and telephony have long played a significant role. Weheliye described the phenomenon a decade ago, drawing on the notion of "hypersoul," in which "soul's religious and spiritual undercurrent is often pushed aside in favour of brazen aspirational materialism (aka the 'playa' culture). Many tracks flaunt an obsession with hi-tech consumer gadgetry, especially mobile phones."[14] Weheliye concurs with this assessment:

> To say that communications and other technologies are leading actors on the stage of contemporary R&B would amount to an understatement of gargantuan proportions; lyrically, hardly a track exists that does not mention cellular phones, beepers, two-way pagers, answering machines, various surveillance gadgets, e-mail messages, and the Internet, stressing the interdependence of contemporary interpersonal communication and informational technologies.[15]

Moreover, Weheliye identifies a few relevant songs of the new hypersoul genre, including Missy Elliot, 702, and Magoo's "Beep Me 911," Destiny's Child's "Bug-a-Boo," Ginuwine and Aaliyah's "Final Warning," and the Firm's "Phone Tap" as all incorporating mobile communication and surveillance device topoi and sounds. Likewise, Weheliye notes the prevalence of the vocoder in 1980s and 1990s R&B and notes the appearance of similar effects in more recent songs, like K-Ci and JoJo's "Crazy," which seem to be relatively early examples of the fad in 2000s R&B for the Pro Tools Auto-Tune plug-in used prominently in Cher's 1998 dance-club hit "I Believe"—a fad strongly associated with the singer T-Pain.[16] As Weheliye notes, this and other alterations to the voice could often be understood as part of a "cell phone effect," one that "marks the performers' recorded voices as technologically embodied."[17] More recent R&B and hip-hop songs featuring cell phone effects include Soulja Boy Tell Em's "Kiss Me Thru the Phone" (2008), in which a synthesizer line tracking the chorus vocal line sounds like touch tones or a low-fidelity ringtone; Beyoncé and Lady Gaga's "Video Phone" (2008), which also begins with what sounds like a monophonic ringtone underneath its Morricone/spaghetti western-esque opening, and with ringtone sounds continuing through the rest of the song (cell phone voice effects appear at certain moments, too); or Trey Songz's "LOL Smiley Face" (2009), which features iPhone "Marimba"-esque sonorities in its chorus.[18]

Weheliye's 2002 article marshaled R&B in the service of an attempt to counter oversimplifications in the discourse of posthumanism that either largely ignored race (as in N. Katherine Hayles' *How We Became Posthuman*) or oversimplified the problem of black diasporic posthumanism (as in Kodwo Eshun's *More Brilliant Than the Sun*). Eshun in particular delineates a black posthumanist music aesthetic (especially in "jazz, funk, hip-hop, techno, and jungle") as part of Afrofuturist discourse, in contrast to a black humanist musical lineage (especially in R&B) and thereby downplaying the ways in which the humanist and posthumanist lineages overlap.[19] At that time, R&B's revived fetish for new sound-production technologies was recent and coterminous with the rise of black cell phone use, during which mobile phones, Blackberrys, pagers, and other technologies were portrayed in R&B and hip-hop videos as relatively elite commodities associated with wealth, particularly wealth acquired illicitly in the drug trade or the sex trade. Although such imagery has not disappeared from the visual languages of those music genres, the elite status of mobile technologies within the black diaspora, at least in the United States, *has* disappeared. Mobile phones and mobile Internet use, as was noted above, are rapidly becoming ubiquitous for blacks in ways that have registered them as a valuable market—particularly where ringtones are concerned.[20] The mainstreaming and banalization of such vocal effects and technologies suggests less that mainstream popular culture is pervasively posthuman in ethos and more that the technological forms that seem to provide their objective correlates are now so common as to provide the pretext for *generic conventions*—as has been the case for many technological adoptions in popular music applied to voices, such as the electrical microphone, reverb unit, overdriven amplifier, multi-track recorder, and delay processor—rather than mere novelty effects.

One more article of evidence comes from R. Kelly's "hip-hopera," *Trapped in the Closet* (2005–2007), a narrative and musical tour de force featuring a hypnotic groove and a plethora of different voices, all sung by R. Kelly himself, corresponding to the production's different characters. The R&B-based genre of the rap opera or hip-hopera, said to be inspired by rock opera and musicals, seems to signal a desire for the genre to expand beyond the confines of the single song or even album, but *Trapped in the Closet* seems strange and ambitious even by the genre's own standards; Chuck Klosterman describes it as "a parody of musical theatre, a soap opera from the late 1970s, and a BET version of *The Red Shoe Diaries*"[21] and features a bewildering number of characters and plot twists, all inaugurated by a man, Sylvester (played in the video by Kelly himself), who awakes to find himself in a strange bedroom and having slept with a married woman, and then hides in the closet when her husband comes home. *Trapped in the Closet* hangs together because of its music, featuring a strange mixture of monotony and amplified tension: Sung over the same, insistently unstable $ii^{ø4/3}$–I^6 groove, the music builds up to a fevered pitch at the end of each of its twenty-two episodes or "chapters," which in every case famously ends with a cliffhanger.[22] The complex and twisted logic of *Trapped in the Closet* is not easily grasped, but, as Klosterman has noted, one of its central concerns is telephonic communication:

> It certainly seems like Kelly is trying to make some kind of point about the fragility of modern communication. He is pathologically obsessed with the telephone: much of chapter 17 centres on Sylvester's inability to get quality cell reception while driving under an overpass. Chapter 22 is exclusively comprised of phone conversations. Within the world of TITC, there is no situation that necessitates voicemail; characters will still answer a ringing phone while actively threatening their adversaries at gunpoint. The irony of this omnipresent phone banter is that the people in this narrative still can't understand what's really happening to them; much like the characters on Lost, they're blindly oblivious to the interpersonal connections they already share. What the rest of us are supposed to take away from that revelation remains unclear. It's possible that R Kelly believes the acceleration of technology erodes the authenticity of human interaction; it's equally possible that this thought has never occurred to him and never will.[23]

As Klosterman notes, the scene that appears at the very end of *Trapped in the Closet* is central to this thesis, but its mode of representation is particularly striking. It features a complex series of overlapping telephone calls (a mix of mobile phones and land-line phones) concerning a rumor about "the package," which has come into the possession of the pastor, Rufus, who, it turns out is gay or bisexual and whose former lover, Chuck, is now in the hospital with HIV/AIDS. In representing this series of phone calls, two features are noteworthy: first, each caller is presented in a floating bubble that moves through an otherwise black screen; second, by the end of chapter 22 (1:23:26–1:23:51), all of the callers are amassed together on the screen and the familiar backing groove has disappeared, leaving only the sung chatter of the conversational voices—again, all sung by Kelly (see figure 9.1).[24]

In scholarly discourse on mobile communication and music, one of the most arresting metaphors is that of the "bubble." It features prominently in the work of Michael Bull, one of the best-known scholars of mobile music devices. In *Sounding Out the City*, Bull draws on interviews with personal stereo users, who themselves mention the word, and he offers the general claim that "Personal-stereo users each inhabit their own cultural 'bubble' facilitated by culturally produced messages from the 'dream pool' of culture."[25] Although Bull tends to emphasize personal stereo users' isolation from their environs and other people (and their communing with the

Figure 9.1
Video still from *Trapped in the Closet*, chapter 22, 1:23:26.

products of the culture industry), he does not oversimplify the problem to one exclusively of isolated monads and discusses communal acts of personal stereo listening—such as when two listeners share the same Walkman and thereby "produce a collective bubble that distinguishes users from non-users."[26] Nonetheless, fundamental to Bull's conception of personal stereo listening is that users, whether collectively or individually, tend to divide themselves from their immediate present in favor of an (imaginary) elsewhere. Bull's conception of stereo listening resonates with writings on mobile phones, which also are said to produce an isolating bubble around their users. In the words of Joachim Höflich, "The mobile phone affects—even changes—the meaning of place. Once more this creates small (more or less) private spheres—'bubbles'—in the public sphere; 'there is a thin layer of private space around the bodies of the people with whom we are sharing non-private space.'"[27] Again, the concept is the same: the bubble isolates one from the immediate world (more or less), even as the occupants of those bubbles might be connected to individuals in distant locations (probably in their own bubbles). At an extreme, one might see in this discourse an opposition between the immediate present and the communicative or listening domain.

In comparison, R. Kelly's bubbles not only highlight the connections between communicating members, showing the interactions between the bubbles, as if fleeting nodes in an ephemeral network; taken in the aggregate, they provide a picture of a relatively geographically proximate community that uses (mobile) telephonic communication to bind itself closer together, in this case through the sharing of an unexplained rumor.[28] The story, thus, is less about the inauthenticity of human interaction as caused or exacerbated by modernity, but rather that, in the midst of confusion, there is a broader, contentious, and ultimately harmonious (even comic) community that is bound together by the telephone. But, we might go further in remarking that nearly all of Kelly's interlocutors—a large proportion of the characters featured in *Trapped in the Closet*—are black. While it is, at one level, clearly an aesthetic product about a black community, the inclusion of white characters makes it self-evident that race is not the exclusive factor in Kelly's conception, which is instead guided by a picture, however fantastical and whimsical, of a collectivity that is, at heart, working-class.[29] That the dynamics of working-class, multi-ethnic mobile communication are somewhat different than the dominant view of its theorization is not surprising, insofar as mobile communication technologies have diffused quickly but not universally to working-class populations, whose distinctive, if necessarily variable, stamp on those technologies may be different from the middle-class and elite-class uses that preceded it historically.[30] That Kelly decided to focus his narrative within such a class context—avoiding the most lavish depictions of the "Don"/"playa" lifestyle—is to his credit and gives *Trapped in the Closet* an unexpected realism amid the almost surreal goings-on.

Ringtone Charts and "Ringtone Rap"

An important corollary of the rise of black mobile telephone use in the United States is the role of ringtone consumption within African American communities, in which the ringtone often functions as a complement to other modes of public music listening. In this regard African Americans and other minorities are not unique, but black working-class Americans developed as a prime market for ringtones during the heyday of ringtone sales (the mid 2000s)—a reality confirmed by ringtone producers' casual descriptions of the primacy of "urban" ringtone consumption. In-depth statistics are hard to come by—and are often produced as for-profit information of questionable accuracy sold in corporate research reports by mobile marketing firms (see chapter 1 for more on this)—but publicly available demographic indicators are telling. In a press release on one report from late 2005, figures indicated that 37 percent of African Americans owning mobile phones downloaded ringtones, while 28 percent of Latinos and 26 percent of American cell phone owners overall did the same.[31] From these numbers alone, it suggests that the remainder—mostly white Americans—download ringtones with significantly lower frequency: 24 percent, working backward from these data with a few assumptions.[32] According to another report, focusing on Latinos in 2004, "29 percent of Hispanics with cell phones downloaded ring tones, compared to 9 percent of the general online population."[33] Such widely varying figures for the general US population may be attributed to differences in definitions of "online" or variable and faulty data collection methods, but nonetheless it seems plausible that minorities were out-buying white Americans during the peak of ringtone sales, and that African Americans were at the vanguard of ringtone consumption.[34]

Residues of this trend can be found in a few locations within the domain of American popular music. To some extent it is reflected in the charts devoted to ringtone sales. The history of such charts is relatively recent and is tied to the changing status of the ringtone industry itself. For some time, individual mobile entertainment firms would list in-house top-10 or top-20 ringtones in sales, which often were indicative of broader popularity of some ringtones but subject to the uneven distributions of their own customer bases. In Europe, the Mobile Entertainment Forum—billed as a "global trade association on behalf of the mobile media industry," though it is primarily based in London[35]—conceived of a UK-based ringtones hit chart in 2002 and began seriously investigating possibilities sometime in mid 2003. An announcement was made in January 2004 that the "MEF UK Ringtones Top Twenty" would help to make the industry better coordinated with the music industry and that the ringtones chart itself would achieve two of the MEF's goals, "to provide the entertainment and mobile industry with a formal chart, reflecting the growing revenue and copyright importance as music and mobile telecommunications converge."[36] The chart came online in late May of that year;[37] in the hopes of seeing similar growth in the mobile entertainment market in the United States, in late October 2004 the music-industry trade publication *Billboard* began publishing a similar chart, which was in turn endorsed by the Mobile Entertainment Forum and by the Cellular Telephone Industries Association and developed with the aid of an analysis firm called Consect.[38] Both charts tracked polyphonic ringtone sales, but attention soon turned to audio-file ringtones. In December 2006, *Billboard* partnered with Nielsen Entertainment's Nielsen Mobile division, whose new tracking system Nielsen RingScan would provide the basis for a new "Hot RingMasters" chart tracking sound-file ringtone sales, since, as *Billboard*'s charts director Geoff Mayfield noted, "the majority of sales have shifted from Polyphonics to master recordings."[39]

Ascertaining why a particular song makes the ringtone charts, let alone predicting what songs might appear on them in the future, is not necessarily a simple matter. In November 2007, the music critic Tim Page was asked about this in a short National Public Radio piece. He seemed to equivocate: "If I could choose what pieces were going to be more popular than the next, I'd be a multi-millionaire. It's so hard to tell what's going to become a hit."[40] But Page's response seemed to misunderstand the query, or at least a more useful formulation thereof—namely, what makes a ringtone a hit, particularly in contrast to other kinds of hit songs? Examining the *Billboard* charts themselves proves instructive here. Take, for example, one of the publication's earliest ringtone charts (table 9.1). Some trends are immediately apparent. First, the number-1 ringtone, "My Boo" by the R&B artists Usher and Alicia Keys, was also number-1 song on *Billboard*'s "Hot R&B/Hip-Hop Singles & Tracks" and "Hot R&B/Hip-Hop Airplay" charts in the same issue.[41] It was also number one on the "Hot 100" and "Hot 100 Airplay" charts, but it only reached number 7 on the "Mainstream Top

Table 9.1
Hot Ringtones Chart, *Billboard,* November 6, 2004.

This Week	Last Week	Weeks On	Title	Original Artist
1	1	2	My Boo	Usher and Alicia Keys
2	2	2	Lean Back	Terror Squad
3	4	2	Drop It Like It's Hot	Snoop Dog, feat. Pharrell
4	6	2	Locked Up	Akon, feat. Styles P
5	5	2	Halloween	John Carpenter
6	12	2	Sunshine	Lil' Flip, feat. Lea
7	3	2	Goodies	Ciara, feat. Petey Pablo
8	10	2	Freek-A-Leek	Petey Pablo
9	–	1	Game Over (Flip)	Lil' Flip
10	15	2	Yeah!	Usher, feat. Lil Jon & Ludacris
11	11	2	Headsprung	LL Cool J
12	8	2	Breathe, Stretch, Shake	Mase, feat. P. Diddy
13	–	1	Ice Ice Baby	Vanilla Ice
14	19	2	Big Pimpin'	Jay-Z, feat. UGK
15	7	2	Get Low	Lil Jon & The East Side Boyz, feat. Ying Yang Twins
16	9	2	Super Mario Brothers Theme	Koji Kondo
17	24	2	She Will Be Loved	Maroon 5
18	17	2	Bad Boys	Inner Circle
19	23	2	The Pink Panther Theme	Henry Mancini
20	–	1	Turn Me On	Kevin Lyttle feat. Spragga Benz

40" chart during that week.[42] The connection between the R&B/Hip-Hop genres and the ringtone charts becomes somewhat clearer when examining the top four hits at this time: in order, "My Boo," Terror Squad's "Lean Back," Snoop Dogg's "Drop It Like It's Hot," and Akon's "Locked Up." The respective chart positions of these tracks on the "Hot R&B/Hip-Hop Singles & Tracks," "Hot R&B/Hip-Hop Airplay," and "R&B/Hip-Hop Singles Sales," respectively, were 1-11-2-10, 1-11-2-10, and 3-13-1-18; the proximity of the ringtones chart to the Top 100, which includes the R&B/hip-hop hits mentioned above, is equally striking, at 1-6-4-10. On the "Mainstream Top 40," in contrast, the positions were 7-18-N/A-N/A. Most remarkable is that of the top three hits on the "Mainstream Top 40" chart (Nelly's "Over and Over," Maroon 5's "She Will Be Loved," and Kelly Clarkson's "Breakaway"), the latter two of which are not R&B/Hip-Hop songs and the first of which is a crossover hit by an R&B star with the country star Tim McGraw, are either very low on the ringtones chart ("She Will Be Loved" is at number 17) or do not appear on the ringtones chart at all.[43] Similarly, most of the hit songs on the "Hot 100" not ranking highly on the R&B/hip-hop charts were likewise not represented on the ringtones chart (including "Over and Over" at number 5, the white rapper Eminem's "Just Lose It" at number 8, and "Breakaway" at number 9). Clearly, then, ca. late 2004, the R&B and hip-hop genres, which are most directly associated with African American music, which feature the highest percentage of black artists when compared to other genres, and sales and airplay of which probably are driven by black consumers and audiences, were clearly dominant on the ringtones hit chart—especially black music that was popular more generally (i.e., as appearing on the "Hot 100").[44] The ringtone-market dominance of these genres, if not the demographic consequences thereof, did not go unnoticed by mobile market research groups and was on occasion mentioned in the business press.[45]

The synchronicity between black music and ringtones, which might be understood particularly for African Americans as being a kind of "new radio," was not complete, however.[46] Other ringtone hits seem to have a widely divergent provenance. For example, John Carpenter's music for the 1978 film *Halloween*—particularly the main theme for the film's psychotic killer, Michael Myers—featured as a popular ringtone on November 6, 2004, surely on account of Halloween's appearance on the calendar one week earlier. In other cases, the ringtone chart features nostalgic revivals of former hits, including Vanilla Ice's "Ice Ice Baby" (1985, 1987), Jamaican reggae group Inner Circle's song "Bad Boys" (1987, 1993), which featured on the Fox television show *Cops*, Koji Kondo's theme to the video game *Super Mario Bros.* (1985), and Henry Mancini's "Pink Panther Theme" (1963).[47] Surely some of the asymmetries and inconsistencies in the chart have to do with the success of rendering a particular musical excerpt in a polyphonic ringtone format, but we might also further speculate that these nostalgic ringtones are less clearly racialized—in the sense that their consumer demographic is less clearly racially marked, or, conversely, may be marked as *white*—when compared to the R&B and hip-hop hits on the ringtone chart (which would include at least thirteen of the twenty listed above).

The clear popularity of ringtones among R&B and hip-hop fans and, more broadly, black cell phone owners, in the 2000s led to a curious and much-debated phenomenon within the world of hip-hop disparagingly known as "ringtone rap." The term is not especially precise, but it seems to indicate a new approach to producing simple, repetitive popular songs (not exclusively rap) that featured catchy choruses—or, more accurately, *perpetual* choruses[48]—designed for low-fidelity phone speakers, rather than strong, innovative songs and albums as a whole. The critic Brian McManus blames the rise of digital culture and "shrinking attention spans" for the appearance of songs such as J. Holiday's "Bed" (a "woeful, repetitive and ploddingly dull shitstorm of a song").[49] The disposable character of these songs, which treat lightweight subject matter, is inseparable from the apparent disposability of the artists themselves, who have not been able to sustain successes for long periods of time with a series of hits. Instead, artists such as Mims (who had a number-1 hit with "This Is Why I'm Hot") seem destined to be one-hit wonders. Symptomatic of the musical world of ringtone rap are lowered expectations for a song or musical offering. In October 2007 an executive at EMI Music Publishing was quoted as saying "About one or two weeks ago, one of the saddest things happened to me, when (an artist) played me a record and said, 'This would make a hot ringtone.'"[50]

Perhaps the most important figure of the ringtone rap era is DeAndre Cortez Way (b. 1990), who goes by the stage name Soulja Boy Tell 'Em. Unlike some of his contemporaries, Soulja Boy was able to follow up his number-1 hit "Crank That (Soulja Boy)" (from 2007) with "Kiss Me Thru the Phone" in 2008–09. His 2010 hit "Turn My Swag On" follows the same formula. These songs are marked by an extreme degree of repetition, much more so than is typical of the contemporary R&B or hip-hop song. The lyrics reiterate over and over (especially "Crank That") on top of an unvarying groove, making the songs less about a narrative or series of thoughts than is typical of most songs. Instead, at his best, Soulja Boy produces a distinctive if not particularly original "feel," which is sustained throughout its three-minute duration and which captures a charmingly naive, juvenile sensibility.[51] Such a feel is easily excerpted for ringtones, highlighting their best moments and providing the necessary exposure to plant them as earworms without overstaying their welcome. When heard in their full-length versions and compared with music by Snoop Dogg, Beyoncé, or Jay-Z, these songs encourage distraction and function better as background music or perhaps dance music.

In part, the divide is generational. Jay-Z's Grammy-winning "D.O.A." (standing for "Death of Auto-Tune") takes these younger artists to task.[52] The first verse begins with the lines "This is anti-Auto-Tune, death of the ringtone / This ain't for iTunes, this ain't for singalongs / This is Sinatra at the opera / Bring a blond / Preferably with a fat ass / Who can sing a song." After aligning himself against digital, disposable culture and with old-fashioned images of opera and Sinatra, Jay-Z exhorts the new generation of rappers to stop singing ("y'all sing too much") and to be much more violent and masculine ("Put your skirt back down, grow a set, men"), even throwing in a dig at skinny jeans ("You niggas' jeans too tight / Your colors too bright, your voice too light"). The chorus, a half-hearted, mocking rendition of the chant from Steam's 1968 song "Na Na Hey Hey Kiss Him Goodbye," is self-consciously invoked as dismissive valediction for the insufficiently "hard" newcomers. Moreover, the song's distorted electric guitar riffs, wailing soprano saxophone, atmospheric electric organ chords, active rock-soul bass line, and loose rock beat (sampled from a 1970 recording by Janko Nilović and Dave Sucky, the jazz/psychedelic-rock fusion album *Psyc Impressions*) provide signifiers of both masculinity and maturity (as historical throwbacks), reinforcing the points made in the lyrics. Jay-Z is not alone in his disdain for ringtone rap; in his song "C'mon Baby" (2007), the far less popular rapper Saigon presciently bade farewell to the ringtone rap phenomenon: "Finally I've arrived / So we can say our goodbyes / to the ringtone rapper / That crap'll never survive."[53] Ultimately, ringtone rap may have been a form of rebellion against the (post-)gangsta ethos and the societal violence it portrayed (and, on occasion, reproduced). Perhaps it figured a productive poetics of (blissfully willful) regression for a generation of African American youths marked by the split between a relatively comfortable middle class, on the one hand, and a languishing working class ravaged by welfare reform, ballooning incarceration rates, and cutbacks in education funding and in need of the emotional and material support of consistent parenthood and mentorship.[54]

The heyday of ringtone rap had—like the growth of ringtones sales in general—passed by early 2010. The surest sign of that was the appearance of an almost academic project by the pop culture blogger Kyra Kyles on a blog called Chicago Now. The task was to create a ringtone rap with great speed and efficiency. Kyles sought the help of Antongenius, a hip-hop producer and beatsmith in Indianapolis, and produced the "Shamrock Shake," a silly musical effort performed by Kyles (a.k.a. Naomi Rambles) and other musicians in homage to McDonald's green, mint-flavored, St. Patrick's Day seasonal milkshake, the name of which is then turned into a dance move. It had some of the distinctive features of ringtone rap—particularly a juvenile chorus ("I got that Shamrock Shake, that Shake, that Shake") repeated many times. The low-grade production values that Antongenius created point to his willingness to produce out-of-character music for the experiment. Perhaps the most interesting aspect of the project was Kyles' account of it after it was completed: the song took a little more than two hours for Antongenius to "make and mix," and the singsong chorus chant performed by the Chicago-area rapper Nikki Lynette required numerous takes in order to get the right effect, which for Kyles demonstrated Lynette's "ability to make it more marketable."[55] The resulting track, a passable imitation of the style, drew a surprising response: "Some people contacted me since its debut and asked me to rap on their projects," Kyles reported. "Others told me I had a hit on my hands. That latter sentiment made me a little sad."[56] To finish the project, Kyles created a video, although it never was released—possibly for copyright/trademark reasons; she did, however, use St. Patrick's Day 2011 as an opportunity to promote the song once again.[57]

The Failure of Ringtone Autoreferentiality

Attempts to exploit the popularity of ringtones among black consumers cannot be confined to the much-vilified pseudo-genre of ringtone rap. More targeted interventions can be delineated, concerning songs or albums that directly refer to ringtones—particularly those incorporating the word "ringtone" into their title. The history of references to ringtones in (mostly) English-language popular music extends at least back to 2003, when the word "ringtone" itself was becoming better known as a distinct label (as opposed to "cell phone ring," "cell phone melody," etc.). The earliest songs are, not surprisingly, European, the first being the London-based band Athlete's "Shake Those Windows" (2003). The song nostalgically recalls "old school" "songs and beats and rhymes in life" that "Live in Edinburgh / Just across the road / Sometimes my stereo and in my aching bones / It's on the sugar hill and in your ringtone." The first song to have the title "Ringtone," however, is not in English but in Dutch. Using both the English and Dutch words for ringtone (*beltoon* in Dutch), the Netherlands-based pop/soul singer Alain Clark opens his first album (2004) with a repetitive, minor-blues chant-style song, introduced with a "cell phone effect" on his voice. Unlike Athlete's song, Clark's chorus features a trope that would become extremely common in future songs about ringtones: that the (male) singer's love interest, who is very often the (female) addressee in the song lyrics, possesses or should possess a ringtone of that very song. This device creates a rather complex, self-referential relationship between the characters of a song—i.e., the song heard by listeners *is* the ringtone mentioned in its lyrics—and perhaps between song and listener[58]:

Mijn ringtone beltoon voor je telefoon gaat
Neem m op, neem m op, neem m op, neem op voordat
Mijn ringtone beltoon voor je telefoon is
Zodat je weet dat ik 't ben
[My ringtone, ringtone rings for (or on) your telephone
Get it, get it, get it, get it before (it's too late)
My ringtone, ringtone is for your telephone
So that you know it's me]

As far as the US market is concerned, however, Clark's modest hit in the Netherlands is irrelevant, if certainly fascinating and pointing to trends in the narration of ringtone use in popular song lyrics. More important is the hit by the group Gym Class Heroes, whose "Cupid's Chokehold/Breakfast in America"—a collaboration with guitarist/vocalist Patrick Stump of the popular rock band Fall Out Boy—became a top-five hit in the United States, the United Kingdom, and Canada in 2006–07. Built upon a sample of the chorus of Supertramp's 1979 song "Breakfast in America," which is combined with a wordless vocal riff by singer Roger Hodgson from the song's bridge, the Gym Class Heroes' hit transformed its source's tale of an alienated, lonely immigrant to the United States into a pedestrian song about romance and its difficulties, adding a pop-ska groove to the song's verses that seems to lower the affective temperature of the music. The following lines in one verse are of interest here: "I love it when she calls my phone / She's even got her very own ringtone / If that ain't love, then I don't know what love is." All of the song's verses are marked by the narrator's optimism at having found "the one," which is then contrasted by the Supertramp chorus' more pessimistic "Take a look at my girlfriend / She's the only one I got / Not much of a girlfriend / I never seem to get a lot." In a sense possessed or controlled by his love interest, the male narrator demonstrates his love by having a special ringtone for his girlfriend—a gesture demonstrating that he has performed the kind of labor necessary to maintain a relationship as well as providing a perhaps sentimental souvenir, and a public token, of his romantic interest.[59]

Following on the heels of the Gym Class Heroes' hit song, a number of R&B and hip-hop artists produced songs or albums with the word "ringtone" in the title, starting in 2007. As table 9.2 demonstrates, the predominant genres for ringtone songs are hip-hop and R&B, including numbers by Miss B./Yung Joc, R. Kelly, Just-A-Girl, Naeto C, Esham, L. L. Cool J., Charles Hamilton, Soulja Boy Tell 'Em, Big Boi, Dorrough, and Warren G; additionally, a Christian rap/hip-hop number by Canton Jones confirms perceived appeal of ringtones within the African American con-

Table 9.2

Songs (or albums) with "ringtone" in the title. Release dates were obtained from iTunes, Amazon.com's MP3 store, and other sites.

Song title	Artist	Release date	Album/release	Genre
Ringtone	Alain Clark	2004	2004 (Netherlands)	Dutch, blue-eyed soul, pop
Ringtone	Miss B., featuring Yung Joc	March 2007?	DJ Smallz, *Smokin R&B*, volume 4 (mix tape), March 2007?	Hip-Hop/ Rap
Ringtone	R. Kelly	5/29/07	*Double Up*, 5/29/07	R&B
Ringtone	Canton Jones	2/5/08	*Kingdom Business*, 2/5/08	Christian & Gospel: R&B/ Hip-hop/Rap
Ringtone	Just-A-Girl	2/6/08	N/A (single)	R&B
Ringtone	Naeto C	8/1/08	*U Know My P*, 8/1/09	Hip-hop/Rap
Ringtone	Esham	8/28/08	*Sacrificial Lambz*, 8/28/08	Hip-hop/Rap
Ringtone M . . . [Ringtone Murder]	L.L. Cool J.	9/9/08	*Exit 13*, 9/9/08	Hip-hop/Rap
Ringtone Rap	Charles Hamilton	11/11/08	*Sonic the Hamilton*, 11/11/08	Hip-hop/ Rap (Hipster Rap)
[Album: Mixtape] *Lord of the Ringtones*	Soulja Boy Tell 'Em	3/3/09 (?)	*Lord of the Ringtones*, 3/3/09	Hip-hop/Rap
Ringtone/Theme Song	Big Boi	3/6/09	Leak/outtake from *Sir Lucious Leftfoot: The Son of Chico Dusty*, 6/6-9/10	Hip-hop/Rap
Ringtone	Without a Face	7/14/09	*Worst Debut Album Ever*, 7/14/09	Alternative (indie), Acoustic
What's My Ringtone	Dorrough	8/4/09	*Dorrough Music*, 8/4/09	Hip-hop/Rap
Ringtone	"Weird Al" Yankovic	8/25/09	*Internet Leaks* EP/8/21/09	Pop/Rock, Comedy
Ringtone	Warren G	9/29/09	*The G-Files*, 9/29/09	Hip-hop/Rap
I Miss Your Ringtone	Lady D	8/11/10	*She's Nobody*, 8/11/10	R&B
Ringtone	Shawty Lo, featuring Soulja Boy Tell 'Em	10/18/10	*I'm da Man, Pt. 3* (mix tape), 10/18/10	R&B/Hip-hop
Ringtone	Yung Dre Dre	1/17/11	single, 1/17/11	R&B

suming public. But although Jones encourages you to put the (musical and textual) phrase "'Taking over for the Kingdom' . . . on your ringtone," by and large the conception of ringtone use in these songs is not explicitly religious, instead concerning itself with romantic relationships and sexual entanglements. Musically speaking, the most successful of these was produced by the R&B artist R. Kelly on his lengthy album *Double Up* (2007), produced at the peak of his legal troubles concerning his alleged statutory rape and pornographic video recording of himself with a 14-year-old girl. The album initially suggested, on the one hand, that R. Kelly would be taking on his critics: the first track, "The Champ," having a Hollywood-soundtrack production quality and combative lyrics. But, after an interlude in which the singer says "All right, now we got that out the way, let's *double up*," proceeding with a characteristically lascivious song about the male singer's penchant for threesomes with two women. The hypercharged libido (or simulation thereof) on display in the album (and noted by reviewers) is typical of Kelly, and it pervades almost the entirety of *Double Up*, including a relatively little-known bonus track at the end of the album, simply titled "Ringtone."[60] The track is reminiscent of the Atlanta-based hip-hop style known as snap music, a style understood as a slowed-down version of crunk, which originated in Memphis and migrated to Atlanta, which derives from the Miami Bass style (itself based on electro), and which is characterized by a relatively minimal production method involving a heavy bass drum (or sample), relatively uncomplicated synth layers, sometimes a bass part, vocal parts, and finger snapping.[61] Produced by the guest artist Polow da Don, who is based in Atlanta and intimately familiar with (if somewhat disdainful of) snap music,[62] Kelly's track is rather distant from the Gym Class Heroes' "Cupid's Chokehold" in terms of its social location, which is close to the black strip club and dance club scenes of the Georgia metropolis,[63] and its ethos is likewise much less sentimental and much more explicitly sexual in tone. The narrative of the song continues the threesome theme from "Double Up," and describes Kelly himself picking up two women off the street, taking them to "the club" and eventually back to his mansion for sex; one can tell that "the sex is major / 'cause only one day later / they got Kells on they ringtone."[64] Here, the ringtone is treated as a demonstration of sexual conquest and, as such, plays into themes typical for both Kelly and for US R&B and hip-hop in the 2000s.[65] Moreover, the "club" topos of the minimalist snap music style mixes with more evocative, even realistic effects, such as the siren sound (or strained yell) distantly placed in the mix (on A4, in the first two measures of figure 9.2), which is then reiterated and foregrounded by Kelly's lead vocal line

Figure 9.2
R. Kelly, "Ringtone" (2007), 0:16–0:25.

on the words "red light" and "hitch hike"—now possibly figuring a car horn or sonically mimicking a flashing light, signaling the girls to get into the car. This siren effect foreshadows both the appearance of police sirens that later frequent the song, as well as the chorus' reiterations of the word "ringtone" sung as response vocals overdubbed by Kelly himself. (See figure 9.3.) The effect is to unite or inflate Kelly's own subjectivity—he is often accused of being egotistical—with the sounds of the nighttime environment in and around the vicinity of the club itself, including its dangers (police sirens, threatening men) and noisemakers and images associated with the narrative subject and his romantic targets (car horn, stoplights, ringtones). In this way, the sound of the ringtone itself is overdetermined—in mutating with other sounds and sense data, it effectively pervades the whole of the song's texture—and the reiterations of the word "ringtone" in the chorus itself arguably reinforce this impression. The song is what

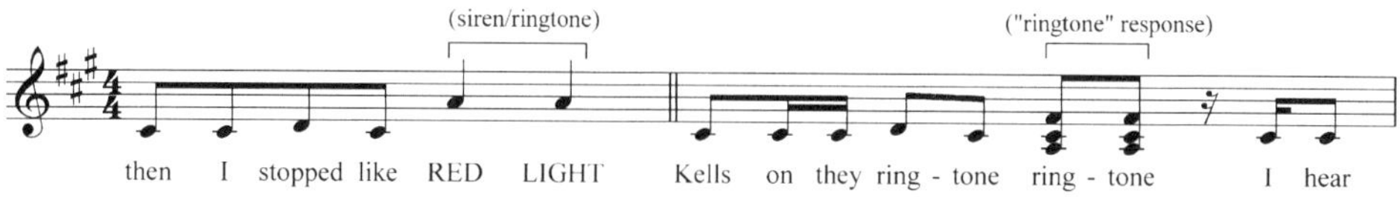

Figure 9.3
R. Kelly, "Ringtone" (2007), 0:20–0:22, 0:45–0:47.

one reviewer described as "an unexpectedly infectious bonus track that doubles as a transparent plug for R. Kelly ringtones"; everything within it sonically and imagistically doubles back onto Kelly himself, transforming the black, nighttime, urban environment he depicts into an advertisement for the R. Kelly brand. But, its production quality and signal-topos thematic would also lend themselves well to the function of a phone signal—that is, the song seems intended, in part, to *be* a ringtone, rather than simply promote them.[66] It may be therefore understood as an exercise in ringtone autoreferentiality.

Although "Ringtone" was one of the better tracks on *Double Up*, Kelly decided not to release it as a single. Perhaps he was right to do so: a self-referential ringtone that itself says the word "ringtone" over and over may quickly become tiresome, its postmodern novelty wearing off after a few incoming calls, but of course popular taste is unpredictable and there was no way to know how the song would have been received had it been promoted heavily.[67] Be that as it may, it is safe to say that ringtone autoreferentiality did not generate huge windfall profits for the ringtone industry and its beneficiary, the music industry. And although few of the songs with "ringtone" in the title feature the kind of wit and creative intelligence demonstrated in R. Kelly's "Ringtone," the fact that most of them used the word "ringtone" far too frequently and none of them were a hit (as a ringtone or otherwise) is telling.[68]

An instructive example comes from the group OutKast and the Atlanta hip-hop luminary Big Boi (Antwan André Patton). Produced during sessions for his debut solo album, *Sir Lucious Left Foot: The Son of Chico Dusty* (2010), "Ringtone" was in part a casualty of Big Boi's conflicts with OutKast's label, Jive Records, which deemed the album insufficiently commercial and requested that he record a song stylistically similar to Lil Wayne's hit "Lollipop."[69] After more than two years of disputes, disagreements, and delays by Jive in releasing the album, Big Boi was permitted to shop the album elsewhere; Island/Def Jam released it in July 2010. One of the results of this music-business conflict was that the earlier, pre-released singles were either not made available by Jive for the new recording or deliberately withheld by Big Boi himself, including the much praised "Royal Flush," which featured his partner from OutKast, André 3000. A rare exception was, in fact, "Ringtone," but its own history is rather convoluted. The song was released as a single more than one year in advance of the album itself in March 2009, and received critical attention quite early on: for example, a reviewer in *Rolling Stone* remarked: "True love travels on a cellular level apparently, as OutKast's shorter one blesses his sweetie with her 'very own' tone in his phone. But once this talkbox-laced club track starts sounding from cells 'round the world, will she still feel special?"[70] Clearly, the song was assumed to be a future hit. Reviewers weren't the only figures paying attention; the Frito-Lay Company gained the rights to use the song in an advertising campaign for the new "Late Night" flavor of Doritos in June 2009, and a part of the campaign involved an "augmented reality" viewing experience that would, with a webcam and a bag of the chips, make it seem as if a concert were taking place on one's own bag. Although one reviewer noted "Unless you're four years old or really, really love Doritos, this'll be interesting for about 10 minutes,"[71] the commercials promoted the song—as did its modest radio play and fans' curiosity about the artist's newest album project. Likewise, the online mobile content provider Dada Music used the song in promoting a joint venture with Sony Music. By sometime in 2010, the song had been heard 1.5 million times on Big Boi's MySpace page.[72] However, the conflicts with Jive Records seem to have prevented the artist from including the song on his album as it was originally named; hence, the original title was scrubbed from iTunes, from Amazon.com's MP3 store, and from other digital sales vehicles. Renamed "Theme Song," it was re-released only as a bonus track on the deluxe edition of *Sir Lucious Left Foot*. Its potential for popularity probably was stymied by its renaming.

Or is it that simple? Certainly, some features of the song were in tune with prevailing trends. For one, the sound world of the song is far outside of the crunk/snap stylistic constellation discussed above, which by 2009–10 would have been a favorable development: the crunk/snap lineage reached its apex of popularity in 2007–08 and would become gradually displaced on the pop charts by various styles, including "electro pop," a musical practice influenced by 1980s synth pop and related styles. The *Village Voice* having consigned crunk to the "microtrend graveyard" of the 2000s, the style's zenith had passed by the time Big Boi's "Ringtone" appeared.[73] "Ringtone," like many of the songs from the *Sir Lucious* sessions, instead is laced with retro synths, talk-box effects, and other markers of musical

Figure 9.4
Big Boi's "Ringtone/Theme Song" (2009) and Miami Sound Machine's "Conga" (1986).

styles from more than 20 years earlier, particularly evoking the sound of synth-funk of the 1980s. Tom Breihan of the webzine Pitchfork captured this sensibility well: "Musically, the album drips with 1980s synth-funk signifiers. The keyboards glimmer as they roam, and talkboxes mutter and blurt."[74] Yet the album combines a dizzying number of musical and textual references, making the early digital effects only a central component of a widely ranging aesthetic statement. (Indeed, a snap-style beat is found on the appropriately erotic "Tangerine," although it is mixed in with uncharacteristically grungy distorted-electric-guitar chordal arpeggios, using a Phrygian/quasi-Spanish i ♭II ♭III progression, and numerous other contrasting elements.) Unlike much of *Sir Lucious*, however, "Ringtone" is more straightforwardly nostalgic for the 1980s, featuring typical synth layers and a near-quotation of the 1980s hit "Conga" by Miami Sound Machine embedded in Big Boi's flow (see figure 9.4).[75]

Far more damning to the cause of Big Boi's potential hit was its unsuccessful chorus, with its clumsily sung line "You got your very own theme song / very own ringtone," in contrast to the rather smooth and assured delivery of the rapped verses by Big Boi himself. Though it begins with a rather promising country-blues slide guitar riff and may generally be reminiscent of his 2003 number-1 hit "The Way You Move," the song devolves into sentimentality and nostalgic clunkiness and neither conveys the innocence of a Soulja Boy tune nor displays the conceptual sophistication or catchiness characteristic of Big Boi's best work.[76] Indeed, few people seemed excited by the song; much more typical was one reviewer's reaction, "it won't knock you out on the first listen," and then followed by the wan claim that it would "grow on you."[77] The song's weaknesses haven't stopped ringtone-hit hopefuls from imitating it, however. In October 2010, in a blatant rip-off combining lines from R. Kelly's "Ringtone" and Big Boi's "Ringtone/Theme Song," the Atlanta-based rapper Shawty Lo released a song called "Ringtone," which featured Soulja Boy singing "She got me on her ringtone / She said it was her theme song." Appearing as part of

Shawty Lo's mix tape *I'm da Man, Pt. 3*, with DJ Scream (2010), the song is a dedicated contribution "for the ladies," which the rapper intones with booming echo/delay effects, making for an incongruous presence in an otherwise saccharine offering. Soulja Boy's extremely repetitive, Auto-Tuned vocals provide its primary, and rather thin, content.

Ringtones, Racial Commentary, and Black Modernity

The arguments outlined above can be recapitulated as follows: Ringtones sold in the US market underwent a remarkable transformation in market demographic in the 2000s, and that change was registered in the content of the English-language popular songs that served as their basis. That demographic shift involved the incorporation of an African American telephonic culture into the world of mobile telephony, the ringtone seemingly driving black popular musical production and the thematic of the ringtone itself spawning an unpopular subclass of R&B and hip-hop songs.[78] If the foregoing demonstrates the effective racialization of the ringtone within the United States, we should note that the phenomenon was already being registered by proxy in the United Kingdom in the early to mid 2000s, and that when white American artists began to comment on the ringtone phenomenon later in the decade they only alluded elliptically to a racial process that was already well underway. In the former cases, the term "ringtone" was initially articulated with the longstanding imaginative affiliations between white British youth and black music that extend at least as far back as the "British Invasion" of the 1960s. The previously mentioned example by Athlete, "Shake Those Windows," attests to this with references to hip-hop music (for example, A Tribe Called Quest's classic 1996 album *Beats, Rhymes, and Life*) and to windows rattled by booming speakers (see lyrics above). Hip-hop may have served, for the UK equivalent of Generation X, the role that electric and country blues did for the generation of Eric Clapton, Jeff Beck, and Jimmy Page; the ringtone here evokes not only the geographical horizons of the UK, but also, unintentionally, the nascent black ringtone culture of the United States.

But the relationship between the American black working class and white Britons has also generated a new phenomenon in the British Isles known as "chav" culture. To oversimplify, chavs are largely working-class, white British youth who imitate aspects of black American hip-hop culture (such as wearing thick gold jewelry, athletic training clothes, and sneakers), but have fashioned their own cultural codes in a peculiarly British way (such as through the adoption of the once-elite Burberry tartan-patterned fabric). One particularly distinctive element of chav culture, ca. 2004–05, was the centrality of the mobile phone and ringtones—chavs presumably constituting a significant market demographic for the UK ringtone industry at the time.[79] Whether as understood as a vilified replacement for the notion of the "underclass," as a means of demonizing the white working class in general, or as part of a multi-racial "convivial culture," the chav phenomenon, including the public display of ringtones, was mentioned by the band Arctic Monkeys in a 2006 song titled "A Certain Romance."[80] Set to an up-tempo ska-punk style after a kind of punk "jungle" music introduction, the song criticizes chavs, and possibly other youth subcultures, describing youths with "classic Reeboks / or knackered Converse / or trackie bottoms tucked in socks."[81] Significantly for present purposes, the song includes the couplet "And over there, there's broken bones / There's only music, so that there's new ringtones," which contemplates the priority of musical production for excerpted hits on mobile phones. Fitting within the broader critique of chavs as a youth subculture significantly oriented around consumption, the chav phenomenon is reminiscent of the UK subcultures of the postwar period, particularly from the 1970s on, but with one difference: on account of the political repression of the British working class from Thatcher onward, the highly mediated consumption of chavs is generally figured not as oppositional to the system but as a sign of being seduced by it, even as that system turns its back on working-class youth as part of the de-industrialized reserve army that has accumulated in fits and starts in the UK in the past few decades.[82]

If one could say that, via a certain kind of mediation into British youth culture, the equivalence between a racialized cultural practice and ringtones was cemented *before* the ringtone concept took root in African American consumer culture itself, this was doubtless on account of the lag in US mobile culture behind Europe and Asia. However, by the mid to late 2000s, American forms of ringtone commentary also appeared. Absent the filter of an indigenous white youth culture mediating American black culture that could serve as a convenient object of critique, treatments of black cultural practices by white Americans were surely less comfortable prospects, and songs by white artists that endeavored to tackle the subject seemingly deflected particularly racial associations with ringtones, or at best treated them obliquely. In other words, lacking chavs to scapegoat for the ills of the ringtone, white American songwriters exercised care in criticizing ringtones, in view of their prevalent use by African Americans, so as not to risk being labeled racists. Two examples from the same year range widely in terms of popularity, but both can roughly be understood as comedy songs, while transcending the genre to a degree. The first, released in July 2009, is by a singer-songwriter named Henry Dillard who records under the name Without a Face. His self-deprecating album *Worst Debut Album Ever* received a lot of local attention and was ranked number 2 among the top 10 albums of 2009 on the *Houston Chronicle*'s "Peep" blog. It features some striking material, including a bizarre and touching tribute to a KHOU news anchorwoman named Lucy Noland.[83] Dillard's song "Ringtone" reflects his preoccupation with illegal downloading and music careers, but also traffics in a degree of ironic self-promotion that seems to reference the comparable habits of hip-hop and R&B artists, including some of those discussed above.[84] In the song's chorus, Dillard intones, in a howl reminiscent of Gordon Gano, "Hey you're listening to Without a Face ringtones / from the studio, straight to your cell phone / No more room for three-minute songs / I only write ones twenty seconds long." Given Dillard's awareness of the ringtone's racialized aspects, the problem of careers in music and the economics of ringtones—and the fantasy that they could provide stable careers—remains the prime focus of the song's lyrical content, and stylistically or thematically it never seems to devolve into racial parody or mimicry. In contrast to the relatively obscure Dillard, the well-known parody artist Weird Al Yankovic produced a Grammy-nominated EP titled *Internet Leaks*, released a month after the Without a Face recording, that included a song also titled "Ringtone." Composing the song as a nearly pitch-perfect style parody of the band Queen, Weird Al's "Ringtone" takes on the subject of ringtones as being socially obtrusive and annoying—hence, uniquely giving voice to "ringtone rage" within the canon of ringtone songs, with the chorus lyrics "Everybody in the whole wide world really hates my ringtone." By the song's end, the universality of ringtone rage is represented, with the people hating the protagonist's ringtone including "Chinese factory workers," "Muslim women in burqas," "starving kids in Angola," "folks with ebola," "nuns and nannies," "welfare mothers," "Pakistanis," and the "Wayans brothers," ultimately claiming that "everyone on the land, everyone on the sea / every single person everywhere unanimously" hates his ringtone. The inversion of a kind of feel-good, "We Are the World" universalism is turned on its head in favor of a universal hatred of ringtones.[85] The song's insistence on this universalism, even in jest, obscures the ringtone's racialization in the United States, which by 2009 was certainly evident, if somewhat in decline.

That brings us to the final point of this narrative—namely, that in the United States ringtones no longer seem to be purchased as extensively by

African American consumers as was once the case. Today's ringtone charts hew more closely to the general pop and radio charts, as opposed to exclusively focusing on particular genres—though a healthy dose of R&B and hip-hop are certainly present. Arguably, other music-consuming demographics are catching up, including country music fans—Jason Aldean and Kelly Clarkson's "Don't You Wanna Stay" was a top-ten ringtone seller for a few weeks. The larger point is that the peak of the ringtone industry, roughly the years 2007 and 2008, coincided with the core US demographic of ringtone sales still being largely African American (as judged by genre), the economic downturn surely hitting disposable incomes quite hard beginning in 2008. Still, one might make the tentative argument that in the 2000s African American consumers helped to support the music industry disproportionately within the United States, during a period in which it claimed that its overall sales decreased massively—an alleged 31 percent global decline from 2004 to 2010.[86] Perhaps this is not entirely dissimilar to the way in which African American consumers supported the film industry during the late 1960s and the early 1970s—a time of general decline in moviegoing and arguably in content. So-called blaxploitation films, which were inexpensive to produce, helped Hollywood to become highly profitable again with minimal expenditures.[87] Likewise, the ringtone, which inflated prices 200–300 percent when compared to digital sound files, was a massive if temporary boon for the recording industry, as it relied on the mobile phone's relatively closed architecture—although this changed somewhat with the rise of smartphones, the iPhone, and the increased availability of ringtone creation software and apps. As disproportionate consumers of ringtones, blacks in the US (and perhaps working-class people in the UK) thereby disproportionately contributed during a critical period to "saving" the music (that is, recording) industry, which the ringtone itself was supposed to do. Whether the recording industry has actually been "saved" from its claimed digital crisis/debacle is difficult to discern, given that lucrative synchronization royalties are not always clearly reported or accounted for in recording-industry revenue figures. In light of this, ringtone revenue either picked up the slack over the 2000s or provided an extra boost in profits.[88]

Of course, the foregoing is only a tentative hypothesis. Be that as it may, the racialization of the ringtone and the sorts of technological congruence found in black consuming demographics more generally point to a remarkable and potentially exploitable modernity within African America—one that fractures in part along class lines but still surely demonstrates trends that cross them as well. To what extent, in the age of Obama, do African

Americans (and, to a lesser degree, Latinos), rather than represent a distinct "countermodernity" or "counterculture of modernity," stand in as *bearers* of modernity? Among those in support of the modern state, capital investment in development and infrastructure, rational efficiency (including state and, to a degree, corporate bureaucracies), and the expansion of sensory experience through new technologies, African Americans now constitute an especially important electoral bloc buttressing once-centrist political positions and an equally important body of consumers drawn to new technological forms and practices—particularly as large segments of the white working class appear, significantly if not exclusively, invested in an incoherent, anti-statist, and even millenarian politics motivated in part by racism and, in turn, encouraged by an almost nihilistically conservative fraction of the capitalist class.[89] Of course, it seems indisputable that modernity still disproportionately benefits dominant class and racial subjects, but the completion of modernity as a *project* seems to have been largely rejected by many white, working-class Americans, particularly conservative Christians and conservative suburban whites. In that case, it would seem that minority populations in the United States, who have by and large not gone in for this kind of anti-statist politics, would also disproportionately share the burden of advocating for a functional state, let alone the democratic redistribution of the privileges of modernity in the short to medium term.[90]

The extent that the ringtone is a symptom of this process is noteworthy. Ringtones and blackness certainly seem to be intertwined in much of the music examined above, even when their intermingling is studiously avoided. Yet if the sphere of inquiry is expanded to include working people and national minorities more broadly, might the burden of inflated ringtone prices have been disproportionately borne by much of the developed world's working class? If so, one might say that, although the occupants of lower economic strata remained trapped in the metaphorical "walled garden," their disposable incomes redirected toward horticultural upkeep, the benefits of the "celestial jukebox" and the "cloud," such as they are, were reaped by relative class elites with particular class-conditioned forms of digital/mobile access and skill, paying, as far as ringtone expenditures go, *less* than those with far smaller incomes. Remarkably, the political economy of the ringtone may provide yet another demonstration of how expensive it is to be poor.

Epilogue

"More Than Ringtones"

With the decline and the particularization of the ringtone industry's cultural products, we glimpse the last turn of the dialectic in its final degrees of rotation. There is, of course, a fundamental incompleteness to the story of the ringtone, whose afterlives in the wake of being downgraded from leading to trailing edge will be numerous and, in part, unforeseeable. Nonetheless, the ringtone is said to be over and done with, and the mobile entertainment industry's attempt to ennoble it with the term "mobile music"—first as the ringtone industry emerged in the late 1990s, then more vigorously with that industry's precipitous rise in the mid 2000s—has foundered as well.[1] As early as June 2009, during the Mobile Music Now conference in London, Alex Vlassopulos (Head of Business Development, Digital at Sony BMG) stated the "mobile music" concept never worked, in contrast to that of ringtones, and that digital music and mobile music no longer needed to be distinguished as separate entities.[2] But even if mobile music was dead on arrival as a marketing rubric, that hasn't stopped those within the mobile entertainment business from distancing themselves from ringtones—once lucrative but now failing, tinged with an air of disreputability from the outset, and tainted by unethical business practices. Claims in mobile marketing press and reports that the latest product is "more than (just) ringtones" or goes "beyond ringtones" attest to anxieties (even guilt) about the industry's association with its formative product and, perhaps, to the lack of comparable success for more recent mobile music products—especially apps, which have been dominated by the Apple App Store (with roughly $4 billion earned by its developers in 2011, it has six times the market share of its closest competitor Google Android Market) and among which profitable music apps seem to be relatively rare.[3] For example, the CEO of the Australian mobile software firm BigTinCan, David

Keene, claimed that BigTinCan's new BuzzMe Beats app "can deliver special beats for phone calls, emails and texts with matching LED colour patterns and vibrations. . . . [It is] *more than ringtones* on a mobile phone—it [is] an 'inside skin' that changes the ringtone to a beat personally created by [R&B musician and producer Roy] Hamilton."[4] In February 2011, a music industry and mobile/social media conference featuring the artist Chamillionaire as a panelist was held in Los Angeles during the Grammy Awards week; its title, "The Social Media Rock Star Summit Music and Mobile: *Beyond the Ringtone*," was telling.[5]

If the industry's attempt to rebrand the ringtone as "mobile music" was in retrospect doomed to failure, it may be because it adopted an extremely limited, profit-driven, and time-specific meaning of the term, concocted simply because the mobile phone and the MP3 player began to converge. Of course, the broader history of mobile music—the idea that you can take your music with you—long predates the appearance of the mobile phone or the compressed digital audio file. Even in its most recent variant, which is rooted in the history of sound reproduction, it is more than a century old. Witness the longstanding appeal of portable music devices: portable gramophones, automobile radios, portable transistor radios, the Sony Walkman and Discman, and the MP3 player, but also gadgets like the musical lighter in *Xiao Wu*. It is perhaps with this longer history and its future development partly in mind that experimental musicians producing new musical forms and practices, such as those connected to the Mobile Music Workshop projects, have also adopted the term "mobile music." Indeed, the expansive socio-aesthetic ambitions harbored by such artists invoke not only recent fads like ringtones, but also the once-radical (and now rather traditional) ambitions of participatory art and Situationist politics from the 1960s.[6]

It seems safe to say that the "synergistic" conjunction of the mobile entertainment industry and the recording industry that produced the ringtone phenomenon is not likely to be reproduced in the near future. The recording industry, still peddling digital sound files, is slowly recovering from the decade-long shakeout caused by its delayed transition to downloadable digital formats. Traditional music consumption is still declining, as Michael DeGusta has demonstrated, although the rate of decline may be tapering.[7] There seem to be numerous reasons for this, including decreasing sales of albums in favor of single tracks, increasing competition from other forms of cultural production (especially video games, which in 2008 surpassed film revenues[8]), and the youngest consumers' coming of age in an era of normalized digital file sharing and not wanting to pay for recordings.[9] All these reasons add up to difficulties in securing sufficient numbers of digital music sales and subscriptions—for example, four-fifths of the subscribers to Spotify, the most successful service to date, are free members, theoretically supported by ad revenue.[10] In contrast, mobile media capital has placed most of its eggs in the app basket. More ambitious music apps push for the familiar goal of interactivity, as well as combining interactivity with music-based social networking to create what one company calls "social music."[11] Many music apps (e.g., recording studio packages for the iPhone) are intended for musical specialists, at least relatively so, or involve more complex interaction than passive music listening—for example, Björk's 2011 album *Biophilia* was accompanied by for-sale apps that allowed listeners to remix and transform tracks in various ways. Although such app projects may appeal to certain subsets of fan bases, the required demands on consumers' time and attention spans and the necessary cultivation of new (and active) media-cultural habits impose tangible limits on mass sales and use.[12]

The mobile music app has not yet taken off in the way that music-industry entrepreneurs have hoped. Although the app economy provides a steady stream of income to innovators in the mobile entertainment business (as well as to Apple and other retailers), the mobile music app is not and probably will not become the proper successor to the ringtone for the recording conglomerates.[13] Entrepreneurial energies and monies will continue to be invested in music and sound in mobile apps, but these probably will continue to be variants of existing forms—mobile gaming, mobile broadcast, mobile viewing, and so on—and, aside from some licensing, probably will not generate such profits as the music industry saw in the days of the ringtone. Having created a false partition in access to consumers' own music collections and the broader spheres of music broadcasting that partly intersect with those collections, the ringtone industry encouraged the partial reproduction of already-purchased or freely available music on mobile phones at great expense, and faddishness provided the momentum necessary to overcome this basic hurdle. This economic arrangement was highly irrational for most consumers, who repeatedly asked, "Why should I pay for what I already have or can freely hear and share?" Mobile entertainment corporations are now trying to do something rather different and more valuable—create new kinds of listening habits and musical activities—but this task is far more difficult than the capitalizing on the cheap that the ringtone once offered. That the ringtone industry—essentially a product of a convenient delivery system and a captive market—existed to the degree that it did is perhaps the more remarkable fact, its very redundancy being all the more improbable in retrospect.

Revisiting the Temporalities of the Ringtone Industry

From beyond the precipice of the ringtone's fall, how should we analyze the ringtone as a relatively closed case study? During its peak, the story of the ringtone amounted to a particular moment of capture of consumers, capital, and media press by a wide-ranging phenomenon that was inextricable from the broader, overlapping histories of global cultural consumption. Previously, I attempted to place the ringtone within a three-level temporal framework for the analysis of capitalism inspired by the work of Fredric Jameson, Antonio Gramsci, Michael Denning, and Fernand Braudel.[14] To recapitulate: The ringtone could be viewed from within a short-term, a medium-term, or a long-term horizon, with each temporality understood through the categories of event, conjuncture, and epoch/*longue durée*, respectively.[15] At the outer levels, the older analysis seems quite stable. The ringtone as an event was situated within the dynamics of fads, including their moments of reification and utopia, and the mobile phone's cannibalistic remediation or repurposing of earlier cultural content and convergent absorption of preexisting media devices (including hand-held game consoles, digital cameras, and portable music players) provided familiar sounds that helped to propel the novel form to mass popularity. Of course, the ringtone was never a single event: rather, it was the conjunction of numerous events of moment, at least within its own domain, and ranged from the technological advances making them possible (such as Nokia's tethering of its RTTTL to SMS), initial efforts to promote and sell uploadable ringtones, the publication of best-selling ringtone books, and numerous other occurrences described in this book. Although the rhythms of these events are almost seasonal, recalling what Guy Debord described as the consumable "pseudo-cyclical time" of industrial capitalism's commodity cycles,[16] they add up to a longer temporality, namely that of the conjuncture (discussed below). When viewed together, they can be seen as a constellation of historical instances that made the ringtone the formative event for the mobile entertainment industry. At the other end of the scale, the epochal shifts in contemporary capitalism, particularly with the rise of China as "the new workshop of the world," have led to speculations on the possibility of the new global hegemon and geographical center of another long century of productive accumulation transferring to East Asia.[17] The ringtone, from this vantage point, appears as a signal of shifting cultural practices and priorities whose formative production in Japan mark them as being in some sense historically Asian, although not essentially so. However, since the prospects of a new Asian hegemony are far from certain, a slide into a Smithian "stationary state" being predicted in some quarters, the ringtone's earliest prospering in East Asia and Europe and not the United States seems to offer a cultural registration of the much more apparent decline of American global power and a corresponding rise in power of the two other central regions of the present tripolar economy.[18]

The most critical temporal level of analysis today, where the nearly obsolete ringtone is concerned, is at the conjuncture. Occupying a variable, rather than absolute, duration—Braudel saw it as differentially comprising "a decade, a quarter of a century, and, at the outside, the half-century"[19]—the conjuncture is attuned to historical blocs within nation-states, to specific phases within the economic cycle, and to the broader dynamics of social movements, all of which overlap with evental and epochal temporalities in crucial ways. In Stuart Hall's words, a conjuncture is simply "the fact that very dissimilar currents, some of a long duration, some of a relatively short duration, tend to fuse or condense at particular moments, into a particular configuration. It is that configuration, with its balance of forces, which is the object of one's analysis or intellectual inquiry."[20] Its two main components, then, are periodization and the analysis of forces. The "ringtone conjuncture," one might argue, extended from approximately 1998 to 2010 and hence roughly overlapped with the coming to power of New Labor in the United Kingdom (1997–2010) and the presidency of George W. Bush in the United States (2001–2008), as well as the period between the last phase of the 1990s tech bubble and the recession of 2008. The condensation of forces within the ringtone industry, as I perceived it in 2005, was easily summed up by the concept of *rent*. Extractions of rents of numerous kinds are operative in the process of ringtone production and exchange, including intellectual-property rents owned by publishers (songs and other compositions) and record labels (recordings); tolls exacted on the proprietary "information highways" of wireless service providers[21]; rental of wireless billing company software, which handles ringtone transactions; and even the rental of ringtones themselves, which were not easily transferred in the process of upgrading phones and some of which were purchased on a temporary basis. As an accumulative procedure, rent extraction has the effects of raising prices grossly in excess of ideal-condition market rates and reproducing existing monopolies and oligopolies in the process.[22] As a moment of effective collusion between wireless entertainment firms, music-industry conglomerates, wireless corporations, business-to-business services (e.g., billing firms), and even, implicitly, handset manufacturers, the ringtone marked an important historical moment that saw the appeal of inflated profits within the mobile sector, as opposed to the digital sector as concretized in the personal computer and PC-based Internet access. However, with the arrival of the mobile Internet and "off-deck" access to goods and services (including ringtones), the previous fixed arrangement came undone, effectively destroying the necessary uniqueness or scarcity as a basis for rent accumulation through ringtone sales.[23] Consequently, the mobile entertainment industry moved into apps, the handset industry continued to advance smartphone sales, and the recording industry continued to founder.

In retrospect, the bid for replacing the Internet with a more constricted mobile Internet failed. The latter has more or less converged with the former—with most aspects of the decreased control over consumer behavior in the former being reproduced in the latter. But the effect of the ringtone—a commodity that, despite its inflated expense, was treated as relatively disposable and, at its most extreme, did not remain in its owner's hands—served to produce a new habit of consumption—one in which the purchase of a commodity need not result in its physical possession.

Digital Information, the Commons, and Property Rights

Within the context of self-avowedly liberal societies, it is easy to take the fact of ownership rights for granted. Aren't the Lockean natural rights of life, liberty, and property (mystified in the US context as "the pursuit of happiness," thanks to Jefferson) the essential tenets of such political orders, as conservatives love to point out?[24] Yet the extensive social impact of an access-based model of consumption with respect to the use of digital information has progressively crept into the mundane habits of billions of Internet and communications network users all over the world, including more than 230 million Internet users in the United States alone (as of 2008 according to the World Bank) and more than 5 billion mobile phone subscribers globally.[25] The most "collective" aspects of this practice, increasingly apparent in the past 20 years, would seem to signal a radical possibility of a new "digital commons" or even John Barlow's "dot-communism" through a decline in ownership practices as a result of the reproducibility of digital information. Although visions of a digital commons based on an increasing collectivization of data use are attractive to Marxists like me, it seems that reality and ideal are confused all too frequently; in contrast to the generally optimistic tone taken in discourse on the digital, it may be advisable to ring out a modest warning about the digital commons, and to advocate, perhaps paradoxically, for individualized forms of property possession.

The notion of the commons is not often associated with that of individual property ownership, precisely because the two stem from entirely different conceptions of ownership, one being at heart individual and the other collective. Yet both arguably depend on the privileged presence of what Marx called "use value" over the role of "exchange value."[26] In the case of individual ownership, by and large the *scene* of the marketplace is merely one brief moment in the life of a commodity that is typically used to reproduce the labor power of the worker-consumer, though of course the principle of ownership leads to the possibility, in theory, of the creation, circulation, and accumulation of capital—particularly in the case of small-business owners or individual proprietors.[27] (Some of the most significant capital transactions, in contrast, take place not as a result of individual ownership but as a result of the collective ownership of stockholding.[28]) Likewise, a commons, by its very definition, is a commonly owned (or better, unowned) resource that is accessed collectively and openly; its very basis exists in its utility, its usefulness, and the availability of that utility. As Marx put it, "The usefulness of a thing makes it a use-value. But this usefulness does not dangle in mid-air."[29] This is even true of information, the precious and highly mutable entity that serves, in digital form, as a primary class of commodities within the global economy today. For while information would itself seem to "dangle in mid-air," particularly when conveyed via the telecommunications network, its use value is "only realized in use or consumption," in Marx's words, which is to say that its material effects as image, sound, executable code, etc.—are plainly evident, as when you play back a downloaded music file or ringtone.

The material effects of information are, of course, highly valuable. For if it is the case that use values generate exchange values, the magnitude of that utility, to judge from trade figures, is astonishing. Already in 2007, global trade in information and communication technology (ICT) goods (total exports and imports) was worth US$3.7 trillion, according to the Organisation for Economic Co-operation and Development, which doesn't even account for domestic consumption of such goods.[30] The magnitude of this value is inseparable from the progressive expansion of capital accumulation all over the world, and from the fact that this expansion has been made possible by the pervasiveness of information commodities. Immanuel Wallerstein's dictum that "the historical development of capitalism has involved the thrust towards the commodification of everything" rings true here, with the added possibility that the leading-edge status of ICTs suggest a more limited corollary in the tendency toward the informatization of everything.[31]

Yet the condition of information as a commodity poses a unique problem for the exchange process, in that digital information is very easily reproducible, with the marginal cost of reproduction tending to almost nothing, limited only by computation time and energy expenditure in distribution, as well as the extremely gradual wear and tear on the physical communications network infrastructure. What this does in practice is to create the possibility of digital information serving as what Elinor Ostrom calls a highly nonsubtractive form of commons, rather different from traditionally subtractive forms of commons like water, land, forests, fisheries, etc., whose use can quickly lead to their depletion.[32] However, this same fact also makes productive capital accumulation, based on the production process, far less profitable than the rent derived from privileged access to the information resource—especially when that access is made scarce, as it is in the case of a monopoly. The "danger" for information capitalism is that, without controlling the scarcity of this information, very little in terms of either monopoly rent or productive capital accumulation can be guaranteed without introducing forms of access control over that information. For advocates of "free software," such as Richard Stallman, this is a powerful aspect of information, one that can be used to produce a more egalitarian world.[33] For content-owning firms, such as the big media conglomerates, this feature of information is to be battled at all costs, lest free exchange and piracy run them out of business—or so they claim.

The story here is familiar to consumers and newsreaders. It is one of MP3 and YouTube-video sharing lawsuits, of digital rights management (DRM) protection schemes, of "walled gardens" through limited Internet access via mobile telephones—a story in which, time and again, control over intellectual-property rights was visited upon the digital information commodity. The conflict explains why it can be difficult to extract video from DVDs, why until 2009 you couldn't burn more than seven copies of the same iTunes playlist if it included an AAC file purchased via iTunes, and why for a long time people paid 200–300 percent extra for ringtones representing less than 20 percent of the sound file from which they were derived. It also explains why hackers and consumers have devised ways of working around these DRM schemes and limited-access portals.

The impasse depicted here, which came to a head in the mid 2000s, has led to a kind of "third way" solution that squares the circle of free culture and monopoly rent through access-based services in which content is no longer owned by users but is merely leased or streamed to them, which amounts to a variation of the rent discussed earlier. Although one might call this a "commons" of content, it seems not to function exactly in this way. A media conglomerate solution to content delivery has long been in the works and has existed under a number of different guises: in the late 1980s, Vincent Mosco described it as the "pay-per society," evocative of content delivery schemes like "pay-per-view" or "per-minute" telephone use, and he used the notion to make an early attempt at theorizing neoliberal capitalism and its negative impact on state regulation of public services.[34] By the mid 1990s, a new term within the media industry had begun to take hold: "the Celestial Jukebox," articulated by Paul Goldstein in 1994 and transformed into a corporate catchphrase by the early 2000s, as discussed in chapter 1. As Patrick Burkart and Tom McCourt describe it, the Celestial Jukebox is "a tollbooth into a web of privately owned and operated networks where traffic in intellectual property is carefully monitored and controlled, a walled garden of closed networks with restricted access and tightly circumscribed activities."[35] This model is predicated on subscription supplanting sales, and streaming replacing downloading (though the difference is something of a fiction). In the case of digital music, it is worth noting that—although DRM has been removed from all compact discs and most digital audio sales (including those by iTunes)—Spotify, Rhapsody, and other subscription services still include DRM as part of "all you can eat" streaming-access models, effectively discouraging you from archiving music; these services are now available as smartphone apps that can produce streaming ringtones. In a rather different example, library subscriptions to academic journals have discouraged libraries from archiving physical copies of journals, leading to potential access problems (particularly if servers are down or if libraries stop paying subscription fees).[36] In these cases, the most important issue isn't the success or failure of such services; it is that a new set of media habits is being formed—habits that favor instant access and the perception of permanence, thereby eliminating perceived needs for ownership of digital files.[37]

The most recent reincarnation of this kind of conceptualization would appear to be "cloud computing." Conceptually, cloud computing dates back to the 1960s, but only recently has it been gaining traction as a business model. Essentially, the "cloud computing" model imagines all major computing services to take place off site, via the distributed servers and networks constituting the Internet (or "cloud"), instead of happening on your own computer. The information-access concept of the Celestial

Jukebox is retained, and essentially all computing information *and* software are to be located in the cloud. As the media evangelist Peter Fingar writes in his book *Dot.Cloud*, the motto of the new, perfectly globalized, collaborative business system will be "one shared world; one shared computer; one shared information base."[38] The rhetoric around cloud computing is eerily reminiscent of discourse on the digital commons. Both offer utopian visions of information and resource access, and both generate resistance to these visions. According to a book serving as the mouthpiece of Google, one of the major advocates of cloud computing, "The *cloud* is a controversial buzzword that still frightens some companies, especially large operations that want, need, or think they need full control over their data. They worry that if the provider company goes belly up their data may disappear along with the provider. There could be hackers, and what if the system goes down at a critical time for their business? They wonder if their intellectual property or proprietary information is safe online."[39] The critique of fears of hackers, control, and, implicitly, *trust* places Google and other ICT firms in the vanguard, leading capitalism into a post-recessionary future.[40] However, users' habits are still crucial here: Web-based e-mail programs such as Gmail, online photo album services such as Flickr and Google's Picasa, searchable text archives like Google Books, and other sites are examples of cloud computing in practice to which users have already been habituated, and new examples keep appearing (e.g., Apple's iCloud system for storing sound files on Apple's own servers).[41] These file-storage and file-management systems are far less dependent on single computer terminals owned and used by single individuals, but through individual users' accounts they still track user information for the purpose of customizing advertising. More disturbing, data uploaded to the cloud typically isn't subject to consumer-protection laws or to citizenship rights, such as the Fourth Amendment's search and seizure restrictions in the United States. Indeed, in compiling an unprecedented degree of user data and by working directly with surveillance technology firms and with federal intelligence officials, Google itself would appear to be at the epicenter of a massive transformation in corporate and state security surveillance, the consequences of which are difficult to foresee and may be profound.[42]

What of the ringtone in this newly clouded world? That the cloud will present new opportunities for listening and experiencing music in the ringtone format goes without saying: the "treat" proposed by Licoppe may be transformed into a "surprise" through unpredictable streaming ringtones via mobile radio/subscription services, with the drawback of ringtone misrecognition compensated by auditory novelty. Concerns about the extremely low remuneration for artists by subscription services such as Spotify, however, make the new development seem less than rosy—one effect of the massification of widely available databases may, in the end, serve to lower the value of music, to the detriment of its producers.[43] Of course, the older, more profitable ringtone was arguably one of the vehicles for inculcating cloud-like consumption habits—as disposable information commodities, ringtones weren't archived for long-term storage, and they were initially accessed through a centralized, cloud-like network. Even in the new conjuncture, the discrete, for-sale ringtone, which for a while represented the maximally profitable intersection of the histories of digital music and mobile music, will continue to have a residual life. The consequences are, however, mixed. If, on the one hand, cloudy ringtone consumption means the end of overpaying for needlessly duplicated sound files, the shrinking market for them gives great power to the largest distributors, who are most capable of surviving and profiting during the shakeout. In some cases, dominant distributors expressly pursue strategies to minimize competition from smaller entities. In a disturbing example, iTunes recently claimed that within its own organizational system the word "ringtone" means only 30 seconds of an existing song or recording, which forced Brice Salek of Ringtone Records to remove all of his label's tracks from iTunes—the largest distributor of digital music in the world.[44] Whether the product of a computing error or of an intentional effort to increase ringtone creation through iTunes itself, the move could theoretically bolster ringtone sales slightly for Apple and its music-industry partners. But it also discourages and in some cases prevents independent labels from selling ringtones or ringtone albums on its premises.

Google and Apple offer the image of a pseudo-commons to Internet users. That image recalls Nick Dyer-Witheford's claim that, in light of the structural failures of neoliberal policies, capital could "turn to a 'Plan B', in which limited versions of commons, pollution trading schemes, community development and open-source and file sharing practices are introduced as subordinate aspects of a capitalist economy, where voluntary cooperation subsidizes profit. One can think here of how Web 2.0 re-appropriates many of the innovations of radical digital activists, and converts them into a source of rent."[45] Indeed, with the rise of the trademarked Digital Commons software platform (licensed by the Berkeley Electronic Press and used by universities and other institutions for the archiving of publications), and with the proliferation of university-based digital and media commons (which are typically limited to fee-paying and/or employed university community members), the very concept of the digital commons appears to be one of these reappropriations.[46] But if, as part of what James Boyle describes as the "Second Enclosure Movement," this very rhetorical move signals the temporary defeats of the alter-globalization and radical hacker movements that claimed the language of the commons, perhaps the advocacy for ownership of digital wares (or at least a form of unalienable, absolute possession, whether individual or communal) would provide a strategic ballast against the proprietary control of large swathes of information by apparently benevolent corporations and institutions.[47] While still dangling in mid-air, the information commodity's consumption might thereby be placed more solidly on common ground.[48]

Notes

Notes to Introduction

1. On the long history of mobile music devices, see Sumanth Gopinath and Jason Stanyek, eds., *The Oxford Handbook of Mobile Music Studies* (Oxford University Press, forthcoming).

2. See "The Mobile Ringtone Is Dead: Apps Define the New Decade" at http://techinstyle.tv; saliquincer, "The Ringtone Is Dead, Long Live Music!" at http://forum2.mobile-review.com.

3. Quotation from the "M-Commerce: Demystifying the World of Mobile Commerce" panel at CTIA Wireless 2010 conference, Las Vegas, 2010. In 2008 Bud resigned from the board of PhonepayPlus, a UK telecommunications regulator, because mBlox partnered with firms illegally duping customers into unwanted ringtone subscription plans. (See chapter 5 below.) See "Andrew Bud Resigns from the Board of PhonepayPlus" at, "The Scream!" (chat forum), http://www.the-scream.co.uk; Andrew Penman, "Mblox Fined Again over Premium Rate Phone Lines" at http://blogs.mirror.co.uk.

4. See Andrew Bud, comments (18 June 2008) on Andrew Penman, "Mblox Fined Again over Premium Rate Phonelines," Penman & Sommerlad Investigate blog, 12 June 2008; "Mobile Entertainment Market to Cross $50 Billion" at http://news.accuracast.com.

5. One marker of that industry's global emergence was the term "ringtone" appearing in languages besides English. Language-specific terms for ringtone include *chaku-mero* (ringtone) and *chaku-uta* (sound file ringtone) in Japanese, *Klingelton* (ringtone) in German, *le ton mobile* or *sonnerie téléphonique* (ringtone or telephonic ringer) in French, and *tonos moviles* and *tonos reales* (ringtones and mastertones) in Spanish.

6. See Dallas Smythe, "Communications: Blindspot of Western Marxism," *Canadian Journal of Political and Social Theory* 1, no. 3 (1977), 1–28.

7. See Julia Angwin, "The Web's New Gold Mine: Your Secrets," *Wall Street Journal*, 30 July 2010; Eli Pariser, *The Filter Bubble: What the Internet Is Hiding from You* (Penguin, 2011), p. 15.

8. See Marc Perlman, "Consuming Audio: An Introduction to Tweak Theory," in *Music and Technoculture*, ed. R. Lysloff and L. Gay (Wesleyan University Press, 2003).

9. Meghan Daum, "Cellphone Psychology: You Are Your Ringtone," *Los Angeles Times*, 7 July 2005. The trope seems to be a journalistic version of genre-listener typologies found in music sociology, as on pages 4–14 of Theodor Adorno's *Introduction to the Sociology of Music* (Continuum, 1976), or of the commonsensical and often reductive mapping of genre and identity. On genre and identity, see Eric Drott, *Music and the Elusive Revolution: Cultural Politics and Political Culture in France, 1968–1981* (University of California Press, 2011), pp. 81–91. Also, the very existence of ringtone "hits" such as Crazy Frog (see chapter 5) suggests the presence of collectivizing, non-individuating forces that cannot be wished away by the interrelated notions of mass customization, the long tail, and personalization.

10. See, for example, Heikki Uimonen, "'Sorry, Can't Hear You! I'm on a Train': Ringing Tones, Meanings, and the Finnish Soundscape," *Popular Music* 23, no. 1 (2004), 51–62; Noriko Manabe, "New Technologies, Industrial Structure, and the Consumption of Music in Japan," *Asian Music* 39, no. 1 (2008), 81–107; idem, "Going Mobile: The Mobile Internet, Ringtones, and the Music Market in Japan," in *Internationalizing Internet Studies: Beyond Anglophone Paradigms*, ed. G. Goggin and M. McLelland (Routledge, 2009); Christian Licoppe, "The Mobile Phone's Ring," in *The Handbook of Mobile Communication Studies*, ed. J. Katz (MIT Press, 2008); idem, "What Does Answering the Phone Mean? A Sociology of the Phone Ring and Musical Ringtones," *Cultural Sociology* 20, no. 10 (2010), 1–21; Imar de Vries and Isabella van Elferen, "The Musical Madeleine: Communication, Performance, and Identity in Musical Ringtones," *Popular Music and Society* 33, no. 1 (2010), 61–74.

11. Manabe's statistical inquiries offer exceptions to qualitative studies of individuals' ringtone use. A more typical example of the latter is Uimonen ("Sorry, Can't Hear You!" p. 54), who interviewed "eleven males (M) and eight females (F)" of high school age. The discussions Uimonen reports focused on the ringtones subjects chose and why (pp. 54–58), but most were not place- or use-specific; Uimonen additionally reflects upon the urban soundscape (pp. 57–60)—suggesting outdoor, public uses of ringtones.

12. Licoppe, "The Mobile Phone's Ring," pp. 144–145, and "What Does Answering the Phone Mean?" p. 12.

13. Licoppe nevertheless draws on Luc Boltanski and Ève Chiapello, *The New Spirit of Capitalism* (Verso, 2005 [1999]), p. xii.

14. Wayne Marshall, "Treble Culture," in *The Oxford Handbook of Mobile Music Studies*; Dan Hickox, "Mobile Disco: How Phones Make Music Inescapable," *The*

Guardian, 12 August 2010, available at http://www.guardian.co.uk; Dick Hebdige, *Subculture: The Meaning of Style* (Routledge, 2002 [1979]); Uimonen, "Sorry, Can't Hear You!" esp. pp. 58–59; Sumanth Gopinath, "Ringtones, or, The Auditory Logic of Globalization," *First Monday* 10, no. 12 (2005), available at http://firstmonday.org. For a brief introduction to chav culture, see chapter 9 below.

15. Paul Vitello, "A Ring Tone Meant to Fall on Deaf Ears," *New York Times*, 12 June 2006; Melissa Block, "Teen Turns 'Repeller' into Adult-Proof Ringtone," *All Things Considered*, 26 May 2006, available at http://www.npr.org; Sarah Lyall, "Rowdies Buzz Off as the Mosquito Bites," *The Age* (Australia), 30 November 2005, available at http://www.theage.com.au; "Anti-Gang Noise Box Switched Off," *BBC News*, 24 March 2006, available at http://news.bbc.co.uk; "Pupils Adapt Mosquito Alarm into Ringtone Unaudible for Over Twenties" at http://compoundsecurity.co.uk; "High-Pitched Mobile Phone Ringtone Gets Dropped" at http://www.prweb.com; Robert Andrews, "MozzyTone, Once a Teen Deterrent, Now an Interactive TV Ad" at http://moconews.net; "The Mosquito," at http://en.wikipedia.org.

16. Louis Menand, "Name That Tone," *The New Yorker*, 26 June 2006; C.A.N., "KILL THE MOSQUITO!" at https://southwalesanarchists.wordpress.com.

17. Block, "Teen Turns 'Repeller' into Adult-Proof Ringtone." The ringtone may have spurred further sales of the Mosquito itself, the fad paradoxically providing good publicity for the device. In fact, sales spiked after a Council of Europe committee called for the device to be banned. See "Sales of 'Mosquito' Youth Disperser Soar After Criticism," *Western Mail* (Cardiff, Wales), 21 July 2010.

18. For Fredric Jameson (*Representing* Capital*: A Commentary on Volume One*, Verso, 2011, p. 131), dialectical contradiction "is identical with the unity of opposites, opposites which no longer need to be identified with and labeled as positive and negative, inasmuch as the dialectic means a perpetual changing of places between them and a perpetual transformation of one into the other."

19. See Robert Heilbronner, *Marxism: For and Against* (Norton, 1980), p. 93 and passim. Göran Therborn argues that in the twentieth century Marxism's "broken triangle"—historical social science, dialectical philosophy, and political modality—was gradually severed as a result of Western Marxism's specialized focus on philosophy, the academic absorption of Marxism, the decline of Marxist politics in the 1980s, and the global decline of the industrial working class. See Therborn, *From Marxism to Post-Marxism?* (Verso, 2008), pp. 116–120.

20. Ibid., p. 153.

21. See John Rule, *Theory and Progress in Social Science* (Cambridge University Press, 1997), pp. 120–143; Manuel Castells, *The Rise of Network Society* (Blackwell, 1996); Castells, ed., *The Network Society: A Cross-Cultural Perspective* (Edward Elgar, 2004); Bruno Latour, "On Actor-Network Theory: A Few Clarifications" at http://www

.nettime.org. For a critique of connectionist thought, see pp. xxii–xxiii and 103–163 of Boltanski and Chiapello, *The New Spirit of Capitalism*.

22. Peter Lunenfeld, "Screen Grabs: The Digital Dialectic and New Media Theory," in *The Digital Dialectic: New Essays on New Media*, ed. P. Lunenfeld (MIT Press, 1999), pp. xviii–xx. Also see Gary Hall, *Digitize This Book! The Politics of New Media, or Why We Need Open Access Now* (University of Minnesota Press, 2008), pp. 24–32.

23. Fredric Jameson, *Valences of the Dialectic* (Verso, 2009), p. 24.

24. But see ibid., pp. 43–48.

25. For a characteristic critique of the superstructural dismissal of culture, see Paul du Gay, Stuart Hall, Linda Janes, Hugh Mackay, and Keith Negus, *Doing Cultural Studies: The Story of the Sony Walkman* (Sage, 1997), pp. 1–5. That textbook's alternative "circuit of culture" model has been much imitated in media studies. Gerard Goggin offers a list of successor studies, including his own, on pp. 7–16 of *Cell Phone Culture: Mobile Technology in Everyday Life* (Routledge, 2006).

26. Raymond Williams, "Base and Superstructure in Marxist Cultural Theory," in Williams, *Problems in Materialism and Culture* (NLB, 1980), p. 34.

27. Williams was critical of focusing on objects and components in the analysis of culture (ibid., pp. 48–49), as opposed to "the conditions of a practice" (p. 48), but a dialectical study of culture must account for both objects and practices, and for both components and processes.

28. Fredric Jameson, *The Political Unconscious: Narrative as a Socially Symbolic Act* (Cornell University Press, 1981), esp. pp. 25–26.

29. See Williams, *Television: Technology and Cultural Form* (Routledge, 2003 [1974]), pp. 3–7. The title of this book is meant as an allusion to Williams' concise masterpiece, with the added effect of "economy" also meaning "brevity" and thereby referencing the short duration of the ringtone.

30. Jameson (*Valences of the Dialectic*, p. 43) attributes the idea that the superstructure "expresses" the base to Walter Benjamin's Arcades Project. See Benjamin, *The Arcades Project* (Belknap, 1999).

31. Jason McGrath, *Postsocialist Modernity: Chinese Cinema, Literature, and Criticism in the Market Age* (Stanford University Press, 2008), pp. 9–13.

32. See Terry Eagleton, "Base and Superstructure Revisited," *New Literary History* 31, no. 2 (2000), 231–240. Novelty forms of the mass-culture era are some hybrid of attractions and gimmicks (some drawn from new historical situations, such as new waves of immigration), adaptations of existing genres from older media, and explorations of the technical possibilities of the new medium (which often is or draws upon leading-edge technology)—with the first and third often being inseparable. See Ronald Rodman, *Tuning In: American Narrative Television Music* (Oxford University

Press, 2010), pp. 236–237; Tom Gunning, "The Cinema of Attractions: Early Film, Its Spectator and the Avant-Garde," in *Early Film: Space, Frame, Narrative*, ed. T. Elsaesser and A. Barker (British Film Institute, 1990), pp. 56–62; Charles Hamm, *Irving Berlin: Songs from the Melting Pot: The Formative Years, 1907–1914* (Oxford University Press, 1997), esp. pp. 22–67.

33. See Karenza Moore and Jason Rutter, "Understanding Consumers' Understanding of Mobile Entertainment," in *Proceedings of Mobile Entertainment: User-Centered Perspectives*, ed. Moore and Rutter (ESRC, 2004), p. 51.

34. See Robert Brenner, *The Boom and the Bubble: The US in the World Economy* (Verso, 2002); idem, "What's Good for Goldman Sachs Is Good for America: The Origins of the Current Crisis" at www.sscnet.ucla.edu. For a critique of Brenner's methodology, see Giovanni Arrighi, *Adam Smith in Beijing: Lineages of the Twenty-First Century* (Verso, 2007), pp. 99–148.

35. Therborn, *From Marxism to Post-Marxism?*, p. 25. See James Katz and Mark Aakhus, eds., *Perpetual Contact: Mobile Communication, Private Talk, Public Performance* (Cambridge University Press, 2002); Mizuko Ito, Daisuke Okabe, and Misa Matsuda, eds., *Personal, Portable, Pedestrian: Mobile Phones in Japanese Life* (MIT Press, 2005); James Katz, *Magic in the Air: Mobile Communication and the Transformation of Social Life* (Transaction, 2006); Goggin, *Cell Phone Culture*; Manuel Castells, Mireia Fernandez-Ardevol, Jack Linchuan Qiu, and Araba Sey, *Mobile Communication and Society: A Global Perspective* (MIT Press, 2007); James Katz, ed., *Handbook of Mobile Communication Studies* (MIT Press, 2008).

36. Michael Denning, *Culture in the Age of Three Worlds* (Verso, 2004), pp. 90–96.

37. Karen Collins, *Game Sound: An Introduction to the History, Theory, and Practice of Video Game Music and Sound Design* (MIT Press, 2008).

38. Consumer surveillance and voluntary labor are treated in greater depth in Sumanth Gopinath and Jason Stanyek, "Tuning the Human Race: Athletic Capitalism and the Nike+ Sport Kit," in *Music, Sound, and Space: Transformations of Public and Private Experience*, ed. Georgina Born (Cambridge University Press, 2013).

Notes to Chapter 1

1. "Apple iPhone Signals Death of Ringtone Industry" at http://www.blackbeardseo.com.

2. However, shortly after the release of the iPhone, Apple announced that the most recent version of iTunes (version 7.4) would in fact allow for making custom ringtones from one's iTunes song library. See Arnold Kim, "Apple Posts iTunes 7.4 with Ringtones Support," at http://www.macrumors.com.

3. See, e.g., Gerard Goggin, *Global Mobile Media* (Routledge, 2011).

4. See Steve Savage, *Bytes and Backbeats: Repurposing Music in the Digital Age* (University of Michigan Press, 2011).

5. See Ellen Coughlin, "A Little Bourdieu Industry: For Pierre Bourdieu Books, a Flurry of New Translations," *Chronicle of Higher Education*, 12 August 1992; Nick Spencer, "Attacked from Within," at http://www.electronicbookreview.com; Joan Ockman, untitled article, *Assemblage* 41 (2000), 61; Norman Finkelstein, *The Holocaust Industry: Reflections on the Exploitation of Jewish Suffering* (Verso, 2000); Jim, "The 9/11 Industry," at http://letters.salon.com; Kenneth Tunnell, *Pissing on Demand: Workplace Drug Testing and the Rise of of the Detox Industry* (New York University Press, 2004); Donald Abelson, "The Business of Ideas: The Think Tank Industry in the United States," in *Think Tank Traditions*, ed. D. Stone and A. Denham (Manchester University Press, 2004); Philip Swanson, "Back to the Boom? Recent Trends in Spanish American Literary Studies," *Latin American Research Review* 36, no. 2 (2001), 202–208, p. 203. A particularly impressive version of this kind of thing includes Mark Nowak's critique of "the American MFA industry" in "Neoliberalism, Collective Action, and the American MFA Industry," in *¡Workers of the Word, Unite and Fight!* (Palm, 2004).

6. Max Horkheimer and Theodor Adorno, *Dialectic of Enlightenment: Philosophical Fragments* (Stanford University Press, 2002), esp. pp. 94–136. The writers also emphasize the self-consciousness of the culture industry, noting that its components "call themselves industries" (p. 95).

7. Sometimes "ringing tone" would refer to the sound one hears while listening to the telephone handset speaker; in current parlance, this is a "ringback tone" or simply "ringback."

8. Stacy V. Jones, "Patents: Electronic Tone for Telephone Users," *New York Times*, 4 July 1981.

9. "New Types of Telephone," Weekly Economic Report, *BBC World* broadcast, 3 December 1982, summarized at http://web.lexis-nexis.com.

10. Julian Macassey, "Understanding Telephones," *Ham Radio Magazine*, September 1985, available at http://www.privateline.com. Also see "Northern Telecom to Introduce Electronic Phones in New Year," *The Globe and Mail* (Canada), 6 June 1978, which mentions electronic handsets in use since 1973 in Belgium and Denmark. Gong ringers were physically unfeasible for use on cell phones because of weight and power requirements, but they never entirely disappeared from land-line telephones.

11. See the 1990 exchange with Julian Macassey on "REN" at http://yarchive.net.

12. A quotation from an article in *The Age* in 1960 suggests that plans for decreasing power expenditures had been on the horizon in the UK well before the digital ringer came into prominence: "The familiar ring will disappear when Britain's new electronic telephone system is installed through the country. *To save the power neces-*

sary to provide the present ringing of telephones, new call sounds, including a bird-like whisper, are being tried out, the post office said." "Telephone Ring Maybe Bird Whistle," *The Age* (Melbourne), 24 September 1960, cited in Gerard Goggin, *Global Mobile Media* (Routledge, 2011), p. 57; emphasis added. Goggin (pp. 54–58) provides a prehistory of the ringtone extending back to the nineteenth century.

13. Driven by military and computer technologies, especially space and missile programs, integrated circuits became commercialized only toward the latter half of the 1960s—in televisions or hand-held calculators—and would first appear in commercial personal devices like telephones in the late 1960s and the early 1970s, such as the Bell "Touch-a-Matic" automatic-dialing phone in 1972. For a diagram that traces the development of silicon integrated circuits in Bell products and systems, see F. M. Smits, ed., *A History of Engineering and Science in the Bell System: Electronics Technology (1925–1975)* (AT&T Bell Laboratories, 1985), p. 108.

14. See "Integrated Circuit" at http://en.wikipedia.org and Ross Bassett, *To the Digital Age: Research Labs, Start-Up Companies, and the Rise of MOS Technology* (Johns Hopkins University Press, 2002), p. 37 and passim.

15. *NEC Corporation: The First 80 Years* (NEC Corporation, 1984), p. 50.

16. See Jon Halliday, "Japan—Asian Capitalism," *New Left Review* I, no. 44 (1967), 3–8; Chalmers Johnson, *MITI and the Japanese Miracle: The Growth of Industrial Policy, 1925–1975* (Stanford University Press, 1982); Bob Johnstone, *We Were Burning: Japanese Entrepreneurs and the Forging of the Electronic Age* (Basic Books, 1999), pp. xi–xxiii; Simon Partner, *Assembled in Japan: Electrical Goods and the Making of the Japanese Consumer* (University of California Press, 1999); "Japanese Post-War Economic Miracle" at http://en.wikipedia.org.

17. See *NEC Corporation*, pp. 63–80; G. C. Allen, How Japan Competes: An Assessment of International Trading Practices with Special Reference to "Dumping," Hobart Paper 81, Institute of Economic Affairs, 1978, pp. 11–14; Jon Halliday and Gavan McCormack, "Japan and America: Antagonistic Alliance," *New Left Review* I, no. 77 (1973), 59–60.

18. Arthur L. Robinson, "Perilous Times for U.S. Microcircuit Makers," *Science*, 9 May 1980, 582; Partner, *Assembled in Japan*, 209–210; Allen, *How Japan Competes*, pp. 17–21.

19. Joel Chadabe, *Electric Sound: The Past and Promise of Electronic Music* (Prentice-Hall, 1997), p. 117.

20. Johnstone, *We Were Burning*, p. 236; Tom Darter, "John Chowning: Part I of an Exclusive Interview with the Father of Digital Synthesis," *Aftertouch* 1, no. 2 (1985), available at http://www.maths.abdn.ac.uk. Paul Théberge notes that the DX-7 and other synthesizers became popular in the mid 1980s through the promotion of preset sounds instead of synthesizer programmability. The digital synthesizer thus

signals a shift in the musical instrument industry's understanding of its target sales market as primarily consumers (rather than specialist creators). See Théberge, *Any Sound You Can Imagine: Making Music/Consuming Technology* (Wesleyan University Press, 1997), pp. 75–77. Also see Trevor Pinch and Frank Trocco, *Analog Days: The Invention and Impact of the Moog Synthesizer* (Harvard University Press, 2002), pp. 316–317.

21. Chadabe, *Electric Sound*, pp. 185–86.

22. Johnstone, *We Were Burning*, pp. 238–241; Karen Collins, *Game Sound: An Introduction to the History, Theory, and Practice of Video Game Music and Sound Design* (MIT Press, 2008), pp. 38, 40. On convergence, see below.

23. See, e.g., Yamaguchi Tetsuo, "Musical Envelope Control Device," US Patent 4478525, filed 30 November 1981, issued 23 October 1984, available at http://www.patsnap.com.

24. Pieter Doensen, *Watch: History of the Modern Wrist Watch* (Snoeck Ducaju & Zoon, 1994), p. 197.

25. Mary Bellis, "History of Pagers and Beepers," at http://inventors.about.com.

26. Bellis, "History of Pagers." By 1982, 1.5 million pagers were in use in the United States, "mainly for business purposes." See Andrew Pollack, "Technology: Lowly Beepers Coming of Age," *New York Times*, 15 July 1982. On the pager's introduction in Japan, see "Portability, Price Give Pocket Pagers the Edge," *Japan Economic Journal* (*Nihon Keizai Shimbun*), 27 October 1990.

27. Pager subscriptions peaked at more than 10 million in 1996. See Tomoyuki Okada, "Youth Culture and the Shaping of Japanese Mobile Media: Personalization and the *Keitai* Internet as Multimedia," in *Personal, Portable, Pedestrian*, ed. Ito et al., p. 44.

28. Okada, "Youth Culture," pp. 42–44.

29. "Pocket Pagers to Be Standardized," *Japan Economic Journal*, 18 June 1988; "Beeps Abound," *Asahi News Service*, 13 April 1995.

30. Noriko Manabe, "Going Mobile: The Mobile Internet, Ringtones, and the Music Market in Japan," in *Internationalizing Internet Studies*, ed. G. Goggin and M. McLelland (Routledge, 2008).

31. Ibid.; Okada, "Youth Culture," p. 55.

32. P. Wilson, "Portable TVs with Earphones Now," *Courier-Mail* (Australia), 19 August 1987; "Portability, Price Give Pocket Pagers the Edge."

33. Vivienne Chow, "Tone of the Future," *South China Morning Post* (Hong Kong), 27 September 2001.

34. See "Nokia" at http://en.wikipedia.org; Dan Steinbock, *The Nokia Revolution: The Story of an Extraordinary Company That Transformed an Industry* (AMACOM, 2001), pp. xxxiv, 11–13.

35. Steinbock, *Nokia Revolution*, p. xxxi.

36. Ibid., p. 92.

37. See "Nokia: Global Launch for Nokia 6100 Series," *M2 Presswire*, 10 November 1997, and the description of the Nokia 6110 at http://www.logitec.ro. Also see Nico Fogelholm, "The Nokia Tune," at http://www.about-nokia.com.

38. "The Nokia 8110 Spearheads a New Generation of Mobile Phones—The Natural Shape of Things to Come!" *M2 Presswire*, 25 March 1996; Annie Turner, "How Many Extras Does a Phone Need?" *The Times* (London), 26 April 1996.

39. The melodies offered on the phone were both newly composed and public domain tunes, and initially followed a uniform marketing strategy globally. For example, the preset melodies did not, for the most part, differ from country to country—though there were exceptions; for example, "Take Me Out to the Ballgame" appeared only in the US.

40. "Nokia: Global Launch for Nokia 6100 Series." This is, by the way, not to assume that other phones (especially in Japan) did not have this feature first. The narrative on offer here is probably, but not necessarily, limited to the European market.

41. Thomas Dolby Robertson, "Improving the Quality of Mobile Ringtones, Quality of Life?" at http://www.newswireless.net.

42. Dan Steinbock, *The Mobile Revolution: The Making of Mobile Services Worldwide* (Kogan Page, 2005), p. 125. On Paananen, see http://www.vesku.com/.

43. Mark Halper, "The Sweet Sound of Success," *Time*, 8 August 2004, available at http://www.time.com.

44. Subrahmanyam Karuturi, "What Is SMS?" at www.funsms.net.

45. Steinbock, *Mobile Revolution*, p. 125.

46. Sasha Frere-Jones, "Ring My Bell," *The New Yorker*, 7 March 2005; Philip Nicosia, "A Tribute to the Inventor of Ringtones," at http://ezinearticles.com.

47. By the end of 1997, Nokia had the greatest market share in the year's handset sales (24%), and by the end of 1998 it was the world leader in handset sales thanks to its newfound dominance in the massive US market. See "Handset Makers Buck Tradition by Releasing Unit Sales Figures," *Mobile Communication*, 5 March 1998; Nicholas George, "Nokia Records 73% Increase in Interim Earnings: Finnish Group Strengthens Position as World's Top Mobile Phones," *Financial Times*, 23 July 1999.

48. See "An Interview with James Winsoar" at http://www.tonez.co.uk.

49. See "Ringing the Changes," *The Economist*, 17 April 2004, p. 62.

50. Sherman Fridman, "Wireless Phones Dial Up Entertainment," *Newsbytes*, 5 July 2000; Tamara Conniff, "EMI Settles Suit on Phone Tunes," *Hollywood Reporter*, 30 January 2001; Winsoar, "An Interview with James Winsoar." The argument about the importance of YourMobile was suggested by Andrew Bud in a brief conversation on 23 March 2010.

51. "First Ever MEF Special Recognition Award Goes to the Pioneer of the Mobile Ringtone Business—'Vesku' Paananen," at http://www.mef.org.

52. See Emily Turrettini, "James Winsoar, First Ringtone Provider," Ringtonia.com, at http://www.textually.org/ringtonia/archives. Without permission, Winsoar used an earlier version of my ringtones research paper to claim that he was the (world's) first ringtone provider—which I never stated (I only made such claims for his priority in the UK, which may yet be incorrect). See his PR announcement of 24 July 2006, "Phat Tones Is Celebrating Seven Years of Excellence," available at http://www.prweb.com.

53. See "Patent Lawyer Settled Out of Court with Siemens," *The Marker*, 30 August 2005, available at http://www.ahkisrael.co; Kobi Ben-Simhon, "The Luzzatto Trials," *Haaretz*, 29 July 2005; Luzzatto et al., US Patent No. 6449359, 10 September 2002, available at http://patft.uspto.gov; Mike Masnick, "Need Another Reason to Hate Ringtones? They're Patented," at http://techdirt.com.

54. See Emily Turrettini, "Ringtone Patent Lawsuit!" Ringtonia.com, 31 August 2005, at http://www.textually.org.

55. Steinbock, *Nokia Revolution*, pp. 125–135.

56. Ibid., pp. 212–216. Also see the press release for the Nokia 3810, which featured Asian-language interfaces and was also designed with some putatively "Asian" aesthetics and consumer preferences in mind. The most amusing (and arguably problematic) of these concerns ringtones: "It also offers very loud ringing tones, a feature preferred by some Asian consumers, and a range of Asian ringing melodies." See "Ultimate Style and Ease of Use for Asia Pacific Consumers—Nokia Introduces the Nokia 3810—the First Specially Made-For-Asia Mobile Phone," 20 March 1997, available at http://press.nokia.com.

57. See Hayden Porter, "Phone It In!" *Electronic Musician*, February 2004, p. 84.

58. Robertson, "Improving the Quality."

59. Porter, "Phone It In!" p. 77.

60. MIDI was developed in the early 1980s by synthesizer manufacturers (especially Yamaha and Roland) in order to coordinate many different synthesizers in rock concerts. Despite its numerous limitations, MIDI became the standard protocol for

the transfer of digital instructions to electronic musical instruments. See Chadabe, *Electric Sound*, pp. 194–197.

61. Okada, "Youth Culture," p. 55. On the East Asian PHS system, see "Personal Handy-Phone System" at http://en.wikipedia.org; Lynette Luna, "Japan's PHS Carriers Take Varying Tacks to Rejuvenate Demand," *RCR Radio Communications Report*, 23 November 1998.

62. Porter, "Phone It In!" p. 84.

63. Ibid.; Okada, "Youth Culture," p. 55. In 2002, Robertson noted in "Improving the Quality" that 32 voices was the norm in Japan; Okada states that 40 voices were available.

64. James Snodgrass, "What's New: Email on Your Phone," *The Guardian*, 13 July 2000.

65. Timothy Clukey, "SoundingOff" blog, available at http://faculty.plattsburgh.edu; Jason Alexander ("Dr. MIDI"), "Interactive WWW Music . . . ," at http://www.monitor.ca.

66. Richard Bloor, "The WDN Symbian DevZone. . . Richard Looks at the Beatnik Audio Engine," at http://www.wirelessdevnet.com.

67. "Beatnik's Rich Audio Technology Available to Enhance Sound for Symbian OS Devices, Including Nokia 9210 and 9290 Mobile Phones," 20 November 2001, at http://www.beatnik.com.

68. Robertson, "Improving the Quality."

69. Ibid.

70. See Joseph Palenchar, "NPD Tracks Rising Adoption Rate of Highly Featured Cellphones," *TWICE: This Week in Consumer Electronics*, 9 February 2004, p. 36.

71. Personal communication with author, 19 August 2007.

72. "Ringtone" and "Truetone" at http://en.wikipedia.org.

73. "Siemens SL45" at http://en.wikipedia.org; "SanDisk Will Supply MultiMediaCards for New Siemens SL45 Mobile Phone With MP3 Player," *Business Wire*, 11 October 2000, available at http://findarticles.com.

74. See "International Mobile Telecommunications-2000" at http://en.wikipedia.org.

75. "3G" at http://en.wikipedia.org; Manabe, "Going Mobile."

76. "KDDI" at http://en.wikipedia.org; "J-Phone Jumping into 3G Market," *Nikkei Weekly*, 9 December 2002.

77. Jan Michael Hess, "Seeing is Believing!—Reporting from Our 1st Mobile Executive Tour to Tokyo in April 2004," at http://www.mobiliser.org. Also see "Advanced Audio Coding" at http://en.wikipedia.org.

78. On the XMF format, see the MIDI Manufacturers Association's description at http://www.midi.org.

79. Robertson, "Improving the Quality."

80. See Nokia press release "Nokia and Warner Music International Take Music Mobile for the Nokia 3300," 2 June 2003, available at http://press.nokia.com.

81. "New Motorola V300 Puts Pictures, Color, Games and Fun at Your Fingertips Exclusively for T-Mobile," *PR Newswire*, 20 October 2003; Sascha Segan, "Nokia Shocks, Amazes with New Fashion Phones," *PC Magazine*, 17 September 2004.

82. The difference in the two signals causes the 40-Hz warbling. See "Ringback Tone" at http://en.wikipedia.org.

83. Ibid. See Akira Takeishi and Kyoung-Joo Lee, "Mobile Music Business in Japan and Korea: Copyright Management Institutions as a Reverse Salient," *Journal of Strategic Information Systems* 14 (2005), 293; Peter J. Howe, "Wireless-Phone Companies Dial for Dollars with Custom Ring Tones," *Boston Globe*, 27 May 2004.

84. Rosemary Barnes, "The Sound of Money," *San Antonio Express-News*, 24 July 2004; Takeishi and Lee, "Mobile Music Business," p. 293.

85. Takeishi and Lee, "Mobile Music Business," p. 294.

86. Manabe, "Going Mobile," p. 320.

87. See Yinka Adegoke, "Music Industry Hopes For, Yet Fears, iPhone Effect," *EWeek*, 28 June 2007. Nonetheless, by May 2010, perhaps driven by ethnoracial minorities, 33% of US mobile phone users listened to music on phones, although frequency of use is not clear. See Aaron Smith, "Mobile Access 2010," at http://www.pewinternet.org. Also see chapter 9 below.

88. For a summary of a report by Research and Markets, see Emily Turrettini, "Identify the Key Forces That Will Shape the New Ringtone Industry," Ringtonia.com, 27 August 2006, available at http://www.textually.org/ringtonia/archives. For the RIAA's "2010 Year-End Shipment Statistics," see http://76.74.24.142/548C3F4C-6B6D-F702-384C-D25E2AB93610.pdf).

89. See Antony Bruno, "The Next Big Rings," *Billboard*, 31 March 2007.

90. Susan Butler, "Ringtone Balancing Act," *Billboard*, 11 November 2006. In 2003 the ringtones market was already worth somewhere between $2.5 billion and $3.5 billion. According to the Yankee Group, based in Boston, the global ringtone market in 2003 amounted to $2.5 billion; the London-based ARC Group estimates that market at $3.5 billion. See Laurie Flynn, "The Cellphone's Next Makeover: Afford-

able Jukebox on the Move," *New York Times*, 2 August 2004; Scott Banerjee, "WMG Inks Mobileway Deal," *Billboard*, 24 July 2004.

91. British Phonographic Institute, *Statistical Handbook 2008* (BPI Limited, 2008), 98. The data from the RIAA (see figure 1.2) suggest a higher figure: US shipment income on "mobile" music was at $880.8 million in 2007. Cited in "2008 Year-End Shipment Statistics," available with subscription at http://www.riaa.com.

92. Antony Bruno, "Who's Counting Mobile Content Sales?" *Billboard*, 1 October 2005.

93. See Jared Moya, "Why Doesn't the RIAA Talk About the $9 Billion USD in Ringtone Sales?" at http://www.zeropaid.com.

94. Bennett Harrison and Barry Bluestone, *The Great U-Turn: Corporate Restructuring and the Polarization of America* (Basic Books, 1988), 32–33; Christopher McKenna, *The World's Newest Profession: Management Consulting in the Twentieth Century* (Cambridge University Press, 2006).

95. Denis Saint-Martin, *Building the New Managerialist State: Consultants and the Politics of Public Sector Reform in Comparative Perspective* (Oxford University Press, 2000), pp. 19–23.

96. For example, a single paper copy of the fifth edition of *Mobile Music 2008* from Informa costs $2,691, and a PDF license costs nearly twice that.

97. Moya, "Why Doesn't the RIAA Talk About the $9 Billion USD in Ringtone Sales?"; IFPI, "IFPI: 07 Digital Music Report," January 2007, available at www.ifpi.org.

98. Adam Webb, "Technology: Hanging Up on Ringtones," *The Guardian*, 28 June 2007; Music Ally, "Ringtone Boom Draws to a Close," at http://musically.com.

99. Data sources for figure 1.1 include the following: "IFPI: 07 Digital Music Report"; "IFPI Digital Music Report 2008," January 2008, at www.ifpi.org; "Gartner Says Consumer Spending on Mobile Music Will Surpass US$32 billion by 2010," 23 January 2007, at http://www.gartner.com; "Mobile Music Searches for Hit Formula," at eMarketer.com.

100. I have not been able to find international ringtone industry figures from 2007 on, but to judge from the US rates alone (as shown in figure 1.2), the industry has already peaked worldwide and is steadily decreasing.

101. Gartner Research's figures were on the extremely optimistic end of the spectrum; the foregoing is necessarily a drastic oversimplification of the marketplace for mobile music data and should be understood as a didactic exercise describing the logic of the industry as viewed from the outside. Still, the fact that the company did not provide periodic, publicly available follow-up reports allowed them not to take responsibility for their inflated predictions.

102. For the Orange shop, see http://www.orange.co.uk; for the Sony BMG "Music Box," see http://www.sonymusicmobile.com.

103. On the "WAP Push" in ringtone sales, see the discussion beginning on 5 May 2004 at http://www.nowsms.com.

104. In Manabe, "Going Mobile." Also see the report by the Commission of the European Communities, Commission Decision of 03/X/2007 Declaring a Concentration to Be Compatible with the Common Market and the EEA Agreement, Case No. COMP/M 3333—Sony/BMG, Regulation (EC) No. 4064/89 Merger Procedure, Article 8 (2), 10 March 2007, available at http://ec.europa.eu.

105. The two major forms of music publishing royalties are mechanical and performance royalties. The former governs the mechanical reproduction of a particular musical composition (in the form of sheet music or recording), whereas the latter covers performances of such music (either scores or recordings) in public spaces.

106. Scott Banerjee, "Ringtone Rumble Brewing," *Billboard*, 22 May 2004.

107. Banerjee, "Ringtone Rumble."

108. Jason Ankeny, "The New Sounds of Music: Ringtones, the Celestial Jukebox and the Mobilization of Media," *Wireless Review*, November 2003, 30–31.

109. Ibid.

110. "Ringtones Left Out of Digital Music Price Wars," *Total Telecom*, 25 May 2004.

111. Perhaps not surprisingly, NCR (National Cash Register) is one of the many firms involved in the present-day wireless billing market.

112. For a helpful explanation of the economics of monopolies, see Ernest Mandel, *Marxist Economic Theory*, volume 2 (Monthly Review Press, 1968), pp. 393–440, esp. pp. 423–426.

113. "Ringtones Left Out."

114. Around 2004, the dominant companies in the US were Zingy, Moviso, and Modtones. See Banerjee, "Ringtone Rumble."

115. "Ringing the Changes," p. 62.

116. Banerjee, "Ringtone Rumble."

117. Scott Banerjee, "Getting Their Cut," *Billboard*, 22 May 2004.

118. See Brian Garrity and Susan Butler, "Sony BMG Sued Over Artists' Digital Rates," *Billboard*, 6 May 2006; *The Youngbloods v. BMG Music*, 07CV2394(GBD), 2008 US Dist. LEXIS 30600 (S.D.N.Y. March 28, 2008).

119. See Serona Elton, "The Ever Changing Legal Landscape of Digital Music," *MEIEA* eZine 6/2 (2009), at http://www.meiea.org. The case is *F.B.T. Productions, LLC,*

Em2M, LLC v. Aftermath Records, Interscope Records, UMG Recording, Inc., Ary, Inc., Case no: 2:07-cv-03314-PSG-MAN in US District Court, Central District of California.

120. See "Ringtones Sales Short of Music Industry Expectations," *Total Telecom*, 13 September 2004; Susan Butler, "Debates Over US Compulsory License Rates Heats Up," *Billboard*, 16 December 2006.

121. Elton, "The Ever Changing Legal Landscape."

122. Ibid. See also "Final Determination of Rates and Terms," issued by Copyright Royalty Judges, 24 November 2008, Docket no. 2006–3 CRB DPRA, available at http://www.loc.gov. Also see Jennifer Mariano Porter, "Compulsory Licensing and Cell Phone Ringtones: The Phone Is Ringtone, a Court Needs to Answer," *Temple Law Review* 80, no. 3 (2007), 907–948.

123. Paul Sexton, "New Chart Calls Up U.K. Ringtone Sales," *Billboard*, 5 June 2004, p. 44.

124. "*Billboard* Bows Ringtones Chart," *Billboard*, 6 November 2004.

125. Jason Ankeny, "The New Sounds of Music"; Antony Bruno, "The Ring King," *Billboard*, 27 October 2007.

126. See, e.g., Dave Laing, "World Record Sales 1992–2002," in "The Music Industry, Technology and Utopia—An Exchange between Marcus Breen and Eamonn Forde," *Popular Music* 23/1 (January 2004), 88–89.

127. Bruno, "The Ring King."

128. But the process is geographically uneven. On polyphonics in Latin America, see Leila Cobo, "The Polyphonic Problem," *Billboard*, 15 March 2008.

129. Banerjee, "Ringtone Rumble."

130. See Rafat Ali, "Zingy Changes Name to Vindingo; Closes Ringtone Divison; Some Layoffs," at http://www.moconews.net; "Vindigo, Inc.," *Hoover's Company Records*, 29 July 2008.

131. Robert Brenner, *The Boom and the Bubble: The US in the World Economy*, paperback ed. (Verso, 2003), pp. 40–43.

132. See "Atrinsic Acquires Ringtone.com," *Business Wire*, 2 July 2008, available at http://www.atrinsic.com.

133. Tim Green, "How to Turn £20 into over £100m," *Mobile Entertainment*, at http://www.mobile-ent.biz. Information on Buckingham was first related to me by Darren Vogel, Business Development Director of Accenture Digital Media Services, at the 2009 Mobile Music Now conference in London.

134. Nic Fildes, "Ringtones Pioneer Monstermob Ousts Founder," *The Independent*, 14 June 2006, available at http://www.independent.co.uk; "Africa's 'Lord of

the Ringtones,'" News24 (South Africa), 22 October 2006, available at http://www.news24.com.

135. Andy Fry, "Hitting the Mark with Customers—The Winners of This Year's APA Awards Had to Show Marketing Effectiveness," *Marketing*, 29 November 2001.

136. "Merger Bells Ring for Mobile Services Groups," *Sunday Business*, 22 July 2001.

137. "Monstermob and Jester Plan Mobile Gaming Venture," *New Media Age*, 9 January 2003.

138. Ben Griffiths, "Monstermob Seeks £10m to Fund Growth," *The Herald* (Glasgow), 30 October 2003, emphasis added.

139. "U.K. Monstermob Quadruples Profit, Pens Dixons Deal," *Total Telecom*, 16 February 2004; Guy Anderson, "Mediaprom Snapped up in £10m Numbers Game," *The Journal* (Newcastle), 10 March 2004.

140. Andrew Penman and Michael Greenwood, "Sorted and the City: A Phortune for Phunky Adam," *The Mirror* (London), 15 June 2004; Ricardo Baca, "Why Don't You Give Me a Jingle? Booming Business of Cellphone Ringing Tones Lets Users Download a Personality," *Denver Post*, 6 October 2002; Rafat Ali, "UK Company Monstermob Buys US Rival 9 Squared," at http://www.moconews.net.

141. See "MonsterMob Group PLC—Acquisition," *Company News Feed*, 25 August 2004.

142. All found in the results of a search for "Monstermob" at http://www.moconews.net.

143. Peggy Anne Salz, "Monstermob Investors Vote for Refinancing; Higginson Brought Back to Run the Business," http://www.moconews.net.

144. Robert Andrews, "Zed Takes $143 Million Loan to Fund M&A," http://www.moconews.net.

145. "NetPlay TV PLC: Acquisitions & Strategic Partnership," *London Stock Exchange Aggregated Regulatory News Service*, 2 January 2009.

146. See Simon Buckingham, "The Consolidation of Global Mobile Content," *New Media Age*, 10 June 2004.

147. Uimonen, "'Sorry, Can't Hear You!'" p. 61.

148. Naomi Baron, *Always On: Language in an Online and Mobile World* (Oxford University Press, 2008), p. 140.

149. Manuel Castells, Mireia Fernández-Ardèvol, Jack Linchuan Qiu, and Araba Sey, *Mobile Communication and Society: A Global Perspective* (MIT Press, 2007), pp. 160–161.

150. Ankeny, "The New Sounds of Music."

151. See Edward Nash, *Direct Marketing: Strategy, Planning, Execution*, fourth edition (McGraw-Hill, 2000), pp. 6–9.

152. See Michael Weiss, *The Clustering of America* (Harper & Row, 1988).

153. See, e.g., Robert Blattberg, Byung-Do Kim, and Scott Neslin, *Database Marketing: Analyzing and Managing Customers* (Springer, 2008).

154. Joseph Turow, *Niche Envy: Marketing Discrimination in the Digital Age* (MIT Press, 2006), p. 128.

155. Fredric Jameson, *The Political Unconscious: Narrative as a Socially Symbolic Act* (Cornell University Press, 1981), p. 76.

156. Turow, *Niche Envy*, pp. 128–133.

157. Kim Veltman, *Understanding New Media: Augmented Knowledge and Culture* (University of Calgary Press, 2006), p. 203; Michelle Hoyle and Christopher Lueg, "OpenSesame! A Look at Personal Assistants," in *Proceedings of the International Conference on the Practical Application of Intelligent Agents and Multi-Agent Technology*, London, 1997.

158. Veltman, *Understanding New Media*, p. 204.

159. Laurie Flynn, "Web Privacy Group to Offer a Seal of Approval," *New York Times*, 3 April 2000. Also see Kevin Featherly, "Personalization," in *Encyclopedia of New Media*, ed. S. Jones (Sage, 2003), pp. 372–374.

160. See Eli Pariser, *The Filter Bubble: What the Internet Is Hiding from You* (Penguin, 2011). Also see Josh Lauer, "Alienation in the Information Economy: Toward a Marxist Critique of Consumer Surveillance," in *Participation and Media Production: Critical Reflections on Content Creation*, ed. N. Carpenter and B. De Cleen (Cambridge Scholars, 2008).

161. Veltman, *Understanding New Media*, pp. 203, 205. "Mass customization" was advocated in management consultant B. Joseph Pine's *Mass Customization* (1992). See Sarah Berry, "Be Our Brand: Fashion and Personalization on the Web," in *Fashion Cultures: Theories, Explorations and Analysis*, ed. S. Bruzzi and P. Church Gibson (Routledge, 2000), esp. pp. 50–54.

162. See "China's First GSM Service Is Based on Schlumberger Subscriber Smartcards," *M2 Presswire*, 20 January 1995.

163. See Okada, "Youth Culture," pp. 42–47, 54–60; Kenji Kohiyama, "A Decade in the Development of Mobile Communications in Japan (1993–2002)," in *Personal, Portable, Pedestrian*, ed. Ito et al., pp. 70–73.

164. "Nokia Introduces New Digital GSM 1900 PCS Phone Designed for Consumers Who Value Ease of Use and Individuality," *Business Wire*, 3 August 1998.

165. "Hand Painted Covers Allow Further Personalization for Nokia 5100 Phones Through Nokia OK Recommend Painter Program," *Business Wire*, 23 September 1998.

166. See Sue Marek, "Raising the Bar on Ringtones," *Wireless Week*, 15 May 2004, 25; Yuki Noguchi, "Teens Ring Up Market Share," *Washington Post*, 25 April 2004.

167. Oliver Burkeman, "Fellowship of the Rings," *The Guardian*, 13 August 2003.

168. "Mobile 2.0" was probably first inspired by Microsoft's MSN Mobile 2.0 software, released in 2000. See "Microsoft Offers New Mobile Services," *CBS Marketwatch*, 28 February 2000.

169. Ankeny, "The New Sounds of Music."

170. Rafat Ali, "InfoSpace Sells Moviso Content Division to FunMobility," at http://www.moconews.net.

171. See Hamish Mackintosh, "Talk Time: Jay Sean," *The Guardian*, 12 August 2004, available at http://www.guardian.co.uk; Lawrence Lessig, *Free Culture: How Big Media Uses Technology and the Law to Lock Down Culture and Control Creativity* (Penguin, 2004), pp. 296–304. Before its 2006 settlement with the recording industry, Kazaa adopted a similar perspective by promoting legal, licensed content available through the website while still defending its facilitation of free file sharing. See www.kazaa.com.

172. Ankeny, "The New Sounds of Music."

173. Patrick Burkart and Tom McCourt, *Digital Music Wars: Ownership and Control of the Celestial Jukebox* (Rowman & Littlefield, 2006), p. 4.

174. See Patrick Burkart and Tom McCourt, "Infrastructure for the Celestial Jukebox," *Popular Music* 23, no. 3 (August 2004), 349–350. The term first appeared in Paul Goldstein's book *Copyright's Highway: From Gutenberg to the Celestial Jukebox* (Hill & Wang, 1994).

175. Brad Smith, "Multimedia Unplugged," *Wireless Week*, 15 May 2004.

176. Ibid.

177. Kevin O'Brien, "Mobile TV Spreading in Europe and to the US," *New York Times*, 6 May 2008.

178. See Ryan Block, "Say Hello to the Motorola E790 Apple iTunes Phone," at http://www.engadget.com.

179. Flynn, "The Cellphone's Next Makeover."

180. Joe Bant, "Music Downloads Sans PC," *Wireless Week*, 15 August 2004.

181. Dan Hickox, "Mobile Disco: How Phones Make Music Inescapable," *The Guardian*, 12 August 2010, available at http://www.guardian.co.uk.

182. "iTunes Store Top Music Retailer in US," Apple press release, 3 April 2008, at http://www.apple.com.

183. David Pogue, "Cellphone Surprise (iPhone Aside)," *New York Times*, 5 July 2007; Martin Fackler, "Chasing the iPhone," *New York Times*, 2 July 2007; Katie Hafner, "Waiting for the Latest in Wizardry," *New York Times*, 27 June 2007; David Pogue, "The iPhone Matches Most of Its Hype," *New York Times*, 27 June 2007.

184. Jeremy Peters, "Gave Up Sleep and Maybe a First-Born, But At Least I Have an iPhone," *New York Times*, 30 June 2007; John Markoff, "For iPhone, Some Shortages and Activation Problems," *New York Times*, 2 July 2007.

185. Brad Stone, "With Software and Soldering, AT&T's Lock on the iPhone Is Undone," *New York Times*, 25 August 2007; David Barboza, "After China Ships Out iPhones, Smugglers Make It a Return Trip," *New York Times*, 18 February 2008.

186. See Dianne See Morrison, "iPhone to Launch 22 August in India, Estonia, and the Czech Republic," at http://www.moconews.net.

187. Burkart and McCourt, *Digital Music Wars*, p. 6.

188. Henry Jenkins, *Convergence Culture: Where Old and New Media Collide* (New York University Press, 2006), p. 8; Howard Rheingold, *Smart Mobs: The Next Social Revolution* (Basic Books, 2002), p. 12.

189. Burkart and McCourt, *Digital Music Wars*, p. 6.

190. Michael Denning, *Culture in the Age of Three Worlds* (Verso, 2004); Alfred Meyer, "Theories of Convergence," in *Change in Communist Systems*, ed. C. Johnson (Stanford University Press, 1970), p. 320; Isaac Deutscher, *Russia After Stalin* (Hamish Hamilton, 1953), esp. pp. 13–17, 72, 159–74; John Kenneth Galbraith, *The New Industrial State* (Houghton Mifflin, 1967), esp. pp. 389–92; Herbert Marcuse, "The Question of Revolution," *New Left Review* I, no. 45 (1967), 4–5; Daniel Boorstin, *The Republic of Technology: Reflections on Our Future Community* (Harper & Row, 1978), p. 6; Jenkins, *Convergence Culture*, p. 10; Ithiel de Sola Pool, *Technologies of Freedom* (Belknap, 1983), pp. 1–10, 226–251.

191. For an example of the term's use in the business press, see Robert Duckett and R. Bradley Jude, "Atlanta: Telecommunications Terminus for All Southeast," *Atlanta Business Chronicle*, 21 April 1986. The mobile phone first appeared in convergence discourse in the English-language press around 1990. See, e.g., Ken Young, "Mobile Muddle; The Message Is Clear; Mobile Telephone," *British Telecom World*, 1 June 1990.

192. See, e.g., Thomas Baldwin, D. Stevens McVoy, and Charles Steinfield, *Convergence: Integrating Media, Information and Communication* (Sage, 1996), esp. pp. 1–4.

193. Veltman, *Understanding New Media*, pp. 26, 15.

194. Jenkins, *Convergence Culture*, p. 11.

195. See Chris Smith, "Apple iPod Sales Down 17 Per Cent," at http://www.techradar.com. Jenkins describes the ideology of convergence in terms of the "Black Box Fallacy," with rampant device clutter belying the idea that we are gradually approaching an era in which one device will solve all needs (ibid., pp. 13–16).

196. Jenkins, *Convergence Culture*, p. 12. On Sony, see Mark Singer, "Stringer's Way: Can Sony's C.E.O. Save an Ailing Brand?" *The New Yorker*, 5 June 2006.

197. "Tingalingalingaling!" *New York Times*, 18 January 2004; BPI *Statistical Handbook 2008*, p. 98.

198. Alec Foege, "Going Gold? Maybe if Enough Cellphones Ring," *New York Times*, 31 August 2003.

199. Scott Banerjee, "Xingtone Gains New Financing," *Billboard*, 19 June 2004.

200. Ironically, on March 15, 2012 the site was being used to sell ringtones.

201. Zutaut has been portrayed as having created a "Napster of the mobile." (See "Xingtone's Ringtones Zing Labels," *Online Reporter*, 15 March 2003.) But, unlike Napster, Xingtone software is designed to be used by only one computer and one phone, presumably owned by the same person. Much of the information here derives from an interview with Zutaut in May 2004.

202. The company was apparently founded in late 2003, marketing itself as selling software for converting MP3 files to ringtones. See "ToneThis Loads MP3 onto Cellphone" at http://www.tonethis.com.

203. See, e.g., John Davitt, "Schools Report: Information Technology Flowering in a Walled Garden," *The Observer*, 25 June 1995.

204. "Ericson(LM)—Re Co-operation Agreement," *Regulatory News Service*, 26 June 1997. SMS, used in the earliest uploading of ringtones as discussed above, was part of the WAP protocol.

205. "Alcatel Brings Online Internet Access to GSM Mobile Phones," *Business Wire*, 16 February 1998.

206. Paul Broekhuyse, "WAP Critics Have Got It Wrong: Guru," *The Australian*, 8 August 2000.

207. Misa Matsuda, "Discourses of *Keitai* in Japan," in *Personal, Portable, Pedestrian*, ed. Ito et al., p. 32.

208. See Vijay Kumar, Srinivas Parimi, and Dharma Agrawal, "WAP: Present and Future," *Pervasive Computing* 2, no. 1 (2003), 79–83, esp. p. 82.

209. Ben Hammersley, "Information Limitations," *The Times* (London), 6 October 1999.

210. Dan Duchamp, "The Discrete Internet and What To Do About It," in Proceedings of the Second New York Metro Area Networking Workshop, Columbia University, 2002, available at http://www.cs.northwestern.edu.

211. See Rheingold, in *Smart Mobs*, pp. 1–7; Geoffrey Elliott and Huong Tang, "The Wireless Mobile Internet: An International and Historical Comparison of the European and American Wireless Application Protocol (WAP) and the Japanese iMode™ Service," *Information Technology and Management* 3, no. 2–4 (2004), 268–281.

212. Cited in "NMA Mobile; Coming of Age," *New Media Age*, 20 September 2007.

213. "M:Metrics Reports Growth in Mobile Music Adoption," *Market Wire*, 15 January 2008.

214. Cited in Webb, "Technology."

215. Ibid.

216. See "Open-Specification Connectivity Envisioned for 2.4 GHz Band," *Wireless Today*, 20 May 1998; "The History of Bluetooth" at http://www.gsmfavorites.com; "Online Security; Wireless Security," *New Media Age*, 11 January 2007; Jaya Narain, "'Bluejacking' Terror of Texters Who Stalk You Via Your Mobile," *Daily Mail* (London), 30 March 2006.

217. "Study: Few Phone Buyers Purchase Accessories," *PC Magazine Online*, 23 July 2008.

218. "Have You Gone Barking Mad?" *Sunday Tribune* (South Africa), 10 August 2008.

219. Elena Pereira (Venezuela), e-mail interview with author, 6 June 2008; Yasmin Eid (Egypt), e-mail interview with author, 19 July 2008.

220. "The Benefits of Bluetooth," *Music Week*, 22 October 2005.

221. Charlotte Goddard, "Mobile Offers Kiosks a New Role," *Revolution*, 1 January 2004; Cherise Fong, "Mobile Phones Say 'Beam Me Up,'" *CNN.com*, 25 July 2008; Jason Tan, "New Kiosks Ring Up a Storm," *New Straits Times* (Malaysia), 24 March 2008.

222. Antony Bruno, "Mobile Content Moving Out," *Billboard*, 4 April 2008.

223. Ibid.

224. Ibid.

225. Bruno, "MTV Networks," *Billboard*, 12 March 2005; Justin Oppelaar, "Future Looks Wired," *Variety*, 13–19 September 2004.

226. See "Ringtone Music Piracy Flourishes in Asia" at http://sify.com.

227. See Steven Patrick, "RIM Out to Eradicate 'Truetone' Piracy Disease," *The Star*, 17 December 2004, available at http://star-techcentral.com; introduction to Joe Karaganis, ed., *Media Piracy in Emerging Economies* (Social Science Research Council, 2011), available at http://piracy.ssrc.org, pp. i–ii.

228. "Mobile Piracy to Cost Europe EU3.5bn in 2006" at http://www.netimperative.com.

229. James Quintana Pearce, "Japanese Mobile Bits: iPhone; Piracy; Yahoo," at http://www.moconews.net.

230. Stokely Baksh, "Ringtones—And Their Theft—On the Rise," United Press International, 23 August 2005.

231. "IFPI Digital Music Report 2008," p. 18.

232. See Karaganis, "Rethinking Piracy," in *Media Piracy*, pp. 13–18.

233. See Keith White, "The Killer App: *Wired* Magazine, Voice of the Corporate Revolution," *The Baffler* 5 (1993), reprinted in *Commodify Your Dissent: Salvos From the Baffler*, ed. T. Frank and M. Weiland (Norton, 1997).

234. See Eliot Van Buskirk, "Stop Paying for Ringtones," *Wired*, 6 February 2006, available at http://www.wired.com.

235. Scott Gilbertson, "How to Make Custom iPhone Ringtones Without Paying Apple $2," *Wired*, 12 September 2007, available at http://www.wired.com.

236. See the *Wired* How-To Wiki page "Make Custom iPhone Ringtones Without Paying Apple $2" at http://howto.wired.com.

237. Sinead Carew, "Sprint to Cut Mobile Phone Song Price to 99 Cents," Reuters UK, 26 March 2007, available at http://uk.reuters.com.

238. Manabe, "Going Mobile."

239. See, e.g., Burkart and McCourt, *Digital Music Wars*, pp. 101–119.

240. BPI, *Statistical Handbook 2008*, p. 10.

241. DRM-based subscription services, wherein consumers rent rather than own music tracks, had failed in the marketplace: in 2007 and 2008, MTV, Virgin, AOL, and Yahoo all shut down their own music subscription services; even the much-touted Spotify has not been profitable. See Colin Gibbs, "Carriers Still Trying to Find Right Notes in Mobile Music: VZW Latest to Alter Strategy," *RCR Wireless News*, 7 July 2008; and Paul Resnikoff, "Spotify CEO: Profitability Is Absolutely Not a Priority . . . ," at http://www.digitalmusicnews.com.

242. Tricia Duryee, "Global Ringback Tone Sales to Hit $4.7 Billion by 2012: Report," at http://www.moconews.net.

243. See the excellent work on M-commerce in Kenya, including ringtones, by Andrew Eisenberg—e.g., "M-Commerce and the (Re)Making of the Music Industry in Kenya: Preliminary Notes and Findings," presented at meeting of Association of Social Anthropologists, New Delhi, 2012.

244. Walter Benjamin, *The Arcades Project* (Belknap, 1999), p. 64.

245. See Dan Schiller, *Digital Capitalism: Networking the Global Market System* (MIT Press, 1999), esp. pp. 1–36.

246. See Brenner, *The Boom and the Bubble*, for a very powerful treatment of this economic dynamic; for the US-fed international recovery, see ibid., pp. 204–206.

247. See Peter Golding, "World Wide Wedge: Division and Contradiction in the Global Information Infrastructure," *Monthly Review* 48, no. 3 (1996), 70–85; Markos Mamalakis, "The Theory of Sectoral Clashes," *Latin American Research Review* 4, no. 3 (1969), 9–46; idem, "The Theory of Sectoral Clashes and Coalitions Revisited," *Latin American Research Review* 6, no. 3 (1971), 89–126.

248. Data drawn from the Recording Industry Association of America (RIAA), "2008 Year-End Shipment Statistics," "2009 Year-End Shipment Statistics," "2010 Year-End Shipment Statistics," and "2011 Year-End Shipment Statistics," available with subscription at http://www.riaa.com. The RIAA's inconsistent adjustment of figures may be corrections or alterations via fixed-dollar reporting (readjustment based on most recent dollar values); latest amounts were used in figure 1.2, but the numbers should be taken as approximate.

249. RIAA, "2009 Year-End Shipment Statistics," "2010 Year-End Shipment Statistics," and "2011 Year-End Shipment Statistics."

250. RIAA, "2011 Year-End Shipment Statistics."

251. See, e.g., Michael Hardt and Antonio Negri, *Empire* (Harvard University Press, 2000), pp. 290–294.

252. Immanuel Wallerstein, *World-Systems Analysis: An Introduction* (Duke University Press, 2004), pp. 26–32. On dialectical reversals, see the introduction to part III below.

Notes to Introduction to Part II

1. Sasha Frere-Jones, "Ring My Bell," *The New Yorker*, 7 March 2005.

2. On "attractions," see Louis Marin, "Disneyland: A Degenerate Utopia," *Glyph Textual Studies* 1 (1977): 50–66, esp. pp. 62, 65. Also see Fredric Jameson, "Reification and Utopia in Mass Culture," *Social Text* 1, winter 1979, 130–148; idem., "Of

Islands and Trenches: Naturalization and the Production of Utopian Discourse," *Diacritics* 7, no. 2 (1977), 2–21.

3. See, e.g., Jukka Gronow, "Fads, Fashions and 'Real' Innovations: Novelties and Social Change," in *Time, Consumption and Everyday Life*, ed. E. Shove et al. (Berg, 2009), esp. p. 134.

4. See W. Edward Steinmuller, "The Economics of ICTs: Building Blocks and Implications," in *The Oxford Handbook of Information and Communication Technologies*, ed. R. Mansell et al. (Oxford University Press, 2007, p. 202); Manuel Castells et al., *Mobile Communication and Society* (MIT Press, 2007).

5. See Joseph Schumpeter, *Capitalism, Socialism and Democracy* (Routledge, 1992 [1943]), pp. 81–86; Richard Nolan and David Croson, *Creative Destruction* (Harvard Business School Press, 1995), esp. pp. 38–67; Lee McKnight, Paul Vaaler, and Raul Katz, *Creative Destruction* (MIT Press, 2001); Vance Packard, *The Waste Makers* (David McKay, 1960), esp. pp. 53–77; Hartmut Rosa and William Scheuerman, eds., *High-Speed Society* (Pennsylvania State University Press, 2009); William Scheuerman, *Liberal Democracy and the Social Acceleration of Time* (Johns Hopkins University Press, 2004); Thomas Hylland Eriksen, *Tyranny of the Moment* (Pluto, 2001); James Gleick, *Faster* (Pantheon, 1999); Giovanni Arrighi, *The Long Twentieth Century* (Verso, 1994), esp. pp. 6–7, 216–217, 330.

6. See Walter Benjamin, "The Work of Art in the Age of Its Technical Reproducibility: Third Version," in *Selected Writings*, volume 4, ed. H. Eiland and M. Jennings (Belknap, 2003); Alois Riegl, "The Modern Cult of Monuments: Its Character and Its Origin," *Oppositions* 25, fall 1982, 21–51; Karen Lang, *Chaos and Cosmos* (Cornell University Press, 2006), pp. 136–178.

7. See Adam Sherwin, "It's Abba on the Phone Making a Lot More Money, Money, Money," *Times* (London), 19 April 2006; Emily Turrettini, "Oldies Dialing Up Nostalgia as Ringtones," Ringtonia.com, 22 May 2006, at http://www.textually.org.

8. See, e.g., the polyphonic version of the Super Mario Bros. theme song at www.free-ringtones.ie. On the NES synthesizer, see Karen Collins, *Game Sound* (MIT Press, 2008), pp. 25–28.

9. This meme developed partly in response to the political and social upheavals of the long 1960s. See Fred Davis, *Yearning for Yesterday* (Free Press, 1979), pp. 60, 104–108; Elizabeth Guffey, *Retro* (Reaktion, 2006), esp. pp. 98–117. On generational cohorts, see Karl Mannheim, "The Problem of Generations," in *Essays on the Sociology of Knowledge*, ed. P. Kecskemeti (Routledge & Kegan Paul, 1952); Arthur Schlesinger Jr., *The Cycles of American History* (Houghton Mifflin, 1986), esp. chapter 2.

10. Gronow, "Fads, Fashions and 'Real' Innovations," p. 135.

11. See, e.g., "The NPD Group: Rap/Hip-Hop Ringtones Rule the Market," *Business Wire*, 11 September 2006.

12. Davis, *Yearning for Yesterday*, p. 61.

13. Ibid., p. 133.

Notes to Chapter 2

1. Garry Trudeau, *Doonesbury*, 20 and 22 January 2009.

2. *Doonesbury*, 19 and 24 January 2009.

3. *Doonesbury*, 20 and 22 January 2009.

4. *Doonesbury*, 23 January 2009.

5. See Sumanth Gopinath, "Ringtones, or, The Auditory Logic of Globalization," *First Monday* 10, no. 12 (December 2005), available at http://firstmonday.org.

6. Cited in Kerry Soper, *Garry Trudeau:* Doonesbury *and the Aesthetics of Satire* (University Press of Mississippi, 2008), 34. Also see Walter Blair and Hamlin Hill, *America's Humor: From Poor Richard to Doonesbury* (Oxford University Press, 1978), pp. 511–515; Diedrich Diederichsen, "Long Strange Strip. Gary Trudeaus 'Doonesbury,'" in *Szenarien des Comic—Helden und Historien im Medium der Schriftbildlichkeit*, ed. S. Diekmann and M. Schneider (SuKuLTuR, 2005).

7. The series of six strips featuring Thudpucker and Slackmeyer was rerun from 15 March to 20 March 2010.

8. For example, creative labor often blurs the line between professional and amateur work. See David Hesmondhalgh and Sarah Baker, *Creative Labour: Media Work in Three Cultural Industries* (Routledge, 2010), p. 13. Also see Anthony Guest-Scott, "Categories in Motion: The Use of Generic Multiplicity in Music Store Guitar Lessons," *Ethnomusicology* 52, no. 3 (2008), 426–457.

9. Harry Braverman, *Labor and Monopoly Capital: The Degradation of Work in the Twentieth Century*, 25th anniversary edition (Monthly Review Press, 1998).

10. See David Suisman, "Sound, Knowledge and the 'Immanence of Human Failure'. Rethinking Musical Mechanization through the Piano, the Player-Piano, and the Phonograph," *Social Text* 28, no. 1 (2010), 13–34; James Kraft, *Stage to Studio: Musicians and the Sound Revolution, 1890–1950* (Johns Hopkins University Press, 1996), esp. pp. 193–201.

11. See Hesmondhalgh and Baker, *Creative Labour*, pp. 113–38; "Musicians, Singers and Related Workers," in *Occupational Outlook Handbook 2010–2011 Edition*, available at http://www.bls.gov.

12. Luke Holden, "Lord of the Ringtones," *Electronic Musician*, July 2006, p. 34.

13. "An Interview with James Winsoar" at http://www.tonez.co.uk.

14. Vesa-Matti Paananen, e-mail interview with author, 12 September 2007.

15. Ibid.

16. "An Interview with James Winsoar."

17. Paananen interview.

18. Andy Riga, "Job Has a Nice Ring to It," *The Gazette* (Montreal), 24 May 2003.

19. Eric Bernatchez, e-mail interview with author, 16 March 2010

20. Chia-Chun Chen, e-mail interview with author, 4 February 2007.

21. Seth Porges, "Lord of the Rings," *PC Magazine*, 20 February 2007, 21.

22. Ibid.

23. Keith Nowak, interview with author, May 2004.

24. Arne Lucas, e-mail interview with author, 19 February 2010.

25. Emily Turrettini, "Behind the Ringtone Scene," Ringtonia.com, 25 July 2003, available at http://www.textually.org. Relatively little can be gleaned from the websites of Melodi Ltd. (now Melodi Media) and Soundonweb.

26. See "Arphiola: Corporate Presentation" at http://www.arphiola.com. The corporate website, at http://www.arphiola.com, notes that the company has "20+ employees."

27. "Craze Is a Key to Success," *Express & Star*, 16 August 2003. Also see Melodi's website.

28. Andy Booth, e-mail interview with author, 10 October 2005.

29. Arne Lucas, telephone interview with author, 16 February 2010.

30. Luke Holden, e-mail interview with author, 12 March 2010.

31. Holden, "Lord of the Ringtones," p. 33.

32. Billy Dixon, e-mail interview with author, 5 March 2010.

33. Listing at http://www.tonez.co.uk.

34. Dixon interview.

35. Holden interview.

36. Luke Holden, email interview with author, 13 March 2010. Discrepancies in US-based and UK-based ringtone piecework rates might have something to do with the relative inaccessibility of music education in the US, as well as its relative standardization in the UK. Also, in view of the non-unionized nature of the work, as well as the secrecy of the labor market, ringtone programmers probably are forced to take

whatever payment they can get, rather than bargaining individually or collectively for better wages and benefits, and surely lack adequate knowledge of competitors' compensation rates. Indeed, employers may not even know what their competitors pay for comparable labor.

37. Information here is drawn from "Ring Tone Transfer Language" at http://en.wikipedia.org and from Hayden Porter, "Phone It In!" *Electronic Musician*, February 2004, 76–86.

38. http://www.ringtone-mania.com/Free_ringtones/B/Beethoven_5th_Symphony_ringtone.html

39. Ibid.

40. Hayden Porter, e-mail interview with author, 1 April 2010.

41. On the ensemble of skills required to do this kind of work, see Andrew Hugill, *The Digital Musician* (Routledge, 2008), p. 191. Also see Colby Leider, *Digital Audio Workstation* (McGraw-Hill, 2004); Joseph Schloss, *Making Beats: The Art of Sample-Based Hip-Hop* (Wesleyan University Press, 2004); Louise Meintjes, *Sound of Africa!: Making Music Zulu in a South African Studio* (Duke University Press, 2003).

42. Porter, "Phone It In!" p. 82.

43. Holden, e-mail interview with author, 12 March 2010; "Lord of the Ringtones," pp. 37–38.

44. Andy Booth, e-mail interview with author, 10 October 2005.

45. For a study of a similar moments of technical transition, and resulting unemployment and labor deskilling, within a media industry, see Emily Thompson, "Remix Redux: In the Silent Film Era, the Roots of the DJ," *Cabinet* 35 (autumn 2009), 23–28.

46. Maneesh Kenia, telephone and e-mail interviews with author, 28 and 29 January 2010.

47. Billy Dixon, e-mail interview with author, 16 March 2010.

48. Ibid.

49. Ibid.

50. Ibid.; Booth, e-mail interview with author, 10 October 2005.

51. Dixon, e-mail interview with author, 16 March 2010.

52. Booth, interview with author, 10 October 2005.

53. Emily Turrettini, "Original Ringtones by Martin Plante," Ringtonia.com, at http://www.textually.org.

54. Jason Ankeny, "Lee Oskar," *Wireless Review*, 1 April 2002.

55. See "Deep Purple Plot Album" at http://www.ultimate-guitar.com.

56. See Emily Turrettini, "Jay-Z's 'Black Phone,'" Ringtonia.com, at http://www.textually.org.

57. See "50 Cent Rings Up Zingy" at http://www.tmcnet.com and comments by Emily Turrettini, Ringtonia.com, at http://www.textually.org.

58. Jeff Leeds, "The Guy from Green Day Says He Has Your Mother on the Cell phone," *New York Times*, 18 August 2004.

59. Video titled "Timbaland Releases His First Exclusive Ringtone" and comment, available at http://www.youtube.com.

60. On the labor division between "stars" and "hacks" that seems to characterize the cultural industries, see Michael Denning, *The Cultural Front: The Laboring of American Culture in the Twentieth Century* (Verso, 1996), pp. 110–114.

61. Dixon, telephone interview with author, 4 March 2010.

62. A countervailing trend, however, might include the presence of mobile phone digital recorders, the products of which could be turned into ringtones—hence making the movement from constrained melody production to sound capturing as encouraging greater potential creativity.

63. Braverman, *Labor and Monopoly Capital*. Also see John Roberts, *The Intangibilities of Form: Skill and Deskilling in Art After the Readymade* (Verso, 2007), esp. pp. 82–88; Stephen Wood, ed., *The Degradation of Work? Skill, Deskilling and the Labour Process* (Hutchinson, 1982); Paul Thompson, *The Nature of Work: An Introduction to the Debates on the Labour Process* (Macmillan, 1983).

64. Gopinath, "Ringtones."

65. Braverman, *Labor and Monopoly Capital*, p. 227.

66. Dixon, e-mail interview with author, 16 March 2010.

67. Braverman critiques the term "semi-skilled" on account of its ideological mystification of unskilled labor; see esp. pp. 294–310. Nonetheless, the production of sound-file ringtones is arguably semi-skilled, since skilled musicians are hired to do it. Low-level audio editing is found throughout the media industries—radio journalism, basic sound editing for television production, simple Web-based sound design—suggesting that this semi-skilled labor is not exclusively musical. "Semi-skilled" also indexes the skilling of consumers, as audio editing software and techniques pass from "general social technique" into something like a technology of everyday life, incubating the "productive consumer." Citations in Roberts, *The Intangibilities of Form*, pp. 87, 181.

68. On micromaterialization, see Jonathan Sterne, "The MP3 as Cultural Artifact," *New Media & Society* 8/5 (2006), 831–832.

69. On economic and social acceleration, see the introduction to part II.

70. However, automation is significantly motivated by a desire for managerial control over the workplace, rather than being especially cost-efficient. See David Noble, *Forces of Production: A Social History of Automation* (Oxford University Press, 1984), esp. pp. 144–147, 321, 333–347.

71. Consider, for example, "reconfigurable computing," a more dynamic model of computation in which software itself is much more mutable through object-oriented programming. See "Reconfigurable Computing" at http://en.wikipedia.org; Lingfeng Wang and Kay Chen Tan, *Modern Industrial Automation Software Design: Principles and Real-World Applications* (IEEE Press, 2006), pp. 93–108.

72. See Braverman, *Labor and Monopoly Capital*, pp. 132–161; Richard Sennett, *The Culture of the New Capitalism* (Yale University Press, 2006), pp. 90–94.

73. Martin Lister et al., *New Media: A Critical Introduction* (Routledge, 2003), pp. 2–3, 180.

74. Tiziana Terranova, "Free Labor: Producing Culture for the Digital Economy," *Social Text* 18, no. 2 (2000), 33–58; Luc Boltanski and Ève Chiapello, *The New Spirit of Capitalism* (Verso, 2005).

75. On the informal sector, see Michael Denning, "Wageless Life," *New Left Review* 66 (November-December 2010), 79–97; Mike Davis, "Planet of Slums," *New Left Review* 26 (March–April 2004), 5–34.

Notes to Chapter 3

1. "Mobile Ringtone Saves Bride's Blushes" at http://www.ananova.com.

2. "'With This Ring Tone I Thee Wed'—Bride's Big Day Starts to the Notes of a Mobile Phone Instead of an Organ," *Western Mail*, 17 October 2002.

3. Examples might include the digital watch and the pager as direct precursors of the monophonic ringtone (as discussed in chapter 1) and/or portable gramophones and radios as prefiguring more recent mobile music technologies, such as the Walkman or iPod. On the latter cluster of examples, see the introduction to Sumanth Gopinath and Jason Stanyek, eds., *The Oxford Handbook of Mobile Music Studies* (Oxford University Press, forthcoming); Michael Brian Schiffer, *The Portable Radio in American Life* (University of Arizona Press, 1991).

4. See Gilbert B. Rodman, "The Net Effect: The Public's Fear and the Public Sphere," in *Virtual Publics: Policy and Community in an Electronic Age*, ed. B. Kolko (Columbia University Press, 2003), esp. pp. 34–36.

5. See Matthew Mirapaul, "Arts Online: Why Just Listen to Pop When You Can Mix Your Own?" *New York Times*, 20 August 2001.

6. Tomoyuki Okada, "Youth Culture and the Shaping of Japanese Mobile Media: Personalization and the *Keitai* Internet as Multimedia," in *Personal, Portable, Pedestrian*, ed. Ito et al., p. 55; Noriko Manabe, "Going Mobile: The Mobile Internet, Ringtones, and the Music Market in Japan," in *Internationalizing Internet Studies*, ed. G. Goggin and M. McLelland (Routledge, 2009), p. 319; Akira Takeishi and Kyoung-Joo Lee, "Mobile Innovation and the Music Business in Japan: The Case of Ringing Tone Melody ('Chaku-Mero')," in *Unwired Business*, ed. S. Barnes and E. Scornavacca (IRM, 2006), esp. pp. 3–4. The book's title can be translated as "Mobile Phone Ring Tone Do Re Mi Book."

7. For example, "Entertainment" includes "Sport," "TV Themes," "Film Themes," and "Bollywood," and "Classics" includes "Classical," "Jazz & Blues," "Latin & World Music," and "Xmas/Celebrations." The book states, on p. 5, "You will get the best results from this book if you have a Nokia mobile phone."

8. The series had three installments, *Ringtone Mania!* (2001), *Ringtone Mania! Gold* (2002) and *Ringtone Mania! Platinum* (2002), and they were perceived as competitors to the growing ringtone industry. According to one customer's review at http://www.amazon.co.uk, "u dont have 2 pay bout £3 to order them!"

9. Also see Vicente Rafael, "The Cell Phone and the Crowd: Messianic Politics in the Contemporary Philippines," *Public Culture* 15, no. 3 (2003), 401–408.

10. See, e.g., *More than 2,000 Ringtones: The Ultimate Collection*, Nokia version (Psicom, 2001). Its music genres include "Praise," "Anime," "Disney," "X'mas," "Movies," "Television," "O.P.M." (Original Pinoy Music), and "Foreign" (the largest section [pp. 54–112 out of 120 pages], subdivided into "Top Hits," "Classic Hits," "More Hits," and "Most Requested"). Also see chapter 6.

11. "New Hindi Bollywood ringtones," available at http://www.indianchild.com. The archive.org "Wayback Machine" first finds the site on 25 June 2003, fourteen months after the film's release.

12. In comparison to the example below, the run from E♭4 down to B♭3 in the second measure is simplified, and the arrival on C4 is half a beat too early, introducing syncopatation. The Siemens entry and several of the others drop half a beat; the effect is similar to the above, but with the second measure in 7/8. An alternate version is shown in Sumanth Gopinath, "Ringtone, or, the Auditory Logic of Globalization," *First Monday* 10, no. 12 (2005), available at http://firstmonday.org.

13. A scene from the film featuring a somewhat different version of the song than the one discussed here (it is slightly slower, at 122 BPM, and features somewhat different instrumentation) can be found at http://www.youtube.com.

14. The percussion track sample could be a tabla, but it is more likely to be a conga. The "ethnic" percussion instruments in the film version of the song (entering *before* the house bass-drum beat) are far lower in pitch range, sounding neo-African in character, though they may be South Asian in origin. What is crucial is that the sonority conveys the "ethnic," "tribal," "exotic," "bodily," and "sensual" (though South Asian instruments, more readily identified by acculturated audiences, introduce regional/national, caste/class, and rural/urban inflections). See Timothy Taylor, "World Music in Television Ads," *American Music* 18, no. 2 (2000), 162–192, which refers to "ethnic" or "tribal" tracks (e.g., on p. 175).

15. On hook finding, see chapter 2.

16. In "Ringtones, or, The Auditory Logic of Globalization" I speculated, probably incorrectly, that the error disguised a ringtone produced without copyright permissions and was symptomatic of South Asian music piracy; it nonetheless indexes a larger system of exchange including mundane piracy, which also benefits local music and film industries. See Lawrence Liang and Ravi Sundaram, "India," in *Media Piracy in Emerging Economies*, ed. J. Karaganis (Social Science Research Council, 2011), available at http://piracy.ssrc.org.

17. If the errors originated from a ringtone melody book, its format may have caused them. Specifically, *Ringtones Mania! Platinum* uses a standardized, half-page box for each ringtone, with large buttons showing each note, limiting the length of a melody. For example, its version of the Who's "Baba O'Riley" (p. 58) copies its minimalist synthesizer introduction. But although it captures the opening F♯3-F♯3-C♯4-C♯4-F♯4-F♯4-C♯4-C♯4 ostinato correctly, the accompanying flourishes (using C♯4-D♯4-F♯4), necessarily rendered monophonically, miss the original's E♯ (or F) and patterns of variation. Moreover, there is only enough space to repeat that flourish once in the book (owing to its format), even though the passage accounts for only about eight seconds of the song. Nonetheless, the song remains recognizable.

18. The continued availability of the "New Hindi Bollywood ringtones" site may confirm that the use of secondhand mobile phones is widespread in India, preserving monophonic ringtones in a national market otherwise given over to sound-file ringtones and ringback tones (called "caller tunes" or "hello tunes"). See "Airtel gets 'Hello Tunes' to Mumbai" at http://www.techtree.com. Also see Thomas N. Smyth et al., "Where There's a Will, There's a Way: Mobile Media Sharing in Urban India," in *Proceedings of the SIGCHI Conference on Human Factors in Computing Systems* (Association for Computing Machinery, 2010); Liang and Sundaram, "India."

19. See Eli Pariser, *The Filter Bubble: What the Internet Is Hiding from You* (Penguin, 2011).

20. See Ad Crable, "Trashing Your Cell Phone," *Lancaster New Era*, 15 April 2004; Andrew Ross, "The Flight of Silicon Wafers," in *Low Pay, High Profile: The Global Push for Fair Labor* (New Press, 2004).

21. Announced in "Google Acquires Applied Semantics," Google press release, 23 April 2003, available at http://www.google.com.

22. "Ringtones USA Introduces Fair Pricing for Ringtones, Mobile Games," *Business Wire*, 22 March 2006.

23. Much of the information above was obtained via The Internet Archive's "Wayback Machine" (http://waybackmachine.org/) on 21 March 2011. On 27 October 2012, the site was being auctioned at godaddy.com with a suggested price of $1,277.

24. See Maureen Dowd, "No Kiss from Kate," *New York Times*, 19 March 2011; Alex Ben Block, "Q+A: New Hollywood Lobbyist Dodd Targets 'Looting,'" Reuters, 1 March 2011, available at http://www.reuters.com.

25. See http://www.monophonicringtones.net/.

26. *A Bach Companion: The Life and Work of Johann Sebastian Bach, 1685–1750*, dir. Kriss Russman with Robert Levin, VHS (Kultur, 2000). This predominantly English-language documentary (by a British director and featuring an American scholar-performer as narrator) was produced by EuroArts Leipzig and Zweites Deutsches Fernsehen/Association relative á la télévision européene in association with the European Broadcasting Union, which gives it more of a German aesthetic than might be expected.

27. "What Is the Most Annoying Mobile Phone Ring Tone You Have Heard?" *MX* (Melbourne), 16 November 2001.

28. See http://www.nokiasiemensnetworks.com/news-events/events/mobile-world-congress-2010/speakers/lauri-kivinen; http://www.nokiasiemensnetworks.com/news-events/publications/unite-magazine-spring-2008/context-the-secret-behind-mobile-advertising; http://en.wikipedia.org/wiki/Nokia; Dan Steinbock, *The Nokia Revolution: The Story of an Extraordinary Company That Transformed an Industry* (AMACOM, 2001), pp. 168, 279–285.

29. Nico Fogelholm, "The Nokia Tune," at http://www.about-nokia.com (based on a report published in the September 2005 *Helsingin Sanomat Kuukausiliite*).

30. See Fogelholm, "The Nokia Tune." As of 28 October 2012, an English-language (probably British) Nokia ad from the 1990s featuring the original guitar version of the "Gran Vals" could be seen at http://www.youtube.com. In view of the obscurity of the Tárrega piece, Steven Rings speculated that one or more classical guitarists might be employed at Nokia as ringtone composers and arrangers. He noted that he had heard a ringtone arrangement of a melody from Mozart's *The Magic Flute* that was clearly taken not from the original work but from a blander arrangement of the tune by the early-nineteenth-century Spanish guitar composer Fernando Sor. (That arrangement was part of Sor's *Introduction and Variations on a Theme by Mozart from The Magic Flute*, op. 9.)

31. Burkeman, "Fellowship of the Rings," *The Guardian*, 13 August 2003.

32. Interview with Keith Nowak, May 2004.

33. See Andy Riga, in "Job Has a Nice Ring to It," *The Gazette* (Montreal), 24 May 2003.

34. Heikki Uimonen, "'Sorry, Can't Hear You! I'm on a Train': Ringing Tones, Meanings and the Finnish Soundscape," *Popular Music* 23, no. 1 (2004), 51–62, quotation on p. 52; Gavin Naden, "Real Life: What Your Ring Tone Says about You," *Sunday Mirror*, 13 May 2001.

35. Nowak mentioned that "Take Me Out to the Ball Game" appeared only in US markets.

36. Uimonen, "Sorry, Can't Hear You!" pp. 52–53.

37. See Bruno Nettl, *The Western Impact on World Music: Change, Adaptation, and Survival* (Schirmer, 1985); Norman Lebrecht, *Who Killed Classical Music? Maestros, Managers, and Corporate Politics* (Carol, 1997), esp. pp. 394–414; Alex Ross, "Symphony of Millions: Classical Music in China," in *Listen to This* (Farrar, Straus and Giroux, 2010).

38. See Hayden Porter, "Phone It In!" *Electronic Musician*, February 2004, 76–86; Matthew Mirapaul, "Composer Plans to Strike Up the Cell Phones," *New York Times*, 16 August 2001.

39. On the variability of tuning, see Frédéric Voisin, "Musical Scales in Central Africa and Java: Modeling by Synthesis," *Leonardo Music Journal* 4 (1994), 85–90.

40. Equal temperament either post-dated or coincided with the spread of Western diatonicism and tonal harmony, a veritable "colonizing force" according to Kofi Agawu. See his *Representing African Music: Postcolonial Notes, Queries, Positions* (Routledge, 2003), pp. 8–10. Also see Max Weber, *The Rational and Social Foundations of Music*, ed. D. Martindale, J. Riedel, and G. Neuwirth (Southern Illinois University Press, 1958), esp. pp. 89–103; Owen Jorgensen, *Tuning* (Michigan State University Press, 1991); Kyle Gann, "An Introduction to Historical Tunings," available at http://www.kylegann.com; Ross Duffin, *How Equal Temperament Ruined Harmony (and Why You Should Care)* (Norton, 1997); J. Murray Barbour, *Tuning and Temperament: A Historical Survey* (Michigan State University Press, 1951); Stuart Isacoff, *Temperament: How Music Became a Battleground for the Great Minds of Western Civilization* (Vintage, 2003).

41. Quoted in Ankeny, "The New Sounds of Music," pp. 30–31.

42. David Suisman, "Sound, Knowledge, and the 'Immanence of Human Failure': Rethinking Musical Mechanization through the Phonograph, the Player-Piano, and the Piano," *Social Text* 28, no. 1 (2010), 13–34, p. 23. On recording's effect on "live" performance, see Mark Katz, *Capturing Sound: How Technology Has Changed Music*,

revised edition (University of California Press, 2010); Philip Auslander, *Liveness: Performance in a Mediatized Culture*, second edition (Routledge, 2008).

43. See Ruth Solie, "The Living Work: Organicism and Musical Analysis," *19th-Century Music* 4, no. 2 (1980), 147–156.

44. Brian Hyer, "Tonality," in *The New Grove Dictionary of Music*, second edition (Oxford University Press, 2003).

45. Heinrich Schenker, *Harmony*, ed. O. Jonas (University of Chicago Press, 1954 [1906]), pp. 134–137.

46. See Leonard Meyer, "The Perception and Cognition of Complex Music," in *Music, the Arts, and Ideas* (University of Chicago Press, 1967), esp. pp. 283–293.

47. See David Temperley, "The Melodic-Harmonic 'Divorce' in Rock," *Popular Music* 26, no. 2 (2007), 323–342.

48. Quoted by memory from a Varig Airlines in-flight television program, January 2005.

49. Of course, there are a number of reasons why popular songs sound especially unimpressive as monophonic and even polyphonic ringtones. Among them are the centrality of comprehensible lyrics to popular music, the importance of timbre in defining popular music (as opposed to relatively standardized timbres in classical music), and the already familiar adaptation and bowdlerization of classical music, including in digitized monophonic formats.

50. See Albin Zak, *The Poetics of Rock: Cutting Tracks, Making Records* (University of California Press, 2001), p. 24.

51. David Suisman notes that notation separated "conception from execution" whereas mechanized reproduction (player-piano and gramophone) separated "execution from audition." See *Selling Sounds: The Commercial Revolution in American Music* (Harvard University Press, 2009), p. 91.

52. See John Oswald, "Plunderphonics, or Audio Piracy as a Compositional Prerogative," presented at Wired Society Electro-Acoustic Conference, Toronto, 1985 (available at http://www.plunderphonics.com).

53. Charles White, "iPhone Ringtone Sheetmusic: Let Angelic Tinkling Remixes Begin," at http://gizmodo.com.

54. Several of the iPhone's initial preset ringtones featured a marimba or a xylophone, including "Ascending," "Time Passing," "Trill," and "Xylophone." Apple may have a ringtone-producing percussionist on its staff.

55. Specifically, BlackBerry's "Notifier_BBpro_5" is a very fast C♯4-F♯4-E4, recalling part 2 of *Drumming*'s opening riff, which begins exactly a perfect fourth lower with

a similarly brisk G♯3-C♯4-B3. The more popular "Notifier_BBpro_6," featuring a simple B3-C♯3, also recalls Reich's composition.

56. See Edward Strickland, *Minimalism: Origins* (Indiana University Press, 1993), pp. 1–2; Robert Fink, *Repeating Ourselves: American Minimal Music as Cultural Practice* (University of California Press, 2005), pp. 72, 120–166.

57. Daniel Jackson, *Sonic Branding: An Introduction* (Palgrave Macmillan, 2003), p. 128.

58. On the connection between the *American Beauty* score and "Xylophone," see Elissa Milne, "The Harmonic Language of Ringtones," at http://elissamilne.wordpress.com.

59. Perhaps the sound file enabled a qualitative possibility of non-hegemonic structures while quantitatively drowning them in a sea of hegemonic wares. For similar critiques of postmodern pluralism, see Veit Erlmann, "The Aesthetics of the Global Imagination: Reflections on World Music in the 1990s," *Public Culture* 8 (1996), 467–487; Timothy Brennan, *Secular Devotions: Afro-Latin Music and Imperial Jazz* (Verso, 2008), pp. 15–48. Also see Peter Manuel, "Modernity and Musical Structure: Neo-Marxist Perspectives on Song Form and Its Successors," in *Music and Marx: Ideas, Practice, Politics*, ed. R. Burckhardt Qureshi (Routledge, 2002). On the other hand, the appearance in 2007 of "Booseytones"—the classical music publisher Boosey & Hawkes' foray into the sound-file ringtone market—attests to the older formation's persistence (and transformation) in the new sound-file period. For the ringtone portal itself, see http://www.booseytones.com/ ; for a report, see Marcel Sim, "Classical Music Ringtones" at http://www.coolbusinessideas.com.

60. Richard Taruskin, *Music from the Earliest Notations to the Sixteenth Century* (*The Oxford History of Western Music*, volume 1) (Oxford University Press, 2005), p. xv.

61. Richard Taruskin, *Music in the Late Twentieth Century* (*The Oxford History of Western Music*, volume 5) (Oxford University Press, 2005), p. 509.

62. See Kiri Miller, "Schizophonic Performance: *Guitar Hero*, *Rock Band*, and Virtual Virtuosity," *Journal of the Society for American Music* 3, no. 4 (2009), 395–429, esp. pp. 408–411.

63. Jay David Bolter and Richard Grusin, *Remediation: Understanding New Media* (MIT Press, 1999). On the recapitulated drive toward greater sonic fidelity in numerous video game platforms (especially home consoles and personal computers), see Karen Collins, *Game Sound: An Introduction to the History, Theory, and Practice of Video Game Music and Sound Design* (MIT Press, 2008), passim.

64. See Fredric Jameson, "Reification and Utopia in Mass Culture," *Social Text* 1 (winter 1979), 130–148; Henry Adams, *The Education of Henry Adams: An Autobiography* (Houghton Mifflin, 1918), pp. 379–390.

65. James Rampton, "The Sound of Peace-Making," *The Independent*, 5 December 2003.

66. Also see chapter 6 on the political uses of ringtones. On Wagner in *Apocalypse Now*, see Lawrence Kramer, *Musical Meaning: Toward a Critical History* (University of California Press, 2002), pp. 152–153; Jonathan Pieslak, *Sound Targets: American Soldiers and Music in the Iraq War* (Indiana University Press, 2009), p. 85.

67. For critiques of excessively general claims about music's commodification, see Suisman, *Selling Sounds*, pp. 9–17; Simon Frith, *Sound Effects: Youth, Leisure, and the Politics of Rock 'n' Roll* (Pantheon Books, 1981), pp. 45, 91, 101.

Notes to Chapter 4

1. See Norman Lebrecht, *Who Killed Classical Music?: Maestros, Managers, and Corporate Politics* (Carol, 1997); Chin-tao Wu, *Privatising Culture: Corporate Art Intervention since the 1980s* (Verso, 2002).

2. "Orchestra Incorporates Ringtones" at http://www.ananova.com; "Die grosse Piep-Show: Musikalische Mobiltelefone," *Stern*, 25 April 2002; Nadine Nöhmaier, "Handy hoch!" at *Der Spiegel*, http://www.spiegel.de. Also see Kremling's webpage at http://www.junge-philharmonie-wuerzburg.de.

3. Citation in "'Cell-Phone Symphony' Laughs at Israel's Wireless Craze," Associated Press Worldstream, 1 June 2001.

4. Citation from Jeremy Eichler, "Music Review: A Trove of New-Found Gems Brings P.D.Q. Bach More Kudos," *New York Times*, 30 December 2003. Also see Celia Baker, "Lower Bach Pain," *Salt Lake Tribune*, 12 December 2003.

5. A performance of Hamelin's *Valse* can be seen at http://www.youtube.com (originally from the DVD *Marc André Hamelin: No Limits—The World of the Piano*, volume 2 (Euroarts, 2007)) or at "Marc André Hamelin, Walzing Away 'Irritation'" (http://www.npr.org). A performance of Lo's "Nokia Fugue" can be seen at http://audiomuse.ca. Other Nokia Tune compositions and performances that can be seen on YouTube include electronic dance music-style remixes (such as "nokia tune REMIX !!!"), a cool-jazz improvisation from 2007 by the Danish jazz bassist and harmonica player Jacob Venndt and the jazz pianist Pojken Flensborg, and an impromptu improvisation by the Slovenian violist Lukáš Kmiť.

6. In contrast with a broadcast recording of the piece I've heard, the score makes no mention of a phone and gives the Nokia Tune line to the first countertenor.

7. Industrialization is evoked in Motion's text, via Blake's critique of the factory ("dark, Satanic mills") and capitalism in the poem "And did those feet in ancient time," later set by Hubert Parry in the anthem "Jerusalem." Motion's text reads "The

years flick past and voices change to speak their mind but stay the same/Jerusalem's mills and mountains green refusing to sleep in the head and hand."

8. See Daniel Wakin, "Horns Up. Bows Ready. Cellphones On." *New York Times*, 3 October 2006.

9. Works quoted include Bartók's Concerto for Orchestra, Strauss' *Don Juan*, Dvořák's Symphony No. 9, Tchaikovsky's *1812 Overture* and *Swan Lake*, Schubert's Symphony No. 8, and Shostakovich's Symphony no. 5. I've heard one recording of the *Concertino* by the Indianapolis Chamber Orchestra, with Alex Noppe as soloist.

10. On the former, see Wakin, "Horns Up."

11. Baker (b. 1931) has more than 2,000 credited compositions; see "Open Source—You Heard It Here First," *The Australian*, 26 September 2006. He is best known as a jazz composer and trombonist in the "third stream" tendency, appearing on albums by the jazz pianist-composer-theorist George Russell such as the classic *Ezz-thetics* (1961). See Baker's homepage's biography at http://www.davidbakermusic.org and "David Baker (composer)" at http://en.wikipedia.org.

12. However, Baker regrets being pigeonholed as a "black composer" and prefers to have his music evaluated in more general terms.

13. The same seems true of the ringtone work by Kremling, who is director of the Academy for Music Pedagogy at Würzburg University.

14. Wakin, "Horns Up."

15. See David Lister, "Ringtones Really Can Be Music to Our Ears," *The Independent*, 7 October 2006; Steve Hendershot, "A Little Funky, and Outrageous," *Crain's Chicago Business*, 9 October 2006.

16. When combined, the two readings represent African Americans as bearers of an older modernity in retreat that once often excluded them; see chapter 9. Also see Shannon Jones, "Major Symphony Orchestras in US Face Funding Crisis," World Socialist Website, at http://www.wsws.org.

17. Volker Hagedorn, "Klingeltöne über der Ruine," *Die Zeit* (Germany), at http://www.zeit.de; George Loomis, "Better the Devil You Don't Know?" *New York Times*, 2 March 2010.

18. Sciarrino's music has been compared to that of Helmut Lachenmann. See Gavin Thomas, "The Poetics of Extremity," *Musical Times*, April 1993, p. 196. For criticisms of this comparison, see Ian Pace, "Recent Sciarrino Premières," *Tempo*, April 1997, 49, note 2; David Ryan, "Composer in Interview: Helmut Lachenmann," *Tempo*, October 1999, p. 20.

19. See David Metzer, *Musical Modernism at the Turn of the Twenty-First Century* (Cambridge University Press, 2009), pp. 92–103.

20. Ibid., pp. 65, 92, 102–103.

21. See Thomas, "The Poetics of Extremity," p. 194; Gianmario Borio, "Der italienische Komponist Salvatore Sciarrino," *Neue Zeitschrift für Musik,* May 1991, 33–36.

22. Borio, "Der italienische Komponist," pp. 35–36. Also see Thomas, "The Poetics of Extremity," p. 194.

23. Salvatore Sciarrino, program note to *Archeologia del telefono* (Rai Trade, 2008 [2005]), p. vii; translation by Barbara Mauer and Steven Lindberg in liner notes to *Donaueschinger Musiktage 2005*, volume 3, Col Legno, WWE 1CD 20246, 2006, pp. 21–22.

24. See Thomas, "The Poetics of Extremity," p. 196; Nicholas Till, "Investigating the Entrails: Post-Operatic Music Theatre in Europe," in *Contemporary Theaters in Europe: A Critical Companion,* ed. J. Kelleher and N. Ridout (Routledge, 2006), pp. 44–45.

25. Maria Kostakeva, "'Ich schreibe keine Musik, Ich schreibe eine psychologische Erfahrung,'" *Neue Zeitschrift für Musik,* July–August 2009, 32–33.

26. Ibid., p. 32.

27. "Die kulturelle Situation verschlechtert sich. Mein Vorhaben war, eine Oper gegen diese Situation zu schreiben. Ohne Schreien. Leise." ("The cultural situation is deteriorating. My intention was to write an opera against this situation. Without shouting. Quietly."), ibid., p. 33. Also see Nick Vivarelli, "Berlusconi Slashes Arts Funding Again," *Variety,* 6 March 2009, available at http://www.variety.com.

28. The double bass sounds one octave lower than written.

29. See Allan Kozinn, "Students Interpret Kafkaesque Score," *New York Times,* 26 September 2010; Henry Klumpenhouwer, "Commodity-Form, Disavowal, and Practices of Music Theory," in *Music and Marx: Ideas, Practice, Politics,* ed. Regula Qureshi (Routledge, 2002).

30. See Nicolas Hodges, "'A Volcano Viewed from Afar': The Music of Salvatore Sciarrino," *Tempo,* October 1995, 22–24. The title quotes Sciarrino on his distanced aesthetic.

31. See Sujata Sen, "Cyberklick: The Music of Ringtones: The Last Couple of Years Has Seen Extraordinary Progress in Telecommunications. Danish Birds Confirm That," *The Statesman* (India), 27 May 2001.

32. See Golan Levin, "An Informal Catalogue of Mobile Phone Performances, Installations and Artworks" at http://www.flong.com; "Ringtone Concerts," Ringtonia.com, at http://www.textually.org.

33. See Lev Manovitch, "The Myth of Interactivity," in *The Language of New Media* (MIT Press, 2001). One substitute for interactivity is "multi-user input"; see Chris-

tiane Paul, *Digital Art* (Thames & Hudson, 2003), p. 68. Also see Nicolas Bourriaud, *Relational Aesthetics* (Presses du réel, 2002).

34. Levin, "An Informal Catalog." A description of *Handywolke*, can be found at http://www.museumsstiftung.de.

35. "Die widersprüchliche Aspekte drahtloser Kommunikation" at http://www.museumsstiftung.de.

36. "Music to his Ears," *Bangkok Post*, 21 August 2001. Also see http://www.flong.com.

37. David Dayen and Joanna Lux, "Cellphone Symphony: In Composer's Hands, Annoying Ringtones Have Musical Potential," at http://www.flong.com.

38. Levin et al., "*Dialtones* (A Telesymphony): Final Report," 2003 [?], at http://www.flong.com, p. 26.

39. Frank Sawatzki, "*Dialtones* (A Telesymphony): Interview with Scott Gibbons and Golan Levin," at http://www.flong.com.

40. Levin et al., "*Dialtones* (A Telesymphony): Final Report," pp. 16–18.

41. Golan Levin, personal email communication with author, 15–16 May 2011. Levin's experience suggests a divide between corporate-sponsored new media artworks emerging from in-house marketing ventures and those pitched to corporations after having been developed externally. The latter—surely more common, with university-based and non-profit-based new media art serving as marketing R&D for technology and communications firms—face greater challenges in securing corporate sponsorship, the typical strategy of which is "culturally 'playing safe'" (Wu, *Privatising Culture*, p. 125).

42. Levin et al., "*Dialtones* (A Telesymphony): Final Report," p. 5.

43. From an interview of Levin by Peter Traub for *Contemporary Music Review*, 11–19 March 2005, available at http://www.flong.com.

44. Levin et al., "*Dialtones* (A Telesymphony): Final Report," p. 5.

45. Ibid., pp. 5–6.

46. See Klumpenhouwer, "Commodity-Form, Disavowal, and Practices of Music Theory," pp. 29–31. Reification and personification appear to be important conceptual themes in mobile phone art and music generally.

47. See, for example, an interview with Levin by Tori Tans for *XFUNS Magazine* (Taiwan), 22 June 2004, available at http://www.flong.com.

48. Interview with Alexandra Nemerov in "Within, Without: New Media and the White Cube," *CUREJ—College Undergraduate Research Electronic Journal*, 2007, available at http://www.flong.com.

49. Levin et al., "*Dialtones* (A Telesymphony): Final Report," p. 19.

50. Ibid., p. 7.

51. See Emily Turrettini, "'Ha, Ha' to 'Wish You Were Here': Why Cell Phones Have Killed the Concert Lighter," Ringtonia.com, at http://www.textually.org.

52. Levin mentions this tonal center in the interview by Peter Traub for *Contemporary Music Review*. On entrainment, see Martin Clayton, Rebecca Sager, and Udo Will, "In Time with the Music: The Concept of Entrainment and Its Significance for Ethnomusicology," *European Meetings in Ethnomusicology* 11, 3–142, available at http://www.dur.ac.uk.

53. See Levin et al., "*Dialtones* (A Telesymphony): Final Report," pp. 7, 9, 10.

54. As quoted in Burkeman, "Fellowship of the Rings," *The Guardian*, 13 August 2003.

55. Plant even creates an "allegorical illustration," describing types of users of mobile phones as species of birds—e.g., "the swift talker," "the solitary owl," and "the calm dove." See Sadie Plant, *On the Mobile: The Effect of Mobile Telephones on Social and Individual Life* (Motorola, 2002), available at http://web.archive.org, pp. 66–67. With the sound-file ringtone, such phenomena seem to have disappeared. See "Mobile Phones Change Birds' Tune" at http://www.cellular-news.com.

56. The timbral tinniness is perhaps heightened on the recording, which for Levin doesn't accurately capture the performance. Personal email communication with author, 15 May 2011.

57. Levin et al., "*Dialtones* (A Telesymphony): Final Report," p. 6.

58. See Tom Moody's critique and exchange with Levin at http://www.digital mediatree.com or at http://www.flong.com.

59. See, for example, "How to Silence Phone Calls from Annoying Friends and Spammy Callers," *Business Insider*, http://articles.businessinsider.com. Also see Julian Stallabrass, "Thomson & Craighead," *Evening Standard*, January 2003, available at http://www.thomson-craighead.net. The now-defunct online store's site can be accessed at http://www.ucl.ac.uk. Also see the artists' website, at http://www.thomson-craighead.net/.

60. The work is sometimes called *#Telephony*, following the label in the artists' video documentation, available at http://www.thomson-craighead.net/.

61. See interview with Levin by Nikos Kaimakamis, *Mobile Magazine*, 26 October 2005, available at http://www.flong.com.

62. William Duckworth even claims that *Dialtones* is "plotted in space and time as carefully as a Mahler symphony, or a 3-D ride at Disneyland." See *Virtual Music: How the Web Got Wired for Sound* (Routledge, 2005), p. 108.

63. Quotations from "Mobile Phone Ring Cycle Planned," BBC News, 22 April 2002, available at http://news.bbc.co.uk; "Wireless Update" at http://il.computers.toshiba-europe.com.

64. Levin, "An Informal Catalogue." Also see "Tobias Trier" at http://da.wikipedia.org.

65. Frauke Behrendt, *Handymusik: Klangkunst und 'Mobile Devices'* (Epos Music, 2005), available at http://www.epos.uos.de. (The translation is mine.) My description draws on Levin, "An Informal Catalogue" and on Behrendt, *Handymusik*. Behrendt notes (p. 43) that "Dial the signals!" references the German text of "The Internationale"—the refrain begins "Völker, hört die Signale!"

66. On the Improv Everywhere group, see Charlie Todd and Alex Scoredelis, *Causing a Scene: Extraordinary Pranks in Ordinary Places with Improv Everywhere* (HarperCollins, 2009). The term "flash mob" derives, in part, from the "smart mob"; see Howard Rheingold, *Smart Mobs: The Next Social Revolution* (Basic Books, 2002).

67. All citations about the Improv Everywhere performance are drawn from "Cell Phone Symphony" at http://improveverywhere.com.

68. Available at http://www.youtube.com.

69. Golan Levin, personal email communication with author, 15 May 2011.

70. See, for example, "Ringtone Dancer 1" at http://video.google.com.

71. Frauke Behrendt, abstract for "From Calling a Cloud to Finding the Missing Track: Artistic Approaches to Mobile Music," in *Creative Interactions—The Mobile Music Workshops 2004–2008*, ed. N. Kirisits et al. (Universität für Angewandte Kunst Wien, 2008), p. 104.

72. Anthony Tommasini, "I'm Playing with the Orchestra. Leave a Message After the Beep," *New York Times*, 23 February 2011.

73. Nathan Davis, personal email communication with author, 25 March 2011.

74. For more, see Matthew Everett, "Shameless Plug of the Day: 'Ringtone' (One Weekend Only)," *Twin Cities Daily Planet*, 30 April 2009, available at http://www.tcdailyplanet.net.

75. See the description of Locally Toned at http://locallytoned.wordpress.com.

Notes to Introduction to Part III

1. Fredric Jameson, *Marxism and Form* (Princeton University Press, 1971), p. 309.

2. International Telecommunications Union, *The World in 2010: ICT Facts and Figures*, 20 October 2010 (www.itu.int).

3. Here I bracket, for the moment, cloud-based ringtones—not to speak of music apps on smartphones.

4. The ringtone industry is both the intersection of mobile handset manufacturers, wireless firms, and content providers *and* their union. This industry is rather small in one sense, focusing on mobile media content providers and ringtone wholesalers, but in another sense it is vast, encompassing both the telecommunications industry and the entertainment industry.

5. See John Markoff, "Steve Jobs Walks the Tightrope Again," *New York Times*, 12 January 2007; Randall Stross, "Want an iPhone? Beware the iHandcuffs," *New York Times*, 14 January 2007.

Notes to Chapter 5

1. Cited in Ian Mansfield, "News Corp. Sells Ringtone Subsidiary," at http://www.cellular-news.com.

2. See "Crazy Frog" at http://en.wikipedia.org; Guy Blackman, "Pop's Great Leap Forward," *Sunday Age* (Australia), 11 September 2005.

3. Guy Dennis, "Crazy Frog, Crazy Business," *Sunday Telegraph* (London), 22 May 2005.

4. See Quickie, "The Truth is out 8)," at http://www.insaneheat.com.

5. See the Wikipedia entry on "Axel F."

6. See Mansfield, "News Corp. Sells Ringtone Subsidiary"; Chris Snyder, "News Corp. Buys VeriSign's Stake in Jamster for $200M," *Wired*, 7 October 2008.

7. Cited on p. 81 of David Bell, *Cyberculture Theorists: Manuel Castells and Donna Haraway* (Routledge, 2007).

8. Amanda Andrews, "Crazy Frog Creator Leaps Up UK Ad Spending League," *The Times* (London), 4 January 2006.

9. See "From the Bookshelf of Warwick Cairns, Planning Director, Brandhouse WTS," *Brand Strategy* 5 December 2005. Also see Lee Knuttila, "User Unknown: 4chan, Anonymity, and Contingency," *First Monday* 16/10 (October 2011), available at http://firstmonday.org.

10. Wernquist used LightWave software. See http://www.newtek.com/products/lightwave.html.

11. See pp. 1–6 of Anna Tsing, *Friction: An Ethnography of Global Connection* (Princeton University Press, 2005).

12. Lesley Thomas, "Top of the Hate Parade," *Daily Telegraph* (London), 27 May 2005.

13. See Liz Morrell, "Special Report: Shoppers of the Future—Tomorrow's Shoppers," *Retail Week*, 24 June 2005.

14. See Alison Thomas, "Volles Handy, Leere Taschen," *Times Educational Supplement*, 24 June 2005; Matthew Saltmarsh, "Rethinking the 'Credit Culture,'" *New York Times*, 26 October 2007.

15. See, e.g., Eric Schlosser, *Fast Food Nation: The Dark Side of the All-American Meal* (Houghton Mifflin, 2001).

16. Paul Williams, "Smash Singles Lead Big Leap Forward," *Music Week*, 30 July 2005.

17. Cited in Morrell, "Special Report."

18. Ibid.

19. The "lyrics" to the original Crazy Frog recording, transcribed by the author, are partly drawn from Quickie, "The Truth is out 8)."

20. For a list of novelty frog songs, see DJ Ratso Ruffo, "DJ Cheat Sheet No. 16, Songs About Frogs and Worms," at http://www.ratso.net. Also see Ivan Raykoff and Robert Deam Tobin, eds., *A Song for Europe: Popular Music and Politics in the Eurovision Song Contest* (Ashgate, 2007); Philip Bohlman, *World Music: A Very Short Introduction* (Oxford University Press, 2002), pp. 88–98.

21. On child animism, see Jean Piaget, *La representation du monde chez l'enfant* (Félix Alcan, 1926); Kim Dolgin and Douglas Behrend, "Children's Knowledge About Animates and Inanimates," *Child Development* 55, no. 4 (1984), 1646–1650, esp. p. 1649.

22. Lucy Mangan, "G2: Don't Ban Crazy Frog Just Because He Has a Penis. After All, Nakedness and Sex Aren't the Same Thing," *The Guardian*, 13 July 2005; Adrian Turpin, "Crazy World: How an Annoying Frog Conquered the World," *The Independent*, 26 May 2005.

23. The "Axel F" video, viewed more than 29 million times, was available at http://www.youtube.com.

24. Although the frog's headgear is often described as a flight helmet, it seems to be a combination of old-fashioned aviator goggles and a classic 1950s-style British motorcycle helmet of the kind sometimes called a "pudding bowl."

25. For example, see "The Croak's on Coldplay as Crazy Frog Hits the Top," *Evening Herald* (Plymouth), 3 June 2005.

26. See the "Crazy Frog Human" video at http://www.youtube.com.

27. Possibly confirming my interpretation, the explosion was actually removed in a more recent version of the "Crazy Frog - Axel F" video (uploaded 16 June 2009), at http://www.youtube.com.

28. See Peter Manuel and Wayne Marshall, "The Riddim Method: Aesthetics, Practice, and Ownership in Jamaican Dancehall," *Popular Music* 25, no. 3 (2006), 447–470, esp. p. 457.

29. See the Reggae Vibes website at http://www.reggae-vibes.com.

30. The video is available at http://www.youtube.com.

31. The melody is also sung intentionally out of tune and "apart" from the riddim, often generating striking dissonances (such as the A4 against the B♭2 in the first full measure of figure 5.4). See Manuel and Marshall, "The Riddim Method," pp. 459–460.

32. See http://www.justsomelyrics.com/417283/LOC-Ring-Ding-Ding-Lyrics.

33. See Jumana Farouky, "Unity Begins at Home," *Time*, 15 February 2007; "Tottenham" at http://en.wikipedia.org.

34. Paul Gilroy, *After Empire: Melancholia or Convivial Culture?* (Routledge, 2004), pp. 104–106, 144–149. Also see various entries at the Reggae Vibes website at http://www.reggae-vibes.com. One survey found that British youths saw Crazy Frog as "chav"; see Iain Burchell, "Ringtones Are a Real Hangup," *Daily Star*, 30 March 2006.

35. See Heather Phares, "Crazy Frog—Biography," at http://www.allmusic.com.

36. Available at http://www.crazyfrogfanclub.com.

37. Available at http://www.crazyfrogfanclub.com.

38. "Crazy Frog," Wikipedia.

39. For a veritable carnival of chav-bashing, homophobia, rape fantasies, racism against blacks and immigrants, and much more, mostly directed at Crazy Frog, see Dave Beasant's "Kill Crazy Frog" petition at http://www.petitionspot.com.

40. See "Crazy Frog and His Stunted Penis Cleared by ASA," *The Guardian*, 2 February 2005.

41. Donald MacLeod, "The Big Stuff: That Crazy Frog Is Driving Me Nuts!" *The Mirror*, 1 June 2005.

42. Miles Brignall, "Company Fined £40,000 over Crazy Frog Advertisements," *The Guardian*, 21 December 2005.

43. Mark Odell, "Crazy Frog Leads to £40,000 Fine," *Financial Times*, 21 December 2005.

44. Quoted in Owen Gibson, "Blame the Crazy Frog," *The Guardian*, 5 October 2006.

45. The history of media development is replete with brief successes and total failures, even if the scale of the ringtone's success in this regard may appear unprecedented. However, video games, arguably the most important of the new media

significantly emerging from East Asia, have demonstrated a longer life span and continued viability.

46. See, e.g., Eric Weaver, *National Narcissism: The Intersection of the Nationalist Cult and Gender in Hungary* (Peter Lang, 2006), p. 166.

47. See Perry Anderson, "After the Event," *New Left Review* 73 (January–February 2012), 49–61.

Notes to Chapter 6

1. Rupam Jain Nair, "In Gujarat, Cells Are Ringing to the Sound of Modi," *Indian Express*, 30 November 2004, available at http://www.indianexpress.com.

2. Chaudhary opposed Hindu nationalist attacks on Muslims and was very critical of Modi. On the riots, see Martha Nussbaum, "Genocide in Gujarat," *Dissent*, summer 2003, available at http://www.dissentmagazine.org.

3. Also see Jonathan Sterne's criticism of sensory essentialisms on pp. 14–18 of *The Audible Past: Cultural Origins of Sound Reproduction* (Duke University Press, 2003).

4. Interview with Julie Corwin of Radio Free Europe, 23 May 2006.

5. Ringtones are reminiscent of neo-surrealist photomontage in post-Situationist "culture jamming." See Leah Lievrouw, *Alternative and Activist New Media* (Polity, 2011), pp. 28–97. On mash-ups and politics, see Sam Howard-Spink, "Grey Tuesday, Online Cultural Activism and the Mash-Up of Music and Politics," *First Monday*, 4 July 2005, available at http://firstmonday.org. Also see chapter 7 below.

6. Cited in DJ Yap and Fe Zamora, "'It's Not My Voice Movement' Urged," *Philippines Daily Inquirer*, 27 June 2005.

7. This account draws on numerous sources, including the following: "Hello Garci Scandal," at http://en.wikipedia.org; "Timeline in 'Hello Garci' Scandal," *Philippine Daily Inquirer*, 25 December 2006; Jan Padios, "Can You Hear Us Now? Ringtones and Politics in the Contemporary Philippines," in *The Oxford Handbook of Mobile Music Studies*, ed. S. Gopinath and J. Stanyek (Oxford University Press, forthcoming). Download counts for "Hello Garci" ringtones are from David Rowan, "Political Ringtones," *The Times* (London), magazine section, 18 March 2006.

8. See Vicente Rafael, "The Cell Phone and the Crowd: Messianic Politics in the Contemporary Philippines," *Public Culture* 15, no. 3 (2003), 399–425.

9. See Benedict Anderson, "Cacique Democracy in the Philippines," in *The Spectre of Comparisons: Nationalism, Southeast Asia, and the World* (Verso, 1998); Rafael, "The Cell Phone and the Crowd," pp. 421, 405, 408–409.

10. See Rafael, "The Cell Phone and the Crowd," pp. 422–425.

11. See Giorgio Agamben, *State of Exception* (University of Chicago Press, 2005).

12. See Human Rights Watch, Scared Silent: Impunity for Extrajudicial Killings in the Philippines, 2007, available at http://hrw.org.

13. See Howard-Spink, "Grey Tuesday."

14. This was the first ringtone that TXTPower (see below) uploaded onto its website. See Padios, "Can You Hear Us Now?" The definition of "buking" is from http://www.bansa.org/dictionaries.

15. Also see Robert Fink, "The Story of ORCH5, or, the Classical Ghost in the Hip-Hop Machine," *Popular Music* 24, no. 3 (2005), 339–356.

16. "In da Club" received the first *Billboard* Ringtone of the Year award in 2004. See Sasha Frere-Jones, "Ring My Bell: The Expensive Pleasures of the Ringtone," *The New Yorker*, 7 March 2005.

17. See http://www.pcij.org/blog/wp-files/ringtones.php. For the original wire-tapped conversation recordings, see http://www.pcij.org/blog/wp-files/tapes.php.

18. Padios, "Can You Hear Us Now?" p. 16, note 24. Also see Craig Lockard, "Popular Musics and Politics in Modern Southeast Asia: A Comparative Analysis," *Asian Music* 27, no. 2 (1996), 164–170.

19. Harmonically, this ringtone is clever, incorporating songs in the same key (E major). It begins with the single-line walkup beginning on B3 to the beginning proper of "Do You Want to Know a Secret" and includes a compressed "[Hell]o Garci; Ma'am"—humorously placed to seem like a romantic exchange ("oh, Garci!") immediately after McCartney sings "closer." The song then reaches an important dominant chord (or B major) arrival, on the word "hear" (after "say the words you want to *hear*"), and then the music stops. Suddenly, the conversation quote plays and is repeated once; the opening E major tonic chord of "Don't Let Me Down" enters and resolves the harmonic tension.

20. "I'm Sorry" ringtones are available at http://www.txtpower.org. Another ringtone octave-shifted Arroyo's vocal pitch downward and combined her apology with the *Star Wars* "Imperial March" and familiarly labored breathing sounds, transforming her into Darth Vader.

21. "IFJ: Anti-Wiretapping Law Gags RP Journalists," ABS-CBN News, available at http://www.abs-cbnnews.com.

22. Dean Jorge Bocobo, "The Right to Privacy and the Public's Right to Know," Philippine Commentary blog, 13 December 2005, available at http://philippinecommentary.blogspot.com.

23. See "TXTPower" at http://en.wikipedia.org.

24. Public playback of ringtones at political protests may have had a precedent in music concerts and staged events. "Records" exist for the largest number of ringtones of the same song being simultaneously played, acting as promotional vehicles for wireless companies, musical artists and record labels. One press release claimed that Virgin Mobile UK set a record for "the most mobile phones ringing in unison" in 2001 (254 phones); this alleged record was broken in Toronto on 1 November 2007 by 554 Spice Girls fans playing the 1996–97 hit "Wannabe" simultaneously. See "Spice Girl Mel B Breaks World Record in 'Spice City' Toronto," *Canada NewsWire*, 1 November 2007. The event promoted the Spice Girls' reunion tour, which was to begin in Vancouver a month later.

25. See Emily Turrettini, "From Ring Tones to Car Horns? Only in the Philippines" at http://www.textually.org/ringtonia/archives.

26. Rafael's deconstruction of the rhetoric of text messaging seems to ignore activists' common-sense understanding of it as nothing more or less than a mobilization tool. According to the TXTPower activist Trixie Concepcion, "People sometimes think that organizing a campaign is just an SMS away. Not really. While mobile technology really helps in mobilizing, the initial education work must really be done face to face and in person." (cited in Rudy De Waele, "Women in Mobile 11—Trixie Concepcion," at http://www.m-trends.org)

27. On Estrada's offer, see Fe Zamora et al., "40,000 Cry Out: GMA, Go," *Philippine Daily Inquirer*, 14 July 2005.

28. See Rafael, "The Cell Phone," 402; Sarah Toms, "The Philippines' Mobile Politics," BBC News, 1 April 2006, available at http://news.bbc.co.uk.

29. See World Bank, Philippines: Recent Economic Developments, at http://siteresources.worldbank.org. Also see E. San Juan Jr., "Interrogating Transmigrancy, Remapping Diaspora: The Globalization of Laboring Filipinos/as," *Discourse* 23, no. 3 (2001), 52–74.

30. See "Middle Fatigue," *Philippine Daily Inquirer*, 13 June 2005; Amando Doronila, "Battle of Attrition Looms," *Philippine Daily Inquirer*, 29 June 2005.

31. See Rina Jiminez-David, "A Bi-Polar Nation," *Philippine Daily Inquirer*, 14 June 2005.

32. Amando Doronila, "Nostalgia Trip," *Philippine Daily Inquirer*, 24 February 2006.

33. Paul Hutchcroft, "The Arroyo Imbroglio in the Philippines," *Journal of Democracy* 19, no. 1 (2008), 141–55; Joel D. Adriano, "Politics of Poverty in the Philippines," *Asia Times*, 2 March 2008, available at http://www.atimes.com.

34. Tim Chambers and Rob Sebastian, *Mobile Media in 21st Century Politics* (New Politics Institute, 2006), p. 14 (available at http://www.newpolitics.net).

35. Cited from http://www.mobilactive.org.

36. Interview with Katrin Verclas, 14 January 2007. Also providing support was Martin Kearns, Executive Director of Green Media Toolshed, a non-profit organization devoted to "building and strengthening the communications infrastructure for the environmental movement." See http://www.greenmediatoolshed.org/aboutus/goalsandstratigies.acs.

37. Cited from http://nten.org/supporters.

38. Interview with Eric Gundersen, 5 January 2007.

39. Later MobileActive conferences were held in São Paulo, Brazil (2007) and in Johannesburg (2008).

40. See Dru Oja Jay, "An Interview with Evan Henshaw-Plath," at http://misnomer.dru.ca.

41. Odeo.com, founded by Evan Williams in 2005, once included a website for the distribution of legal, free MP3 files, but it was acquired by an investor in 2007 and transformed into a for-profit company focusing on video management. See "Odeo Enters Enterprise Market with SaaS Solution," PRWEB, 2 April 2009, available at http://www.odeoenterprise.com.

42. See http://anarchogeek.com/?s=riot+tones.

43. Ringtone obtained from http://www.developmentseed.org.

44. This awkward ringtone simplifies the well-composed harmonic progressions of the song and substitutes them for a progression that meanders further and further away from the home tonic (F♯ major) by beginning with a I-V-I progression (up to "City of New Orleans"), then going from V to II♯ (or V/V) (at "fields"), and then moves to iii, in what might appear like part of a tonicization of V (or C♯ major). In the original song, the progressions never move far beyond the tonic for any stretch of time, while the effect of the ringtone is to exacerbate a sense of *movement*, of aimless wandering. That said, these observations might be too generous. Brian Hyer noted (personal conversation, 27 April 2007) that, in the ringtone, it wasn't Bush who gets in the way of the city's recovery, but Gundersen himself—figuratively speaking. Indeed, given his placement within the nonprofit world, Gundersen possibly had, and blew, an opportunity to make a more powerful political intervention with a well-crafted ringtone.

45. Gundersen, interview, 5 January 2007.

46. Ringtones obtained from http://www.x910.com. Also see Alexander Cockburn and Jeffrey St. Clair, "Time-Delayed Journalism: The *New York Times* and the NSA's Illegal Spying Operation," at http://www.counterpunch.org.

47. Also see "They Might Be Ringtones" at http://www.tmbw.net.

48. Harmonically, the ringtone is very effective in that it refuses to land on the tonic harmony (C major), instead beginning with an oscillation between IV and iii, and ending up on V, creating a harmonic tension that both leads the hearer to want to resolve it, in a sense, by picking up the phone *and* signifies the unease the listener should feel at the thought of eternal surveillance.

49. See Katrin Verclas, "US Mobile Activism Coming to Life," at http://www.mobileactive.org; "How To Run a Mobile Advocacy Campaign (And How Not To)," at http://www.mobileactive.org.

50. See Brendon Williams, "Prescott in Ring Again," *The Mirror*, 5 June 2001.

51. See "Mom Is 'Dallas,' Dad Is Beethoven's Fifth," *Ha'aretz*, 5 June 2001.

52. See Anindita Ramaswamy, "India's E-Campaign Is Electioneering via E-Mail, Cellphone," *Deutsche Presse-Agentur*, 18 April 2004.

53. See Yogendra Yadav, "The Big Picture: Radical Shift in the Social Basis of Political Power," *The Hindu*, 20 May 2004, available at http://www.hindu.com.

54. Catharine P. Taylor, "Is It Time for Advertisers to Jump into Cell-Phone Messaging?" *Adweek*, 12 July 2004; Pew Charitable Trusts, "Youth Voter Turnout Up for the Second Major Election in a Row," available at http://www.pewtrusts.com; Scott Keeter, Juliana Horowitz, and Alec Tyson, "Young Voters in the 2008 Election," available at http://pewresearch.org.

55. Ashley Parker, "Ring-Tone Politics," at http://thecaucus.blogs.nytimes.com.

56. See Tariq Ali, *Pirates of the Caribbean: Axis of Hope*, revised edition (Verso, 2008).

57. The referendum, defeated in its first tryout, has a complex history and electoral trajectory that has been discussed at length elsewhere, including James Petras, "Venezuela in the Aftermath," at http://www.counterpunch.org; Federico Fuentes, "Venezuela: Social Tide Activists on the Referendum Defeat and the PSUV," at http://www.venezuelanalysis.com; Alan Woods, "Venezuela: The Referendum Defeat—What Does It Mean?" at http://www.venezuelanalysis.com; Erik Demeester, "Venezuela: The Economic Tug of War Between Revolution and Counter-Revolution," at http://www.marxist.com. For more general treatments of Chávez's politics, see Gregory Wilpert, *Changing Venezuela by Taking Power: The History and Policies of the Chávez Government* (Verso, 2007); Richard Gott, *Hugo Chávez and the Bolivarian Revolution* (Verso, 2005).

58. See Sujatha Fernandes, *Who Can Stop the Drums? Urban Social Movements in Chávez's Venezuela* (Duke University Press, 2010), esp. pp. 19–24; Fernando Coronil, *The Magical State: Nature, Money, and Modernity in Venezuela* (University of Chicago Press, 1997); Perry Anderson, "Jottings on the Conjuncture," *New Left Review* II, no. 48 (2007), 24.

59. See Julia Buxton, "Economic Policy and the Rise of Hugo Chávez," in *Venezuelan Politics in the Chávez Era: Class, Polarization, and Conflict*, ed. S. Ellner and D. Hellinger (Lynne Rienner, 2003).

60. A number of Spanish-language blogs included the popular YouTube video clip and transcriptions of the flare-up; see, e.g., http://www.noticias24.com/actualidad/?p=9650. A longer, less widely available clip from Cubainformación.tv, "¿Quién fue el responsable del incidente entre Chávez y el Rey de España?," is available at http://www.bitsrojiverdes.org.

61. A YouTube search for "por que no te callas" revealed several versions of the news clip being viewed around 800,000 times; see http://www.youtube.com.

62. See "Ahora nadie se calla," BBC Mundo, 14 November 2007, available at http://news.bbc.co.uk and cited in "¿Por qué no te callas?" at http://en.wikipedia.org. A YouTube search for the phrase on 6 July 2008 yielded 1,290 items, more than twice as many as were found a few days after the event.

63. See Simon Romero, "When a Mother Country Tells Its Kid, 'Shut Up,'" *New York Times*, 25 November 2007; "Descubrimientos arqueológicos," at http://www.atravesdevenezuela.com. Also see "La Puerta de Tannhauser," by Aloriel, "'El Rey me hizo millonario,'" at http://aloriel.no-ip.org.

64. See "'Shut Up' Chavez Ringtone a Hit," at http://www.cnn.com. The original article, "1,5 millones de euros gracias al politono '¿Por qué no te callas?'" (*El País*, 19 November 2007) itself cited a BBC article with the estimate: "'Shut Up' Chavez Is Ringtone Hit," BBC News, 19 November 2007, available at http://news.bbc.co.uk.

65. "'Shut Up' Chavez Ringtone a Hit."

66. On 14 June 2011, the jamba.es site was redirected to jamster.es, with most of the same PQNTC ringtones remaining available.

67. Here I refer, anachronistically and non-contextually, to the recordings of Alvin and the Chipmunks, created by Ross Bagdasarian in the late 1950s through the speeding up of slowly recorded voices, which made adult voices sound remarkably high-pitched.

68. For examples, see Nicholas Deleon, "Spanish king to Venezuela's Chavez: Why Don't You Shut Up, Now a Ring Tone; Download Here," at http://www.crunchgear.com. Also see the chat room page at http://vagos.wamba.com and "'Shut Up' Chavez Is Ringtone Hit."

69. See "Oliver Urdaneta" at http://es.wikipedia.org.

70. The original tune, to which Urdaneta's parody closely hews, presents familiar musical topoi. For example, the music in the song's verse follows a typical "Spanish" descending parallel-chord minor-mode progression along the upper-scalar tetrachord from i to VII to VI to V in B♭ minor, and then prolongs V for the remainder of

the verse. But, toward the end of the verse, the music veers into the major mode, providing a celebratory arrival on I for the cheerful, anthemic chorus. On Escobar, see Jorge Peréz, "'¡Hay qué motorizarse!': Mobility, Modernity, and National Identity in Pedro Lazaga's *Sor Citroen*," *Arizona Journal of Hispanic Cultural Studies* 11 (2007), 22, note 16. On the modal implications of this tetrachord, see Peter Manuel, "From Scarlatti to 'Guantanamera': Dual Tonicity in Spanish and Latin American Musics," *Journal of the American Musicological Society* 55, no. 2 (2002), 311–336.

71. The link to the video, posted by "oliverurdaneta," is available at http://www.youtube.com.

72. This link and comments were viewed at http://www.youtube.com.

73. The literature on *300* and the "This is Sparta!" meme is extensive. See Saër Maty Bâ, "'Planetary Humanism' Calling: New Sightings of 'Black' Bodies in Epic Films," *Film International*, 2 March 2011, available at http://filmint.nu; Jamie Dubs et al., "This Is Sparta!" at http://knowyourmeme.com.

74. From DJ Pichu's *Imeem* page, available at http://www.imeem.com.

75. http://www.youtube.com/watch?v=v5qmgdRW4BE.

76. Augusto, "porque no te callas remix @dj devil," at http://yahel.wordpress.com.

77. See http://vagos.wamba.com/archive/index.php/t-255486.html. Also see http://rapidshare.com/files/72159149/callas.rar.

78. See, e.g., http://www.youtube.com/watch?v=g1J2gvjp4fs&feature=related.

79. See Eva Golinger, *Bush vs. Chávez: Washington's War on Venezuela* (Monthly Review Press, 2008), pp. 115–125.

80. Luis Carlos Diaz, "Venezuela: Why Don't You Shut Up?" at http://www.globalvoicesonline.org.

81. For example, note the tone of defiance in the following statement by Chávez: "We don't want this to become a political crisis. Spain has many investments, private companies here and we don't want to damage that, but if they are damaged, they are damaged. Spanish investment in Venezuela is not indispensable." (Cited in Kiraz Janicke, "Spanish-Venezuelan Spat Continues as Both Try to Calm Issue," at http://www.venezuelanalysis.com.) For Chávez's rhetorical question to King Juan Carlos, see Jaime Garcia, "CUMBRE IBEROAMERICANA," *El Mundo*, 12 November 2007. See the mention of Francisco Javier Peréz's *El insulto en Venezuela* (The Insult in Venezuela) in Romero, "When a Mother Country Tells Its Kid, 'Shut Up.'"

82. Gustavo Bueno, "¿Por qué no te callas?" *El Catoblepas* 69 (November 2007) (available at http://www.nodulo.org).

83. Ibid., section 6.

84. See the documentary film *The Revolution Will Not Be Televised* (2003); Wilpert, *Changing Venezuela*, pp. 223–226; Wilpert, "RCTV and Freedom of Speech in Venezuela," at http://www.venezuelanalysis.com.

85. See the show's website at http://www.alopresidente.gob.ve/. Also see Juan Forero, "'Aló Presidente,' Are You Still Talking?" *Washington Post*, 30 May 2009; Rachel Nolan, "The Realest Reality TV Show in the World," *New York Times Magazine*, 4 May 2012; Fernando Buen Abad Domínguez, "Alo Presidente . . . Happy Birthday," at http://venezuelanalysis.com; Cristina Marcano and Alberto Barrera Tyszka, *Hugo Chávez* (Random House, 2007), pp. 192–203.

86. Félix Martin and Pablo Toral, "The Role of Spanish Firms in Latin America's Structural Reforms, 1989–2005," in *Latin America's Quest for Globalization: The Role of Spanish Firms*, ed. F. Martin and P. Toral (Ashgate, 2005), pp. 1–2.

87. Daniel Treisman, "Stabilization Tactics in Latin America: Menem, Cardoso, and the Politics of Low Inflation," *Comparative Politics* 36, no. 4 (2004), 412–413.

88. David Hojman, review of Antonio Luis Hidalgo Capitán, *El pensamiento económico sobre desarollo: De los Mercantilistas al PNUD*, *Journal of Latin American Studies* 32, no. 1 (2000), 290–291.

89. See Forrest Hylton and Sinclair Thomson, "The Chequered Rainbow," *New Left Review* II, no. 35 (2005), 40–64. It should be emphasized, though, that in the case of water privatization in Bolivia, Bechtel (US) and Edison (Italy) were the main instigators, and the gas wars were a result of trade favoring Chilean export processing of Bolivian petrocarbons.

90. The quotation and the Morales citation are from Salim Lamrani, "Hugo Chávez, José Luis Rodríguez Zapatero, and the King of Spain," at http://www.venezuelanalysis.com.

91. Romero, "When a Mother Country Tells Its Kid, 'Shut Up.'"

92. Ibid.

93. Lamrani, "Hugo Chávez, José Luis Rodríguez Zapatero, and the King of Spain."

94. See the previously cited Cubainformación television program, available at http://www.cubainformacion.tv.

95. Beatriz Padilla, "Latin American Immigration to Southern Europe," at http://www.migrationinformation.org; Pilar Marrero, "Immigration Shift: Many Latin Americans Choosing Spain over US," at http://www.imdiversity.com.

96. Cited in "Immigration Rises to Top Campaign Theme in Spanish Elections" at http://www.expatica.com. Also see Romero, "When a Mother Country."

97. "King's 'Shut Up' Is a Ringtone Hit," *Miami Herald*, 16 November 2007, available at http://www.hemisferio.org.

98. See http://www.gobiernoenlinea.ve/servicios/tonos_musicales/tonos.html. The chant "¡Uh! ¡Ah! ¡Chávez no se va!"—perhaps inspired by sporting chants, labor songs and protests, and/or African-diasporic collective vocalizations—emerged spontaneously during the protests against the coup in April 2002, when a million people surrounded the Miraflores palace in Caracas; it became one of the unofficial slogans of the Bolivarian Revolution and was again frequently intoned during the 2006 recall referendum. Indeed, one reporter notes that techno (electronic dance music) versions of the chant were heard in the streets after the 2006 recall referendum victory. See Shawn O'Bryant, "Uh, Ah, Chavez No Se Va," at http://narcosphere.narconews.com.

99. Wilpert, *Changing Venezuela*, p. 27; Elena Pereira, e-mail interviews with author, 31 May and 6 June 2008.

100. See the website of the Ministerio del Poder Popular para la Cultura, available at http://www.ministeriodelacultura.gob.ve/.

101. See Rory Carroll, "Chávez Pours Millions More into Pioneering Music Scheme," *The Guardian*, 4 September 2007, available at http://www.guardian.co.uk; "El Sistema" at http://en.wikipedia.org.

102. Ali, *Pirates*, p. 183.

103. Ibid., p. 184; Fuentes, "Venezuela."

104. See "Venezuelans Vote to Eliminate Two-Term Limit on All Elected Office 54.4% to 45.6%" at http://venezuelanalysis.com.

105. Manuel Castells, Mireia Fernández-Ardèvol, Jack Linchuan Qiu and Araba Sey, *Mobile Communication and Society: A Global Perspective* (MIT Press, 2007), pp. 127–178, esp. 163–167.

106. See ibid., pp. 185–213.

107 See Ron Eyerman and Andrew Jamison, *Music and Social Movements: Mobilizing Traditions in the Twentieth Century* (Cambridge University Press, 1998), pp. 1–2.

108. Guy Debord, *Comments on the Society of the Spectacle* (Verso, 1990), p. x.

109. Some examples post-dating those discussed above suggest an overall decrease, if not disappearance, of middle-class, left-liberal dominance of political ringtone use, as well as an increasing trend of top-down, orchestrated uses of ringtones controlled by political figures themselves. Consider South African president Jacob Zuma's use of a recording of himself singing the Zulu-language anti-apartheid song "Awuleth' Umshini Wami" (Bring Me My Machine Gun) or "Umshini Wam" as a ringtone, first as part of his defense during a rape trial in 2006 and then toward his election in 2009, which seems to have had a more populist and not elite orientation (this ringtone also ended up being remixed, by Zimbabwean musician See4); the

uses of Obama speech remix ringtones during the 2008 primary and presidential campaigns mentioned above; the organized ringtone use of Narendra Modi, mixing speeches with Gujarati folk songs, in 2009, perhaps following a more conservative middle-class trend and closer to the anti-Chávez ringtones; and the recent uprisings in Egypt, which also involved ringtone-use of pro-revolution popular songs, here more probably conforming to the middle-class, left-liberal trend (although I haven't found examples of remix ringtones). See Isaac Mahlangu, "There's No Stopping Umshini Wam," *The Times* (South Africa), 18 December 2007, available at http://www.timeslive.co.za; Ashley Parker, "Ring-Tone Politics"; "Modi's Speeches Turned to Guj Folk Songs, Ringtones," CNN-IBN (Indo-Asian News Service), available at http://ibnlive.in.com; Elizabeth Blair, "The Songs of the Egyptian Protests," National Public Radio, 11 February 2011, available at http://www.npr.org. The planned nature of all of these speech/vocal examples is also striking, in contrast to the examples given above.

110. Ian Baucom, "Globalit, Inc.; or The Cultural Logic of Global Literary Studies," *PMLA* 116, no. 1 (January 2001), 160, cited in Vilashini Cooppan, "World Literature and Global Theory: Comparative Literature for the New Millennium," *symploke* 9, no. 1 (2001), 38.

Notes to Chapter 7

1. Brice Salek, e-mail interview with author, 16 March 2011.

2. See mayurmotgi, "Make Custom Ringtones for Free(any length)" [*sic*], at http://minedge.wordpress.com.

3. The Munich-based group Super Smart released a polyphonic-only ringtone album titled *Panda Babies* on GoFresh in April 2004. See "Album Gets Ringtone Only Release," BBC News, 30 April 2004, available at http://news.bbc.co.uk.

4. See http://tones.wolfram.com/about/ and http://tones.wolfram.com/about/how.html. In many ways, Wolfram*Tones* is essentially a sophisticated polyphonic ringtone remix engine of the sort that was available on polyphonic-equipped mobile phones in the early to mid 2000s.

5. Ambient phone speakers have been increasingly used to publicly broadcast recorded music, as in the UK phenomenon of "sodcasting." See Dan Hancox, "Mobile Disco: How Phones Make Music Inescapable," *The Guardian*, 12 August 2010, available at http://www.guardian.co.uk.

6. Consider Nathan Davis' composition *Bells* (2011), discussed in chapter 4.

7. Here I exclude discussions of video ringtones, music apps, and other multimedia forms that transcend the sonic, even though ringtones triggering the vibrate-mode motor and lights cannot be understood as purely sonic.

8. Christian Licoppe, "The Mobile Phone's Ring," in *The Handbook of Mobile Communication Studies*, ed. J. Katz (MIT Press, 2008), pp. 144–145; Luc Boltanski and Ève Chiapello, *The New Spirit of Capitalism* (Verso, 2005), pp. 103–63, esp. pp. 112–115. Licoppe's term is relevant to all ringtones, not just popular songs.

9. Sometimes, multiple ringtones are made from a single song, each tone using excerpts. Perhaps this suggests that a song can have multiple hooks. On hooks, see Gary Burns, "A Typology of Hooks," *Popular Music* 6, no. 1 (1987), 1–20; Charles Kronengold, "Accidents, Hooks and Theory," *Popular Music* 24, no. 3 (2005), 381–397.

10. However, a properly edited ringtone could articulate a song chorus as a stand-alone section; formal divisions within a song can independently demonstrate some degree of completeness.

11. A 1981 interview with Wendt is cited in Nicholas Zubrugg, "Marinetti, Boccioni and Electroacoustic Poetry," in *Critical Vices: The Myths of Postmodern Theory* (G+B Arts International, 2000), 15; MacKay, personal communication, 22 February 2010. The term appears more frequently in literary discourses (especially on text/sound-poetry) than in musical ones.

12. See Kofi Agawu, *Playing with Signs: A Semiotic Interpretation of Classic Music* (Princeton University Press, 1991), pp. 51–79.

13. See Charles Rosen, *The Romantic Generation* (Harvard University Press, 1995), pp. 41, 48.

14. Douglas Green, *Form in Tonal Music: An Introduction to Analysis*, second edition (Holt, Rinehart and Winston, 1979), pp. 91–93.

15. See Mark Katz's discussion of Stravinsky's *Serenade* for piano on p. 3 of *Capturing Sound: How Technology Has Changed Music*, revised edition (University of California Press, 2010).

16. The hardcore punk band Minor Threat featured songs under a minute in length—for example, their "Straight Edge," which named an entire punk subculture, is only 45 seconds long. On grindcore, Napalm Death, and "You Suffer" (which includes the lyrics "You suffer, but why?"), see Albert Mudrian, *Choosing Death: The Improbable History of Death Metal and Grindcore* (Feral House, 2004), esp. pp. 117–42; "Grindcore," at http://en.wikipedia.org. Many other groups in various styles have also produced short songs since the 1980s, although short skits, brief tracks, and pseudo-advertisements appeared in art and psychedelic rock albums of the 1960s—the fake ads on *The Who Sell Out* (1967), for example.

17. See the 60×60 webpage, available at http://www.voxnovus.com.

18. On sonic branding, see Daniel Jackson, *Sonic Branding: An Introduction*, ed. P. Fulberg (Palgrave Macmillan, 2003). Eno's Windows 95 startup jingle, 3.25 seconds

long, earned him $35,000; see David Sheppard, *On Some Faraway Beach: The Life and times of Brian Eno* (Chicago Review Press, 2009), p. 411.

19. On bumpers, see pp. 55–57 of Ronald Rodman, *Tuning In: American Narrative Television Music* (Oxford University Press, 2010).

20. See, e.g., Scott Burnham, "The Role of Sonata Form in A. B. Marx's Theory of Form," *Journal of Music Theory* 33, no. 2 (1989), 247–71, esp. p. 251.

21. Other examples of original ringtone composition are found on the French label Sonic Object; formed in 2001 by Antonie Schmitt and Adrian Johnson, it seems to be the first label devoted to ringtones, although they are freely available. See http://www.sonicobject.com/. The Ringtone Society, a Dutch-based collective, likewise gives away original ringtones. See http://www.ringtonesociety.com/.

22. Citation in Steven Leckart, "Solid Gold Bits," *Wired*, March 2007, available at http://www.wired.com. For this article in *Wired*, They Might Be Giants produced five more short ringtone songs: "Bite Size," "Untangle the Phone Cord," "Devil Spawn," "Friend or Foe?" and "My Other Phone Is a Boom Car." Also see "Dial-A-Song" at http://tmbw.net. Among the shortest songs by the band are the few-second-long "fingertips" on the album *Apollo 18* (1992).

23. It is specifically reminiscent of the accompanimental figuration toward the end of Bartók's Bagatelle, op. 14, no. 2. Debussy and other modernist composers used similar kinds of piano figuration. In "Phone Phone Phone," the figure is a whole-tone collection.

24. This kind of truncation is not unique to the miniature popular song of the postwar pop/rock era; one can find similar effects in nineteenth-century art music.

25. Karen Collins notes on pp. 133 and 28 of *Game Sound* (MIT Press, 2008) that boss music in classic video games was often "chaotic and frantic" and tended to have "much shorter loops . . . [which] added to the tension."

26. Licoppe, "The Mobile Phone's Ring," p. 143.

27. J. Philip Lambert has discussed the combination of interval cycles in the music of Charles Ives; the upper, descending cycle would be called a "3/4-combination cycle" because it uses two different interval sizes (m3 = 3 half-steps; M3 = 4 half-steps). See "Interval Cycles as Compositional Resources in the Music of Charles Ives," *Music Theory Spectrum* 12, no. 1 (1990), 43–82.

28. The discussion here veers into ecological understandings of listening, which found powerful proponents in John Cage and R. Murray Schafer. See Schafer's book *The Soundscape: Our Sonic Environment and the Tuning of the World* (Destiny Books, 1994).

29. For a discussion of "developing variation," Arnold Schoenberg's concept of thematic transformation, see Walter Frisch, *Brahms and the Principle of Developing Variation* (University of California Press, 1984).

30. See Robert Fink, *Repeating Ourselves: American Minimal Music as Cultural Practice* (University of California Press, 2005), pp. 42–47.

31. The ringtone could also be heard as evoking birdsong: Unit A, in particular, sounds like the chirp of a house sparrow, suggesting the degree to which such signals and sonorities might demonstrate complex processes of determination. Also, the harmonic resolution implicit in unit C confirms the sense of "achievement."

32. On the *grito*, see Daniel Sheehy, *Mariachi Music in America: Experiencing Music, Expressing Culture* (Oxford University Press, 2006), p. 95. The self-aware hipster irony directed at various subcultural and consumer identities in the Ringtone Records catalog—Latinos in *East L.A. Ringtone Ringtones*, young pre-teen girls in *Ringtones for Cloud Girls* (2010), video game fans in *Game Players* (2008), stoners, hippies, and other counterculturally minded persons in *Prophets of Armageddon Ringtones* (2008), and so on—suggests that these ringtones are not necessarily (or only) targeted *to* these constituencies, but rather (or also) create critical or lightly caricatured portrayals *of* them. Here, Ringtone Records effectively undertakes, in its politically incorrect postmodernism, a critique of the ideology of ringtone personalization.

33. See Regina Lynn, "Silly, Porn Ringtones Are for Adults," at http://www.wired.com. The "Sex Drive Daily" blog has moved to http://www.reginalynn.com/.

34. As examined in March 2011.

35. Brice Salek, email interview with author, 25 February 2011.

36. See Tally Deushane, "More 30 Second Songs," at http://www.youtube.com; Sean Cooper, "Just One Hook," *Wired*, March 2007, available at http://www.wired.com. Also see Gerard Goggin's treatment of "mobisodes" on p. 180 of *Cell Phone Culture* (Routledge, 2006).

37. See Alexander Cockburn, "The Myth of Microloans," *The Nation*, 6 November 2006, available at http://www.thenation.com.

38. Daniel Miller, "Sterilizing Cyberspace" (review of Jonathan Zittrain's *The Future of the Internet—And How to Stop It*), *New Left Review* II, no. 51 (2008), 154–160, esp. pp. 155, 154.

Notes to Introduction to Part IV

1. For an argument within discourses on world music that emphasizes the relationship between the discourse of difference and the production of homogeneity, see

Veit Erlmann, "The Aesthetics of the Global Imagination: Reflections on World Music in the 1990s," *Public Culture* 8, 1996, 467–487.

2. Fredric Jameson, *Valences of the Dialectic* (London: Verso, 2009), 16.

3. See Philippe Legrain, "Cultural Globalization Is Not Americanization," *Chronicle of Higher Education*, 9 May 2003. Donald Sassoon notes that not "everyone . . . consumes American culture; only . . . most of the culture that circulates *across national boundaries* originates in the US." See "On Cultural Markets," *New Left Review* 17, September–October 2002, p. 124.

4. By "world music" I mean both the hybridized, international elements of a new quasi-genre of music (formerly "world beat") far smaller in scope than the recorded musics of the world *and* the dominant Western forms of music, especially popular music, coming from the United States. On the former, see Steven Feld, "A Sweet Lullaby for World Music," *Public Culture* 12, no. 1, 2000, 145–171; Martin Stokes, "Music and the Global Order," *Annual Review of Anthropology* 33, 2004, p. 61. On the latter, see Timothy Brennan's argument, on p. 21 of *Sacred Devotions* (Verso, 2008), that "world musics" are "being globally disseminated and popularly, even reverently, internalized almost everywhere." Also relevant are globalized production techniques such as the Auto-Tune plug-in. The ringtone as a type of world music is best exemplified in globally dominant preset ringtones, such as the iPhone's "Marimba" and the Nokia Tune. Of course, the most important "world music" treated in chapter 9, in the sense of a dominant type of Western music, is that of black popular music. Despite its predominantly US-based origins, its transatlantic and Caribbean histories make it an especially global aesthetic in and of itself—even if it has been readily assimilated into the global music industry's economies of scale.

Notes to Chapter 8

1. See http://www.youtube.com/watch?v=XIZVcRccCx0. It was the subject of a minor plagiarism controversy; see Mary Elizabeth Williams, "The Horror of NPR's Copycat Story," at http://www.salon.com.

2. See Steven Pustay, Cell Phones and Cinema: Filmic Representations of Mobile Phone Technology and New Agency, MA thesis, Ohio University, 2007, available at http://etd.ohiolink.edu; Sara Swain, Roaming Across Cinematic Space: The Cell Phone in Narrative Film, MA thesis, Concordia University (Montreal), 2008; Matt Schley, "The Cell Phone in Modern Thriller Cinema," film studies course project, University of Utah, 2009, available at http://mattschley.com.

3. Fredric Jameson, *A Singular Modernity: Essay on the Ontology of the Present* (Verso, 2002).

4. Walter Benjamin, "The Work of Art in the Age of Its Technical Reproducibility," third version, in *Selected Writings*, volume 4 (1938–1940), ed. H. Eiland and M. Jen-

nings (Belknap, 1996–2003). For the original quotation (in translation), see V. I. Pudovkin, *Film Technique and Film Acting*, memorial edition (Grove, 1976 [1958]), p. 143.

5. This does not mean that disparate objects become the same thing; rather, they can be made to have equivalent functions or work in similar ways—in the way that bartered or tender exchange creates equivalences between different things (through a presumed equivalence of value in the marketplace).

6. See Rudolf Arnheim, *Film as Art* (University of California Press, 1957), pp. 138–139; Béla Balázs, *Theory of the Film: Character and Growth of a New Art* (Dobson, 1952), p. 279; André Bazin, *What Is Cinema?* volume 2 (University of California Press, 1971), p. 120.

7. Gilles Deleuze, *Cinema 1: The Movement-Image* (Athlone, 1986), p. 159.

8. See, e.g., Chris Foresman, "Apple Responsible for 99.4% of Mobile App Sales in 2009 (Updated)," *Ars Technica*, 18 January 2010, available at http://arstechnica.com.

9. These terms came from the Institute for Propaganda Analysis during the late 1930s; its categories of propaganda devices included name-calling, glittering generalities, transfer, testimonial, plain folks, card stacking, and bandwagon. Also see Marvin Bressler, "Mass Persuasion and the Analysis of Language: A Critical Evaluation," *Journal of Educational Sociology* 33, no. 1 (1959), 17–27.

10. Other leading-edge commodities arguably functioned in similar ways in the past. For example, during the heyday of Hollywood cinema, in the 1940s and the 1950s, certain makes or models of automobiles were associated with certain characters. In some films of the early 1980s, including *Tron* (1982) and *Wargames* (1983), a personal computer was often associated with a hacker or a geek character.

11. See Walter Benjamin, *The Origin of German Tragic Drama* (Verso, 1998), esp. pp. 132–134 and 197–198.

12. Andrew Sofer, *The Stage Life of Props* (University of Michigan Press, 2003), pp. 11–12; Frances Teague, *Shakespeare's Speaking Properties* (Bucknell University Press, 1991), p. 17, cited in Sofer, p. 12. Mobility is thematized in filmic appearances of mobile phones, which are often contrasted with fixed land-line phones in stalker/slasher movies. See Swain, "Roaming Across Cinematic Space," pp. 37–56.

13. Sofer, *The Stage Life of Props*, p. 27.

14. Sofer (pp. 20–29) provides a number of possible functions of the prop.

15. See Christine Spengler, *Screening Nostalgia: Populuxe Props and Technicolor Aesthetics in Contemporary American Film* (Berghahn, 2009), pp. 96–97.

16. But see Scott Paulin, "Richard Wagner and the Fantasy of Cinematic Unity: The Idea of the *Gesamtkunstwerk* in the History and Theory of Film Music," in *Music and Cinema*, ed. J. Buhler, C. Flinn, and D. Neumeyer (University Press of New England, 2000).

17. See Theodor Adorno and Hanns Eisler, *Composing for the Films* (Continuum, 2007 [1947]), esp. pp. 2–3; Theodor Adorno, *In Search of Wagner* (Verso, 1981), p. 36. On Debussy's criticism, see Marie Rolf, "Symbolism as Compositional Agent in Act IV, Scene 4 of Debussy's *Pelléas et Mélisande*," in *Berlioz and Debussy: Sources, Contexts and Legacies: Essays in Honor of François Lesure*, ed. B. Kelly and K. Murphy (Ashgate, 2007), p. 132. But Debussy did use leitmotifs; see David Grayson, *The Genesis of Debussy's* Pelléas et Mélisande (UMI Research Press, 1986).

18. Adorno and Eisler, *Composing for the Films*, p. 7.

19. Robbert Van Der Lek discusses and rejects the term "prop music" on pp. 34–35 of *Diegetic Music in Opera and Film: A Similarity Between Two Genres of Drama Analyzed in Works by Erich Wolfgang Korngold (1897–1957)* (Rodopi, 1991).

20. See Michel Chion, *Film, A Sound Art* (Columbia University Press, 2009), pp. 365–373.

21. Sofer, *The Stage Life of Props*, p. 20.

22. Bruno Latour, *We Have Never Been Modern* (Harvard University Press, 1993), pp. 91–129.

23. Although this is not an example of schizophonia, as discussed by Steven Feld and R. Murray Schafer, the concept is not irrelevant. See Steven Feld, "From Schizophonia to Schismogenesis: The Discourses and Practices of World Music and World Beat," in Charles Keil and Steven Feld, *Music Grooves: Essays and Dialogues* (University of Chicago Press, 1994). A more appropriate term is "rhizophonia," treated in Jason Stanyek and Benjamin Piekut, "Deadness: Technologies of the Intermundane," *TDR* 54, no. 1 (2010), p. 19. Also see Steven Connor, *Dumbstruck: A Cultural History of Ventriloquism* (Oxford University Press, 2000).

24. Michel Chion, *Audio-Vision: Sound on Screen* (Columbia University Press, 1994), p. 6.

25. NBC acquired the rights to *Coupling* (in 2002) and *The Office* (2003), the latter of which was successful in the US. See "The Buzz," *Advertising Age*, 19 May 2003.

26. Although scholarly studies are lacking, good journalistic treatments of the show (and the American remake) are available. See Caryn James, "Television: Sitcoms Can Be Sexy. Just Ask the British," *New York Times*, 15 June 2003; Alessandra Stanley, "Two Nations Split by a Sense of Humor," *New York Times*, 25 September 2003.

27. The sound consultant Julian Treasure estimates that the Nokia Tune is heard 1.8 billion times per day, or 20,000 times per second (interview, Radio New Zealand

National, *Nine to Noon*, 27 January 2010, available at http://static.radionz.net.nz; cited in "Nokia Tune" at http://en.wikipedia.org).

28. See Cristel A. Russell, "Toward a Framework of Product Placement: Theoretical Propositions," *Advances in Consumer Research* 25 (1998), 357–362, available at http://www.acrwebsite.org.

29. Richard Wray, "In Just 25 Years, the Mobile Phone Has Transformed the Way We Communicate," *The Guardian*, 1 January 2010, available at http://www.guardian.co.uk; Marc van Impe, "No Growth in UK Mobile Market Penetration," at http://www.mobilemonday.net.

30. Maurice Yacowar claims it may be "the best TV series ever made" in *The Sopranos on the Couch: The Ultimate Guide: Analyzing Television's Greatest Series* (Continuum, 2007), p. 9.

31. Glen Gabbard, *The Psychology of The Sopranos: Love, Death, Desire, and Betrayal in America's Favorite Gangster Family* (Basic Books, 2002); David Lavery, ed., *This Thing of Ours: Investigating* The Sopranos (Columbia University Press, 2002); Richard Greene and Peter Vernezze, eds., The Sopranos *and Philosophy: I Kill Therefore I Am* (Open Court, 2004); Christopher Kocela, "Unmade Men: The Sopranos After Whiteness," *Postmodern Culture* 15, no. 2 (January 2005), available at http://muse.jhu.edu; Martin Hipsky, "Post-Cold War Paranoia in *The Corrections* and *The Sopranos*," *Postmodern Culture* 16, no. 2 (2006), available at http://muse.jhu.edu; Dana Polan, *The Sopranos* (Duke University Press, 2009); Alex Schulman, "The Sopranos: An American Existentialism?" *Cambridge Quarterly* 39, no. 1 (2010), 23–38. Also see "*The Sopranos*" at http://en.wikipedia.org.

32. See, e.g., Kevin Fellezs, "Wiseguy Opera: Music for *Sopranos*," in *This Thing of Ours*, esp. p. 162. Some film auteurs privilege source music; see Kathryn Kalinak, *How the West Was Sung: Music in the Westerns of John Ford* (University of California Press, 2007), pp. 90–91.

33. See Fellezs, p. 165 (but note that the argument on returns as earnings is not in the source). Musical puns—wherein text (title, lyrics, etc.) is captured by music and redeployed in this guise—are undertheorized; see, however, Matthew Mihalka, From the Hammond Organ to "Sweet Caroline": The Historical Evolution of Baseball's Sonic Environment, PhD dissertation, University of Minnesota, 2012.

34. Claudia Gorbman notes that in *Eyes Wide Shut* (1999) the director Stanley Kubrick "often uses popular songs—rather, their titles or (unstated lyrics)—to provide apt or ironic commentary." See "Auteur Music," in *Beyond the Soundtrack: Representing Music in Cinema*, ed. D. Goldmark, L. Kramer, and R. Leppert (University of California, 2007), p. 153.

35. Sofer, *The Stage Life of Props*, p. 21.

36. Gorbman, *Unheard Melodies: Narrative Film Music* (Indiana University Press, 1987), p. 25.

37. Much blogging and chat-group commentary on Tony Soprano's ringtone can be found. See http://www.blackberryforums.com.

38. See Claudia Gorbman's characterization of auteur-directors as "mélomanes" (music maniacs) on pp. 149–162 of "Auteur Music." See especially p. 150.

39. Robert Brenner, "What's Good for Goldman Sachs Is Good for America," at [illegible], p. [illegible]

40. As Perry Anderson notes, the real economic story of this conjuncture was "the emergence of China as the new workshop of the world." See "Jottings on the Conjuncture," *New Left Review* II, no. 48 (2007), p. 6.

41. Interview with Julian Treasure, Radio New Zealand National, *Nine to Noon*. The two examples of sonic logos, as well as the concept itself, was discussed by Treasure, in addition to the percentage of mobile phone users with default ringtones. On sonic logos, also see Daniel Jackson, *Sonic Branding: An Introduction* (Palgrave Macmillan, 2003), pp. 141–146.

42. However, it is worth noting that the Nokia Tune has been viewed as Nokia's sonic logo; it was used by the company in advertisements as early as 1992 or 1993. See Sonja Kastner, *Klang macht Marken: Sonic Branding als Designprozess* (Gabler, 2008), pp. 1–2.

43. Jackson, *Sonic Branding*, p. 146. Although one might suspect that sonic logos could, in fact, be too short in duration to warrant copyright—and hence, could be subject to fair use laws—this turns out not to be the case, apparently. See Ivan Hoffman, "FAIR USE: Music Sampling," at http://www.ivanhoffman.com.

44. See http://www.seeing-stars.com/Locations/ILoveYouMan1.shtml.

45. Released on *Moving Pictures* (1981), "Tom Sawyer" is Rush's best-known song.

46. See Abe Sauer, "Why Apple Deserves an Oscar Too," *The Awl*, 4 March 2010, available at http://www.theawl.com.

47. Ibid.

48. Most of this is argued in Sauer, "Why Apple Deserves an Oscar Too," and in "The Significance of Apple's Product Placement" at http://www.edibleapple.com. In one report from 2006, the author noted that "On older episodes of the show '24,' 'good guy' characters used Apple products while the 'bad guys' used Windows and non-branded PCs." See Sara Kehaulani Goo, "Apple Gets a Big Slice of Product Placement Pie," *Washington Post*, 15 April 2006, available at http://www.washington post.com.

49. See Robert Marich, *Marketing to Moviegoers: A Handbook of Strategies and Tactics*, second edition (Southern Illinois University Press, 2009), pp. 98–99 and 114–115.

50. The percentage of market share that Apple commands varies, of course, from year to year and within particular product categories. In the early 2000s, Apple's percentage of the home computing market steadily dwindled, decreasing to 1.98% by 2004. See Jeffrey Cruikshank, *The Apple Way* (McGraw-Hill, 2006), p. 2.

51. Of course, one would also need to account for Apple products' apparent signification as cosmopolitan in design sensibility and luxury-good status, which could disarticulate national associations with the brand. Excepting any possible pretensions to "buy American" when it comes to Apple (whose products are largely made in China), one might argue that Americans generally do not associate American brands with the nation-state. Karl Gerth speculates: "In China, branding is more overtly an issue of economic nationalism than in the United States. Most Americans associate the work of branding with companies and the marketplace, not with government officials and the state. Americans think it is Apple's job to make the iPod brand a household name, not the US government's." In *As China Goes, So Goes the World: How Chinese Consumers Are Transforming Everything* (Macmillan, 2010), p. 114.

52. See Stefan Constantinescu, "Apple's iPhone Is the Best Thing to Happen to the American Smartphone Industry," at http://www.intomobile.com.

53. The text message is actually from his agent (as confirmed by Bryden's commentary on the matter), but it isn't clear that the agent is American when we hear the ringtone.

54. Gorbman, *Unheard Melodies*, p. 24.

55. Ibid.

56. Michel Chion, *Audio-Vision: Sound on Screen* (Columbia University Press, 1994), p. 8.

57. See Michael Fried, *Absorption and Theatricality: Painting and Beholder in the Age of Diderot* (University of California Press, 1980). Chion notes the possibility of music that "is neither empathetic or anempathetic, which has either an abstract meaning, or a simple function of presence, a value as a signpost: at any rate, no precise emotional resonance" (*Audio-Vision*, p. 9); this description is apt for the effect discussed here.

58. Jia Zhangke, "Jia Zhangke On . . . ," on p. 11 of *The World* (press kit, Zeitgeist Films, 2004), 11, available at http://www.zeitgeistfilms.com.

59. "Postsocialism" refers to the post-1989 period in China, but it is also associated with the cultural and socioeconomic effects of the reforms instituted by Deng Xiaoping starting in the late 1970s. See Chris Berry, *Postsocialist Cinema in Post-Mao China: The Cultural Revolution after the Cultural Revolution* (Routledge, 2004), esp. pp. 13–14.

Zhang Xudong describes Chinese postsocialism as "a result of the historical overlap between the socialist state-form and the era of capitalist globalization" (*Postsocialism and Cultural Politics: China in the Last Decade of the Twentieth Century*, Duke University Press, 2008, p. 16). Jason McGrath notes that Jia Zhangke's "aesthetic is postsocialist both in the sense of being a successor and a contrast to the previously dominant aesthetic of socialist realism and in the sense that it is a realism of the postsocialist condition" (*Postsocialist Modernity: Chinese Cinema, Literature, and Criticism in the Market Age*, Stanford University Press, 2008, p. 132). Jia has long been critical of the fifth generation filmmakers but was influenced by the Chen Kaige–Zhang Yimou film *Yellow Earth* (1984).

60. On Jia's understanding of the 1980s, see Stephen Teo, "Cinema with an Accent—Interview with Jia Zhangke, Director of *Platform*," *Senses of Cinema* 15 (2001), available at http://archive.sensesofcinema.com. On Jia and music, see Michael Berry, *Xiao Wu, Platform, Unknown Pleasures: Jia Zhangke's "Hometown Trilogy"* (Palgrave Macmillan, 2009), pp. 124–126. On Jia's realism, see McGrath, *Postsocialist Modernity*, pp. 129–64; Zhang Xudong, "Poetics of Vanishing: The Films of Jia Zhangke," *New Left Review* II, no. 63 (2010), 71–88.

61. On Jia's experience of popular culture in the 1980s, including as a break-dance performer, see Berry, *Xiao Wu, Platform, Unknown Pleasures*, pp. 125–133.

62. See Tonglin Lu, "Music and Noise: Independent Film and Globalization," *China Review* 3, no. 1 (2003), 57–76, esp. pp. 65–66. Indeed, Berry (p. 27) notes that the original sound designer of *Xiao Wu*, Lin Xiaoling, left the shoot because of the director's demands to make the sound quality rougher, fearing it would hurt her professional reputation.

63. See Lu, "Music and Noise."

64. This song appears in both *Xiao Wu* (1997) and *24 City* (2008).

65. Much of this discussion of "Ren xiaoyao" in *Unknown Pleasures* draws on Berry, *Xiao Wu, Platform, Unknown Pleasures*, pp. 106–108.

66. Recoding is not unique to Jia's work—it is common in much Hollywood cinema, in which musical materials are often reused and recontextualized. On "chameleon music," which crosses between diegesis and non-diegesis, see Peter Larsen, *Film Music* (Reaktion, 2007), p. 164. On leitmotivic film music, see James Buhler, "*Star Wars*, Music, and Myth," in *Music and Cinema*, ed. J. Buhler, C. Flinn, and D. Neumeyer (Wesleyan University Press, 2000), pp. 33–57, esp. pp. 41–44. Jia's technique remains only quasi-leitmotivic, in that musical themes or motives do not become treated as leitmotifs that are orchestrated, reharmonized, recontextualized, and precisely cued to specific narrative and affective moments, associated with a particular character or characters.

67. See Berry, *Xiao Wu, Platform, Unknown Pleasures*, p. 124.

68. McGrath, *Postsocialist Modernity*, pp. 157–158.

69. For a politicized view of headphone listening in East Asian studies, see Rey Chow, *Writing Diaspora: Tactics of Intervention in Contemporary Cultural Studies* (Indiana University Press, 1993), pp. 161–164.

70. See McGrath, *Postsocialist Modernity*, p. 150; Zhang Xudong, "Poetics of Vanishing," esp. pp. 84–86.

71. Lu, "Music and Noise"; Berry, *Xiao Wu, Platform, Unknown Pleasures*; McGrath, *Postsocialist Modernity*.

72. This is the sequence in which Xiao Wu steals an apple from a moving cart and then chases another character, Fatty, who in turn steals the apple from him; it is also an interlude during which Xiao Wu mulls over his newly developing feelings for Meimei.

73. Lu, "Music and Noise," pp. 70–72.

74. My comments here reflect my lack of access to (or knowledge of) the original song. However, as Xiao Wu then whistles the song in the next scene, he sings the different sections of the song together (which were split by the cut), confirming that they are part of the same song.

75. See luminaryling, "Loyal Fans in China," at http://www.michaeljackson.com.

76. See Lu, "Music and Noise," pp. 66, 73–75.

77. See Giovanni Arrighi, Terence Hopkins, and Immanuel Wallerstein, "1989: The Continuation of 1968," in *After the Fall*, ed. G. Katsiaficas (Routledge, 2001). Claudia Gorbman, in "Three Moments in *The World*," a talk given at the 2010 Music and the Moving Image conference, reads the appearance of the "Ode to Joy" from Beethoven's Ninth Symphony in *The World* as indexing the conjuncture of 1989 and the Tiananmen protests.

78. Lu, "Music and Noise," pp. 66–68.

79. In a personal communication dated 7 August 2010, Clare Eng states: "The ad starts with a greeting (basically, X company warmly welcomes its associates that are part of Y corporate group). It then goes on to advertise its product—a line of men's and women's leather shoes. It finally closes with the tagline, roughly translatable to 'Spring's a time for new things, buy our shoes!'"

80. Lu, "Music and Noise," pp. 64–65.

81. Jia Zhangke, "Jia Zhangke On . . . ," 10–11. Also see "Lim Giong" at http://en.wikipedia.org.

82. Respectively from the 2008 New Yorker Films DVD release and a separate translation by Sheldon Lu on p. 113 of "Emerging from Underground and the Periphery:

Chinese Independent Cinema at the Turn of the Twenty-First Century" in *Cinema at the Periphery*, ed. D. Iordanova, D. Martin-Jones, and B. Vidal (Wayne State University Press, 2010).

83. All references in this scene were identified by Jason McGrath (personal communication, 27 July 2010). See Hu Mingrong, "*Yearnings, Aspirations*," *Encyclopedia of Contemporary Chinese Culture*, at http://www.bookrags.com; "*The Bund* (TV Series)," at http://en.wikipedia.org. See also Ackbar Abbas, "Poor Theory and New Chinese Cinema: Jia Zhangke's *Still Life*," paper presented at annual conference of Society for Cinema and Media Studies, available at http://www.humanities.uci.edu.

84. The recoding technique appears again in this film. In a scene presenting Shen Hong (played by Zhao Tao), a nurse from Shanxi looking for her estranged husband in Fengjie, a child on a ferry transporting her sings a song *a capella*. In the very next scene, showing Shen Hong crossing a street in a newly developed area, we hear a professional recording of the same song. Although the cut is relatively smooth, the half-step key difference between the two versions is somewhat jarring—perhaps punctuating the gap between the slow, low-tech ferry and the new buildings/development in the two scenes, between the class difference between the indigent child and the middle-class Shen Hong, and, ultimately, between the past and the future of Fengjie itself.

85. Brother Mark's death, if not entirely accidental, seems produced either by the "dog-eat-dog" world he inhabits (as he puts it—during the ringtone conversation, Han Sanming learns that Mark is somewhat mistreated by the dockworkers, though it isn't clear to what degree or in what capacity) or by the payment he and a group receive to "beat up some guy" from Yunyang. Jie Li argues that Mark is "employed by local demolition officials to beat up poorly compensated residents who are not moving out quickly enough" and that "the rubble [of the demolished buildings] is exacting karmic retribution on the very human bodies that laid them to ruins." See Jie Li, "Home and Nation Amid the Rubble: Fei Mu's *Spring in a Small Town* and Jia Zhangke's *Still Life*," *Modern Chinese Literature and Culture* 21, no. 2 (2009), p. 104. On p. 112 of "Emerging from Underground and the Periphery," Sheldon Lu describes Mark's work as that of a "small-time gangster" and notes that his name is a reference to a character played by Chow-Yun Fat in the 1986 John Woo film *A Better Tomorrow*.

86. The lyrics to "Shanghai Bund" can be found in Roman transliteration (from Cantonese) and in English translation at http://www.ampedasia.com.

87. Interestingly, it is only after Mark's death that we hear Han Sanming's ringtone again—this time, relaying a call from Brother Ma that his sister Missy Ma (Han's estranged wife) has returned to Fengjie. Drawing on Tonglin Lu's framework in *Xiao Wu*, perhaps the death of the homosocial bond between Mark and Han produces a space that makes the rekindling of the heterosexual bond between Han and Missy possible.

88. The gap between aspiration and reality in Jia's more recent films would seem to be represented powerfully in the visual domain through CGI effects—including the strange, magical-realist spaceship sequences in *Still Life*. Jason McGrath reads these sequences in *Still Life* as speaking to utopian imaginaries, as an "ideal of freedom that sustains people through such traumatic transformations." The gap between mundane music and idealized musics, as sometimes demonstrated in the director's recoding technique, may function in a similar fashion. See Jason McGrath, "The Cinema of Displacement: The Three Gorges Dam in Feature Film and Video," in *Displacement: The Three Gorges Dam and Contemporary Chinese Art*, ed. Wu Hung with J. McGrath and S. Smith (Smart Museum of Art, University of Chicago, 2008), p. 42.

89. There is a contradiction in a film aesthetic that would purport to be more concerned with acts of communication between characters rather than from characters to spectators. The structural priority of the spectator as a consumer within the cinematic apparatus makes, at one level, any film a communicative act, however opaque or obscure, intended for an audience member. Nonetheless, the characters in Jia's films and their interactions seem to be the focus of what is conveyed to audiences, instead of directorial commentary on those characters and interactions. Jia Zhangke's realist aesthetic initially seems to remove normative editorial commentary in favor of *observation*, even though this observing subjectivity stitches together apparently observed moments into a cinematic totality.

90. In cinema and television, the ideology of ringtone personalization operates at two levels: the mere identification of character with ringtone; and understanding that this identification is intended for a spectator. The correlate of the second level in everyday life would be that ringtone personalization is intended to make others' personalities transparent and legible to bystanding auditors: without this communicative clarity, ringtone personalization is largely rendered ineffective. Ringtone personalization is therefore not merely for the individual consumer but a way of structuring that consumer's relationship to those proximate to him or her.

Notes to Chapter 9

1. Berin Szoka, "We Must End the (Reverse) Digital Divide!" *The Technology Liberation Front*, at http://techliberation.com.

2. Examples include very high levels of black unemployment during the post-2008 recession, disproportionate incarceration within the prison-industrial complex, stop-and-frisk policing, individual cases of racist violence, an Obama-era upturn in white supremacist activity, etc. See, however, Colin Asher, "The Recession Is Closing Private Prisons," at http://criminaljustice.change.org. On the post-Soul era, see Greg Tate, "Introduction: Nigs R Us, or How Blackfolk Became Fetish Objects," in *Everything But The Burden: What White People Are Taking from Black Culture*, ed. G. Tate (Broadway Books, 2003).

3. See Bob Herbert, "America Is Better Than This," *New York Times*, 27 August 2010; Sam Roberts, "Births to Minorities Approach a Majority," *New York Times*, 12 March 2010.

4. Alexander Weheliye, *Phonographies: Grooves in Sonic Afro-Modernity* (Duke University Press, 2005), p. 2.

5. Ibid., pp. 2–3.

6. Eszter Hargittai, "Second-Level Digital Divide: Differences in People's Online Skills," *First Monday* 7, no. 1 (2002), at http://firstmonday.org; "The Second-Level Digital Divide: Differences in Young People's Internet Uses," 2007, at www.indire.it.

7. See David McCarthy, "Tell It What It Needs to Hear: Labor and 'The Machine' in the IVR Enabled Automated Call Center," in *The Oxford Handbook of Mobile Music Studies*, ed. S. Gopinath and J. Stanyek (Oxford University Press, forthcoming). Some reports suggest that black predilections for prepaid mobile plans span the class spectrum. See the summary of the report, "Mobile Phones and the Multicultural Consumer—US—Feburary 2010," at http://oxygen.mintel.com. African Americans are most likely to possess prepaid cell phones (22%, versus 17% of Americans generally)—see New Millennium Research Council, *The First Annual New Millennium Research Council Survey on Consumers, Cell Phones, and the Economy* (2009) at http://newmillenniumresearch.org.

8. See "African-Americans, Women and Southerners Talk and Text the Most in the US," at http://blog.nielsen.com.

9. Cited in John Barber, *The Black Digital Elite: African American Leaders of the Information Revolution* (Praeger, 2006), p. 174.

10. Ibid., pp. 174–175.

11. See Henry Jenkins, *Convergence Culture: Where Old and New Media Collide* (New York University Press, 2006), pp. 59–92.

12. Aaron Smith, "Mobile Access 2010," Pew Internet & American Life Project, at http://www.pewinternet.org.

13. Ibid. Also see Aaron Smith, "Technology Trends Among People of Color," Pew Internet & American Life Project, at http://www.pewinternet.org.

14. In Bat, "What Is Hypersoul?" (2001), at http://www.riddim.ca; cited on p. 32 of Alexander Weheliye, "'Feenin': Posthuman Voices in Contemporary Black Popular Music," *Social Text* 20, no. 2 (2002), 21–47.

15. Weheliye, "Feenin'," p. 33.

16. Ibid., pp. 33–39. On Auto-Tune, see Greg Milner, *Perfecting Sound Forever: An Aural History of Recorded Music* (Faber and Faber, 2009), pp. 293–346, esp. 341–344.

Smule's popular iPhone app I Am T-Pain auto-tunes users' singing in real time and projects the result from the phone's speaker.

17. Weheliye, "Feenin'," p. 34.

18. In "Ring Ring Ring: Mobile Sensations," in *The Oxford Handbook of Mobile Music Studies*, Weheliye lists almost 30 R&B songs from 1995 to 2010 in which "a significant part of their lyrics and sounds" are devoted to "mobile technologies."

19. Kodwo Eshun, *More Brilliant Than the Sun: Adventures in Sonic Fiction* (Quartet Books, 1998), pp. 22–30.

20. But see Navarrow Wright, "Will Mobile Supply Meet Minority Mobile Demand in 2011?" at http://www.blackweb20.com.

21. Chuck Klosterman, "Sex, Pies and Videotape," *The Guardian*, 25 August 2007, available at http://www.guardian.co.uk.

22. The groove is offset by a string introduction that starts many of the episodes, and other musical elements are introduced throughout the series—such as the gospel music that ingeniously recomposes the groove in chapters 18 and 19.

23. Klosterman, "Sex, Pies and Videotape."

24. The music here recalls electronic and tape pieces using only harmonically coordinated voices (e.g., Steve Reich's *Come Out* or Paul Lansky's *More Than Idle Chatter*), as well as madrigals, Meredith Monk's vocal music, and *a capella* vocal hocketing in gospel and black popular music (as in Sly & the Family Stone's "Dance to the Music").

25. Michael Bull, *Sounding Out the City: Personal Stereos and the Management of Everyday Life* (Berg, 2000), p. 183, note 1. The interviewed stereo user is cited on p. 37.

26. Ibid., p. 112. Also see David Beer, "Tune Out: Music, Soundscapes, and the Urban *Mise-en-scène*," *Information, Communication & Society* 10, no. 6 (2007), 846–866.

27. Joachim Höflich, "Places of Life—Places of Communication: Observations of Mobile Phone Usage in Public Places," in *Mobile Communication in Everyday Life: Ethnographic Views, Observations and Reflections*, ed. J. Höflich and M. Hartmann (Frank & Timme, 2006), p. 27. Citation is of Lyn Lofland, *The Public Realm: Exploring the City's Quintessential Social Territory* (Aldine de Gruyter, 1998), p. 12.

28. As Castells et al. note on p. 174 of *Mobile Communication and Society* (MIT Press, 2007), "most of [the mobile phone's] uses are local, even sedentary, and related to close personal interaction."

29. In *Trapped in the Closet*, some characters' wealth indicates home ownership, and a range of domiciles are represented. But no one is extraordinarily wealthy, and wealth itself is largely procured by organized crime, itself parasitically tied to working-class populations.

30. See Castells et al., *Mobile Communication and Society*, pp. 55–64.

31. See "Solutions Research Group Study States African Americans and Hispanics Lead Mobile Culture," Eworldwire press release, 29 November 2005, available at http://eworldwire.com.

32. If we assume that the survey was demographically representative of the United States and assume a population of roughly 12.5% African Americans and 10% Latinos, leaving 77.5% remaining, roughly 65% of whom are white, according to 2009 US Census information. See US Census Bureau, "2009 American Community Survey, 1-Year Estimates, B02001. RACE—Universe: TOTAL POPULATION," at http://factfinder.census.gov.

33. "Dynamic Trends, August 2004," Hispanictrends.com, archived at http://web.archive.org.

34. In view of the similarities in ringtone consumption by Latinos and African Americans, it makes sense to present information on both groups, even though African Americans have outpaced Latinos in ringtone consumption. See Leila Cobo, "Latin Notas: Shakira Rings Up Hot Ringtone Sales," *Billboard*, 2 July 2005. Also, Latino consumers purchased polyphonic ringtones long after most other American consumers had switched to sound-file ringtones—because Latino consumers probably continued using older mobile phones. See Leila Cobo, "Latin Notas: The Polyphonic Problem," *Billboard*, 15 March 2008.

35. "About MEF" at http://www.m-e-f.org.

36. See "Mobile Entertainment Forum Launches the UK's First Authoritative Ringtones Chart" at http://www.m-e-f.org.

37. Paul Sexton, "New Chart Calls Up U.K. Ringtone Sales," *Billboard*, 5 June 2004.

38. See Ian Burell, "Phone Tones Ring £70m Change to Charts Music," *The Independent*, 24 May 2004, available at http://www.independent.co.uk; "History (cont'd)" at http://www.m-e-f.org.

39. "Nielsen RingScan Launched by Nielsen Mobile to Monitor and Compile Mobile Ringtone Sales Data," *Mobile Internet*, December 2006. The earlier chart also relied on Nielsen Mobile data; see Antony Bruno, "Dialing for Gold," *Billboard*, 12 August 2006. In July or August 2009, the two ringtone charts were renamed "Hot Master Ringtones" and "Hot Polyphonic Ringtones," replacing the "Hot RingMasters" and "Hot Ringtones" charts respectively. The "Hot Polyphonic" chart may have been discontinued; the last version I can find is from 17 April 2010.

40. National Public Radio, "What Makes a Ringtone Popular?" *All Things Considered*, 27 November 2007, available at http://www.npr.org.

41. "Hot R&B/Hip-Hop Singles & Tracks" and "Hot R&B/Hip-Hop Airplay," *Billboard*, 6 November 2004. "My Boo" was also number three on the "R&B/Hip-Hop Singles Sales" chart.

42. "Mainstream Top 40," "Hot 100," and "Hot 100 Airplay," *Billboard*, 6 November 2004.

43. "Over and Over" was not successful on the Hot R&B/Hip-Hop Songs chart, reaching a peak position of 51.

44. Also see David Brackett, "(In Search of) Musical Meaning: Genres, Categories and Crossover," in *Popular Music Studies*, ed. D. Hesmondhalgh and K. Negus (Arnold, 2002).

45. See, e.g., "The NPD Group: Rap/Hip-Hop Ringtones Rule the Market," *Business Wire*, 11 September 2006.

46. That the top four ringtone hits on 6 November 2004 seemed more closely linked to airplay figures than sales figures is telling, although ringtone sales registered new hits ("My Boo" and "Drop It Like It's Hot") and demonstrated a persistence of slightly older hits ("Lean Back" and "Locked Up"). Also see Bethany Klein, "'The New Radio': Music Licensing as a Response to Industry Woe," *Media, Culture & Society* 30, no. 4 (2008), 463–478.

47. For more on nostalgia and cycles of accumulation, see the introduction to part II.

48. As Weheliye astutely observes in "Ring Ring Ring," "ringtone rap reverses the absorption of monophonic ringtones into popular music that occurred ten years ago by transmogrifying songs into extended ringtones."

49. See Brian McManus, "Putting It to 'Bed,'" *Philadelphia Weekly*, 5 December 2007, available at http://www.philadelphiaweekly.com.

50. "New Rap Stars Find Success Fleeting," Yahoo! Music news, 18 October 2007, available at http://new.music.yahoo.com.

51. "Chain Hang Low," by the ringtone rapper Jibbs, drew on "Do Your Ears Hang Low?"—a children's song that was based on the nineteenth-century minstrel song "Turkey in the Straw." Weheliye mentions in "Ring Ring Ring" that Trey Songz's "Lol:-)" sounds "like a crunk version of the children's song 'Frère Jacques'"; he also notes that much ringtone rap/R&B features "nonsensical rhymes," which contribute to its juvenile character.

52. The title of a classic noir film from 1950, D.O.A. ordinarily stands for "dead on arrival."

53. Cited in "BLOG: R.I.P. Ringtone Rapper" at http://www.xxlmag.com. Also see Paul Tingen, "Secrets of the Mix Engineers: Young Guru," at http://www.soundonsound.com.

54. See Bob Herbert, "'A Very Deep Hole,'" *New York Times*, 8 June 2010; "Outside the Casino," *New York Times*, 13 July 2010; "Too Long Ignored," *New York Times*, 21 August 2010; "Two Different Worlds," *New York Times*, 18 September 2010; "This

Raging Fire," *New York Times*, 16 November 2010. Also see Jane Collins and Victoria Mayer, *Both Hands Tied: Welfare Reform and the Race to the Bottom in the Low-Wage Labor Market* (University of Chicago Press, 2010).

55. See Kyra Kyles, "Ringtone Rap Project Presents: the Shamrock Shake," at http://www.chicagonow.com.

56. Kyra Kyles, "Shamrock Shake: The Video Coming Soon," at http://www.chicagonow.com.

57. Kyra Kyles, "The Shamrock Shake: The Beverage and the Song for St. Patty's Day," at http://blogs.redeyechicago.com.

58. An allegory of consumption is at work in such songs: a male singer describing a female love interest with a ringtone of his song (or telling her to put his song on her phone) is easily transposed onto the relationship between singer and (female) consumer, who is described as owning (or is encouraged to own) the song herself.

59. See Imar de Vries and Isabella van Elferen, "The Musical *Madeleine*: Communication, Performance, and Identity in Musical Ringtones," *Popular Music and Society* 33, no. 1 (February 2010), 61–74.

60. For reviews, see Tal Rosenberg, *Stylus Magazine*, 6 June 2007, available at http://www.stylusmagazine.com; Robert Christgau, *Rolling Stone*, 29 May 2007, available at http://www.rollingstone.com; Eric Henderson, *Slant*, 31 May 2007, available at http://www.slantmagazine.com; Nick Marx, Tiny Mix Tapes (2007?), available at http://www.tinymixtapes.com; Laron Cue, http://blogcritics.org/music/article/music-review-r-kelly-double-up/; Jackie Im, *Treble*, 26 June 2007, available at http://treblezine.com; Alexis Petridis, *The Guardian*, 1 June 2007, available at http://www.guardian.co.uk; Kyle Anderson, "Our Kelly," *Spin*, August 2007, p. 124. Kelly denied the allegations; see John Norris, "R. Kelly: In His Own Words," at http://www.mtv.com.

61. Snap music hits include Dem Franchize Boyz's "Lean Wit It, Rock Wit It," D4L's "Laffy Taffy," and Ying Yang Twins' "Wait (The Whisper Song)." See Roni Sarig, *Third Coast: OutKast, Timbaland, and How Hip-Hop Became a Southern Thing* (Da Capo, 2007), p. 306. Crunk features collectively shouted vocals and a great deal of musical repetition; it emerged in the work of Lil John and was popularized in his "Get Low" (2003), with The East Side Boyz and Ying Yang Twins. According to the ethnomusicologist Wayne Marshall (personal communication, 30 January 2011), most crunk is in the range of 60–90 BPM in tempo. Crunk and snap are not entirely separable from "ringtone rap"; in fact, some of the songs listed above are associated with the term. Although not synonymous, both crunk and snap involved a much-disparaged simplification of songwriting and allegedly lowered the standards for songwriting craft within the music industry.

62. See the Fawn Renee, "Polow da Don: Talks to Black Women," at http://allhiphop.com.

63. DJs try out potential hit songs at crunk-associated strip clubs in Atlanta. See Elizabeth Blair, "Strip Clubs: Launch Pads for Hits in Atlanta," National Public Radio, 23 December 2010, available at http://www.npr.org. Also see Sarig, *Third Coast*, pp. 302–303; Darren Grem, "'The South Got Something to Say': Atlanta's Dirty South and the Southernization of Hip-Hop America," *Southern Cultures* 12, no. 4 (2006), 55–73.

64. The phrase "in da club," which appears in the song's first verse (and elsewhere on the album), became commonplace in R&B and hip-hop after 50 Cent's hit "In da Club" (2003).

65. "Ringtone" arguably flaunts Kelly's sexuality unrepentantly, not only in the song's subject matter, but also by depicting sexual encounters (perhaps illegal ones) with two fairly young women—the beginning of the song describes a woman who is "four [foot] five, lookin' fly." Since the song is addressed to "haters," one might interpret the term "shawty" (affectionately describing women/girlfriends/female sexual partners in African American English) more literally (i.e., as meaning "short") and, likewise, the ringtone as a teenage youth-culture commodity.

66. Nathan Rabin, review of *Double Up* at http://www.avclub.com.

67. However, on autoreferentiality and modernism, see Fredric Jameson, *Brecht and Method* (Verso, 1998), p. 89. Also see Timothy Brennan, *Secular Devotion: Afro-Latin Music and Imperial Jazz* (Verso, 2008), p. 73.

68. The closest was Jay-Z's "D.O.A.," which peaked at number 15 on the Billboard US Hot Rap Tracks. But only Alain Clark's and Weird Al Yankovic's (discussed below) received much attention or sold significantly.

69. See Alexis Petridis, review of Big Boi's *Sir Lucious Left Foot: The Son of Chico Dusty*, *The Guardian*, 1 July 2010, available at http://www.guardian.co.uk; David Peisner, "A Rapper's Long Battle to Go It Alone," *New York Times*, 2 July 2010, available at http://www.nytimes.com. An extensive release history of the album can be found at http://en.wikipedia.org.

70. Peter Relic, review of "Ringtone," *Rolling Stone*, 5 March 2009.

71. Jason Chen, "Doritos' Late Night Augmented Reality Bags Like a Party at Taco Bell," at http://www.gizmodo.com.

72. See "Dre Allen" at http://en.wikipedia.org.

73. Christopher Weingarten, "The Decade in Music Genre Hype," *Village Voice*, 22 December 2009. Also see "Crunk" at http://en.wikipedia.org.

74. Tom Breihan, review of *Sir Lucious Left Foot* at http://pitchfork.com.

75. Like several other songs on *Sir Lucious*, "Ringtone/Theme Song" was produced by the Atlanta-based production crew Organized Noize, with whom OutKast has long been associated.

76. Burmy, "Big Boi—Ringtone," at http://www.djbooth.net.

77. Ehren Gresehover, "Big Boi's 'Ringtone' Not Nearly as Annoying as Your Average Ringtone," at http://nymag.com.

78. Still, some of these songs—like "Ringtone" on R. Kelly's *Double Up* or "Theme Song" on Big Boi's *Sir Lucious Left Foot: Son of Chico Dusty*—appeared on albums demonstrating strong sales figures. By February 2011, *Double Up* had sold more than a million copies, and Big Boi's solo recording more than 175,000.

79. See Mia Wallace and Clint Spanner, *Chav!: A User's Guide to Britain's New Ruling Class* (Bantam Books, 2004), pp. 89, 147–149.

80. See, respectively, Keith Hayward and Majid Yar, "The 'Chav' Phenomenon: Consumption, Media and the Construction of a New Underclass," *Crime Media Culture* 2, no. 9 (2006), 9–28; Owen Jones, *Chavs: The Demonization of the Working Class* (Verso, 2011), 1–12; Paul Gilroy, *After Empire: Melancholia or Convivial Culture?* (Routledge, 2004), pp. 144–153. Also see "Chav" at http://en.wikipedia.org.

81. But see the forum "Arctic Monkeys—A Certain Romance" at http://www.songmeanings.net.

82. Hayward and Yar, "The 'Chav' Phenomenon," pp. 12, 20–22. Compare Philip Thornton, "Unemployment Level at Highest since 2000," *The Independent*, 13 July 2006, available at http://www.independent.co.uk; Graeme Wearden, "UK Unemployment Inches below 2.5m," *The Guardian*, 19 January 2011, available at http://www.guardian.co.uk. Concomitant with the general disparaging of chavs are the feelings of despair and powerlessness felt by many working-class Britons in recent years. See Jones, *Chavs*, pp. 247–248.

83. See http://www.youtube.com/watch?v=FB7jB-SZHwQ.

84. See Shea Serrano, "Artist of the Week: Without a Face," at http://blogs.houstonpress.com.

85. See the SuperNews! animation group-produced video, available at http://www.youtube.com.

86. This Recording Industry Association of America figure is available at http://www.riaa.com. Also see Steve Knopper, *Appetite for Self-Destruction: The Spectacular Crash of the Record Industry in the Digital Age* (Soft Skull, 2010); Victor Keegan, "The Demise of the Music Industry Is Visible Everywhere but in the Facts," *The Guardian*, 12 March 2010, available at http://www.guardian.co.uk.

87. See Ed Guerrero, *Framing Blackness: The African American Image in Film* (Temple University Press, 1993), pp. 82–84.

88. See Richard Smirke, "IFPI 2012 Report: Global Music Revenue Down 3%, Synch, PRO, Digital Income Up," *Billboard*, 26 March 2012.

89. Weheliye, *Phonographies*, p. 3; Paul Gilroy, *The Black Atlantic: Modernity and Double Consciousness* (Harvard University Press, 1993), p. 36. On modernity, see Ben Singer, *Melodrama and Modernity: Early Sensational Cinema and Its Contexts* (Columbia University Press, 2001), pp. 17–35.

90. For a critical view of identity-political ties to the state, the Democratic Party, and President Barack Obama in the US, see Tariq Ali, *The Obama Syndrome: Surrender at Home, War Abroad* (Verso, 2010).

Notes to Epilogue

1. The London-based "Mobile Music Now" conferences, which began in July 2007, are one index of this industry-wide marketing effort. On "mobile music," also see Gerard Goggin, *Global Mobile Media* (Routledge, 2011), pp. 55–79.

2. Mobile Music Now, 9 July 2009, BAFTA, London, during the Panel Session and Speed Meet discussion. The mobile entertainment industry's looming "death" pervaded the discussion. The reason for the collapsing of "mobile music" into "digital music" is that mobile platforms are now largely dominated by off-deck Web browsing and by apps roughly equivalent to personal computer programs, bypassing the walled garden of WAP.

3. "Apple: Boom Time," *Dailywireless*, 25 January 2012; Hendrik Koekkoek, *Distimo Publication Full Year 2011* (Distimo, 2011), 6, available at www.distimo.com. Most of the popular apps are games; see Wilson Rothman, "iTunes Lists All-Time Top Selling iPhone/iPad Apps" at http://technolog.msnbc.msn.com; See Brian Chen, "Sketchy MP3 Downloader Soars to No. 1 in iPhone App Store" at http://www.wired.com. Some of the biggest mobile music successes, including the apps Ocarina, Leaf Trombone, I Am T-Pain, and Glee Karaoke, have been produced by the company Smule.

4. Fran Foo, "Mobile Firm Leads Musical Revolution," *The Australian*, 24 August 2010, emphasis added. Keene is referring to for-sale "skins" used to personalize the outer appearance of phones. The app would purportedly do the same for the inner workings of the phone.

5. "Chamillionaire to Share Social Media Savvy at Grammy Summit," *AOL Music* (weblog delivered by Newstex), 10 February 2011; emphasis added.

6. Goggin, *Global Mobile Media*, pp. 76–79. For an in-depth treatment of the problem of mobile music, see Sumanth Gopinath and Jason Stanyek, eds., *The Oxford Handbook of Mobile Music Studies* (Oxford University Press, forthcoming).

7. Michael DeGusta, "The REAL Death of the Music Industry" at http://www.businessinsider.com; Richard Smirke, "IFPI 2012 Report: Global Music Revenue Down 3%; Sync, PRO, Digital Income Up," *Billboard*, 26 March 2012.

8. Terry, "Global Videogame Sales Surpass Movie Industry in 2008!" at http://www.videogamesblogger.com.

9. But North American consumers have been chastened by the RIAA lawsuits of the 2000s and are the least likely among global listeners to download and share music illegally (11% had done so in the three months before one inquiry); the next lowest were Europeans (21%); Latin Americans, Asians and Pacific Islanders, Middle Easterners, Africans, and Pakistanis were all upward of 40%. See "The Hyper-Fragmented World of Music: Marketing Considerations and Revenue Maximisation" at http://www.nielsen.com.

10. See Mike Flacy, "Spotify Extends Free, Unlimited Listening Period in the United States," at http://www.digitaltrends.com.

11. On 3 June 2011, at www.smule.com, Smule's trademark was "Experience Social Music."

12. See Stuart Dredge, "Bjork's New Biophilia Album? There'll Be Apps for That. 10 of 'Em," *Mobile Entertainment*, at http://www.mobile-ent.biz. Björk's fans have been posting online remixes of her music at least since 1997; see Matthew Mirapaul, "Why Just Listen to Pop When You Can Mix Your Own?" *New York Times*, 20 August 2001.

13. For a more optimistic view, see the video of Pandora founder and CEO Tim Westergren at http://www.youtube.com. For a critique of the recording industry, see Steve Knopper, *Appetite for Self-Destruction: The Spectacular Crash of the Record Industry in the Digital Age* (Soft Skull, 2010).

14. Gopinath, "Ringtones, or, the Auditory Logic of Globalization," *First Monday* 10, no. 12 (December 2005), available at http://firstmonday.org. Jameson seems to draw on Braudel's "History and the Social Sciences: The Longue Durée" (1958), which can be found in Braudel, *On History* (University of Chicago Press, 1980 [1969]). Many thanks to Keya Ganguly for alerting me to this reference.

15. See Fredric Jameson, *The Political Unconscious: Narrative as a Socially Symbolic Act* (Cornell University Press, 1981), pp. 74–102.

16. See Guy Debord, *The Society of the Spectacle* (Zone Books, 1995), chapter 6 ("Spectacular Time"), pp. 110–17, paragraphs 147–164.

17. Perry Anderson, "Jottings on the Conjuncture," *New Left Review* 48 (November–December 2007), 6. Also see Giovanni Arrighi, *The Long Twentieth Century: Money, Power and the Origins of Our Time* (Verso, 2010 [1994]), especially the new postscript.

18. See Immanuel Wallerstein, *The Decline of American Power: The US in a Chaotic World* (New Press, 2003); Gopal Balakrishnan, "Speculations on the Stationary State," *New Left Review* 59 (September–October 2009), 5–26; Giovanni Arrighi, *Adam Smith in Beijing: Lineages of the Twenty-First Century* (Verso, 2007), esp. pp. 49–50.

19. Braudel, "History and the Social Sciences," p. 29.

20. Stuart Hall, "Epilogue: Through the Prism of an Intellectual Life," in *Culture, Politics, Race, and Diaspora: The Thought of Stuart Hall*, ed. B. Meeks (Randle, 2007), p. 280. Also see Alex Callinicos, "Epoch and Conjuncture in Marxist Political Economy," *International Politics* 42 (2005), 353–363.

21. See Gopinath, "Ringtones," and chapter 1 of the present volume. Publishing has increasingly come under the control of large umbrella organizations, transforming songwriters and composers from rentiers in the early days of the music industry into something closer to proletarians. Thus, the holders of intellectual-property rights should be, on average, conceived decreasingly as individuals and instead increasingly as rights-aggregating firms.

22. See David Harvey, "The Art of Rent: Globalization and the Commodification of Culture," in *Spaces of Capital: Toward a Critical Geography* (Routledge, 2001), esp. pp. 398–399.

23. Harvey, "The Art of Rent," 396. Sales enable free distribution of digital music (through copying), which downgrades the product's rent value.

24. See Luigi Marco Bassani, "Life, Liberty, and . . . : Jefferson on Property Rights," *Journal of Libertarian Studies* 18, no. 1 (2004), 31–87.

25. "Internet Users," World Development Indicators, World Bank, available at http://data.worldbank.org; International Telecommunications Union, The World in 2010. ICT Facts and Figures, available at www.itu.int.

26. See Karl Marx, *Capital*, volume 1 (Vintage, 1977 [1976]), pp. 125–131.

27. But see Robert Reich, "Entrepreneur or Unemployed?" *New York Times*, 1 June 2010.

28. Vincent Mosco, *The Pay-Per Society: Computers and Communication in the Information Age* (Garamond, 1989), pp. 24–25.

29. Marx, *Capital*, volume 1, p. 126.

30. OECD Science and Technology and Industry Scoreboard 2009, section 3.4 ("International Trade in ICT Goods and Services"), available at http://www.oecdilibrary.org.

31. Immanuel Wallerstein, *Historical Capitalism; with Capitalist Civilization* (Verso, 1996), p. 16. David Harvey proposes, on p. 33 of *A Brief History of Neoliberalism* (Oxford University Press, 2005), that neoliberalism is "the financialization of everything."

32. See Charlotte Hess and Elinor Ostrom, eds., *Understanding Knowledge as a Commons: From Theory to Practice* (MIT Press, 2007), p. 5.

33. See Julian Stallabrass, "Digital Commons" (review of Sam Williams, *Free as in Freedom: Richard Stallman's Crusade for Free Software*), *New Left Review* II, no. 15 (2002), 141–146.

34. Mosco, *The Pay-Per Society*, esp. pp. 17–39.

35. Patrick Burkart and Tom McCourt, *Digital Music Wars: Ownership and Control of the Celestial Jukebox* (Rowman & Littlefield, 2006), p. 5.

36. See Nancy Kranich, "Countering Enclosure: Reclaiming the Knowledge Commons," in *Understanding Knowledge as a Commons*, ed. Hess and Ostrom, p. 89.

37. The habits produced here are passive, essentially involving the outsourcing of storage labor to the subscription services; those habits are different from more ambitious, active ones that producers of mobile music apps are hoping to cultivate among consumers.

38. Peter Fingar, *Dot.Cloud: The 21st Century Business Platform* (Meghan-Kiffer, 2009), p. 54.

39. Janet Lowe, *Google Speaks: Secrets of the World's Greatest Billionaire Entrepreneurs, Sergey Brin and Larry Page* (Wiley, 2009), p. 246.

40. The centrality of large tech corporations in managing the "cloud" contradicts claims of its decentralization—true, perhaps, of literal servers, but not of who is managing them.

41. Miguel Helft, "Apple Unveils a 'Cloud' Music and Storage Service," *New York Times*, 6 June 2011.

42. See Christopher Ketcham and Travis Kelly, "The Cloud Panopticon: Google, Cloud Computing and the Surveillance-Industrial Complex," *CounterPunch Newsletter* 17, no. 7 (2010), 1–5.

43. See David McCandless, "How Much do Artists Earn Online?" at http://www.informationisbeautiful.net.

44. Personal communication with Brice Salek, 12 February 2012. The information Salek received from iTunes reads: "1. The most important issue is that 'format' is listed as LP [for Ringtone Records albums]. 2. Ringtone releases are limited to one track only, and these have multiple. If he had selected 'ringtone release' when cre-

ating these initially it would have held him to one track. 3. 'Ringtone' shouldn't be listed in the track titles—iTunes specific style point below: Track titles must not reference 'ringtone,' 'ringtone version,' etc. 4. Ringtones must be exactly 30 seconds long. No longer, no shorter."

45. Nick Dyer-Witheford, "The Circulation of the Common," talk given at University of Minnesota, 2009 (available at www.fims.uwo.ca), paragraph 4.

46. See http://www.bepress.com/ir/ and http://en.wikipedia.org/wiki/Digital_Commons.

47. Hess and Ostrom, *Understanding Knowledge as a Commons*, p. 12. These entities might, in turn, interpret such actions increasingly as a form of digital squatting.

48. Also see Martin Scherzinger, "In Divisible Mobility: Music in an Age of Cloud Computing," in *The Oxford Handbook of Mobile Music Studies*, ed. S. Gopinath and J. Stanyek (Oxford University Press, forthcoming).

Index